Ice Age of the Dimmer Sun in 30 years
Illustrated Science Exploration by Rolf A. F. Witzsche

© Text Copyright Rolf A. F. Witzsche 2018
all rights reserved

This book contains the transcript with images of the exploration video with the above title:
see: http://www.ice-age-ahead-iaa.ca/

Lead in:

Ice Age of the dimming Sun in 30 years? Yes, but how to live with it

Question: How many colors are in the rainbow?

If your answer is, seven, then you confirm that the Sun that lights up the rainbow cannot be a sphere of hydrogen gas heated by nuclear fusion from within, because if it was, you would see only 3 colors at the most, in 3 narrow strips, according to the hydrogen emission spectrum

Since the Sun, according to the evidence that we see every day, cannot be a hydrogen sphere powered from within, what then is it? How is it powered? These questions can be answered. The evidence speaks for itself. In the course of the exploration it becomes apparent that the Sun is not its own master, but is powered from interstellar plasma streams, focused on it by electromagnetic primer fields. The process is delicate. It is fragile to the point that the Sun can become 'inactive.

It is hard to imagine that our Sun - which is widely regarded as an invariable constant, undergoes activity fluctuations, resulting in climate changes on Earth. But this what we happening, in a process that has just begun. The Sun may turn dim overnight with less than a third of its energy output remaining. This may happen in roughly 30 to 50 years from now, with which the next Ice Age begins.

Surprisingly, evermore evidence comes to light, from research in plasma physics and astrophysics, that indicates that the dimming of the Sun is near. With this considered, the already happening astrophysical event will shape the course of humanity more extensively and more profoundly than any other event in known history. Science becomes of critical importance here, to shape the direction.

As human beings, we have the power to mobilize our creative and productive resources, and thereby meet the great challenge before us. As we do this, we will find that in the process of meeting the great Ice Age challenge, the greatest challenge of all times, we will create ourselves a brighter world than we can yet imagine in which the coming Ice Age won't affect us, versus having no future at all.

Of course, we can also do nothing, as is presently the case, and thereby die on the 'default train' when the cosmic phase shift begins without us being prepared for it. In order to inspire some interest in society, for getting itself off the train to tragedy, I have re-organized and expanded my earlier video production, "Ice Age of the Dimming Sun in 30 Years." By popular request after more than 25,000 viewings of the original video, and in response to comments received, I gave the presentation a new form, and included additional leading-edge discoveries in science. This book contains all the transcripts and images of the expanded version.

Because of the extensive nature of the subject, I have divided the exploration into nine parts.

Table of Contents

Part 1 - Introduction to the dimmer Sun 20

    The science-focus of the series 21

    The Primer Fields 22

    The most basic electromagnetic fields 23

    The Primer Fields and their effects on our Sun 24

    Intensely focused and compressed plasma 25

    Our solar system as a single functional unit 26

    The dynamics of the ice ages 27

    The Sun as an electrically powered star 28

    Sunspots on the surface of the Sun 29

    The fusion Sun theory 30

    A minimal threshold of conditions 31

    If the threshold is not reached 32

    The Sun simply turns off 33

    When our brilliant sun is turned off 34

    The Milankovitch theory 35

    The large temperature fluctuations 36

    Primer Field theory brings a huge difference 37

    The Primer Fields theory presents us a great blessing 38

    To avoid the catastrophe of nuclear war 39

    An incentive to get our act together 40

    A single basic principle 41

    Close to the minimal threshold 42

When the assumed value of money becomes uncertain	43
The Glass Steagall law	44
During the interglacial warm period	45
We will cross the cut-off threshold	46
The Sun became inactive 120,000 years ago	47
Ice sheets more than 10,000 feet deep	48
The Sun inactive in 30 to 50 years	49
The chlorophyll in plants	50
Yes, we will do this	51
Truthful science supplies the potential	52
But what is love?	53
Without science standing at the heart of it	54
False science is destructive	55
Society raising itself out of the trap of false science	56
False science to forcefully strangle civilization	57
When honest science displaces constricted science	58
The solar-forced Great Global Warming	59
The Greenland Ice Sheet experienced some melting	60
Humanity never was a climate factor	61
The most horrific holocaust	62
The electric motivator	63
The cold deep currents flow slowly	64
CO2 is coming back to us	65
In the order of a millionth part of it	66
CO2 has no climate-effect	67
Most of the outgoing energy	68
Latent heat that is released	69

That's where CO2 is a factor 70

The solar-forced Great Global Cooling 71

Part of the system that affects cloudiness on Earth 72

A scientist is an economist 73

Part 2 - Effects of the Primer Fields on the Sun 74

The plasma environment in our solar system 75

In the laboratory environment 76

David LaPoint discovered in laboratory experiments 77

The effects are amazingly critical 78

Condensed plasma interacts with the Sun 79

Illustrated in the Red Square nebula 80

The Sun's great brilliance generated at its very surface 81

The plasma concentration process 82

A highly compressed plasma sphere was formed 83

Magnetic fields operating environment 84

In the real world 85

The plasma sphere in which our sun is located 86

Different types of atoms emit light in different bands 87

The color-rich world that we cherish 88

The Sun cannot be as a sphere of hydrogen gas 89

Hydrogen atoms emit light in only a few narrow bands 90

The color-rich white sunlight is clear tangible evidence 91

The surface plasma fusion also emits highly energetic solar cosmic-ray flux 92

Solar cosmic rays have an ionizing effect 93

Increased cloudiness results in colder climates 94

When a pot of water is boiled into steam 95

Clouds cooling latent energy into space 96

When the solar corona is weak...97

That's what we saw in the 1600s 98

Indicated in Carbon-14 measurements 99

The ever-changing climate on Earth 100

Global warming after the Little Ice Age, was not manmade 101

The solar system barely recovered from the Little Ice Age 102

When the Primer Fields collapse, a phase shift occurs..103

It has become evident in lab experiments...104

The solar system is not as robust as is generally believed 105

A colder, darker, yellow sun..106

A deactivated Sun 107

A massive reduction in solar energy 108

A radically different world unfolds 109

We are presently near the phase-shift point..110

NASA's Ulysses spacecraft...111

When the Sun's powered state ends 112

No one is prepared for the consequences...113

The Ice Age consequences promise to be far bigger 114

Part 3 - The digital Ice Age 115

The two long climate cycles that overlap 116

In the last 500,000 years of the resulting glaciation epoch...117

The difference between the two climate states..118

Produced from two different drilling sites...119

At the beginning of the last Ice Age 120

Rapid oscillations in Greenland ice 121

The Sun can alternate on and off states 122

Greenland ice is much more sensitive 123

- Antarctica, being an ice desert 124
- Oxygen isotope O-18 ratio is temperature sensitive 125
- Antarctica the washed out major trends 126
- Rapid fluctuations in the Greenland ice 127
- Dansgaard Oeschger oscillations 128
- Power-off of the Sun 129
- Giant red sprites 130
- After the Sun turns off 131
- Absorption spectrum of chlorophyll 132
- When the Sun enters its off-state 133
- Agriculture afloat on the equatorial seas 134
- Worse than the effect of a nuclear war 135
- Where the sunlight is the strongest 136
- The next deep glaciation to begin 137
- The cut-off level 138
- NASA's Ulysses spacecraft 139
- The brilliant life-giving 'fire' in the sky 140
- The Primer Fields will vanish in the near future 141

Part 4 - The Primer Fields dynamics 142
- Plasma sphere around the Sun 143
- David LaPoint uses two bowl-shaped magnets 144
- The Zeta Pinch effect 145
- Plasma currents in space become compressed 146
- Electromagneticly confined 'high-density' plasma On the platform of the Primer Fields 147
- Interglacial period of the Sun's active time 148
- Let me illustrate now how the process functions 149
- Dense Plasma Focus Device 150

A ring of electrodes 151

A plasma sheet forms 152

The plasma sheet 153

The plasma becomes extremely pinched 154

It becomes unstable 155

The plasma twists itself into a spiral 156

Then the spiral becomes compacted 157

The more unstable the spiral becomes.. 158

The twisting forms a complex knot 159

From a video about the Dense Plasma Focus Device 160

A high-density plasma concentration forms 161

When the plasma streams are too weak 162

When the Sun is powered 163

A number of interesting effects 164

In the narrow space between 165

Two bowl-type electromagnetic fields work together 166

A polarity flip point 167

Three functional magnetic elements 168

Each has a specific function to fulfill 169

The flip ring 170

The magnetic choke ring... 171

The out-flowing stream... 172

A high-power plasma experiment.. 173

Archetypal drawings 174

The plasma jets 175

The 11-year solar cycles 176

Operation of the Red Square Nebua.. 177

Two complementary bowl-type structures in operation ... 178

Observed in a laboratory plasma-flow experiment ... 179

Our galaxy, when it is observed as a whole ... 180

The center of the galaxy ... 181

Large intergalactic plasma streams ... 182

Two long intergalactic connecting streams ... 183

very long electric resonance cycles ... 184

Very long electric resonance cycles ... 185

Both near their minimum point ... 186

Breakdown of the Primer Fields ... 187

The Primer Fields cannot form ... 188

The Sun becomes inactive, dim, and cold ... 189

The Pleistocene Epoch ... 190

The Sun remains not totally shut down ... 191

The Dansgaard-Oeschger oscillations ... 192

Part 5 - The Dansgaard Oeschger oscillations ... 193

During the extremely weak conditions in the galaxy ... 194

During the long glacial periods ... 195

The Dansgaard-Oeschger oscillations ... 196

The snowball-Earth concept is just a theory ... 197

Ocean levels dropped 400 feet ... 198

The Milankovitch Cycles ... 199

The Serbian geophysicist and astronomer Milutin Milankovich ... 200

Theatrical cause, doesn't match the measured ice core data ... 201

When it comes to the big Dansgaard-Oeschger oscillations ... 202

These drilling projects are not small undertakings ... 203

The GRIP project drilled out a 3028-metre ice core ... 204

The drilling was repeated with another 4-year effort 205

The new drilling confirmed 206

When one overlays the Greenland ice core data 207

The coldest, driest, and windiest continent on Earth 208

In comparison with Antarctica, Greenland is a wet place 209

Ice cores go back in time up to 740,000 years 210

Differences in the resolution of details 211

The Dansgaard-Oeschger oscillations are very real 212

Spaced 1470 years on average 213

A reasonable determination three events spaced 1470 years apart 214

The most recent Dansgaard-Oeschger event 215

Preventing the collapse of the Little Ice Age into the next big Ice Age 216

The previous Dansgaard-Oeschger event 217

A two-fold nested system of Primer Fields 218

If the Oort clouds are real 219

Evermore evidence keeps coming to light 220

The cosmic plasma streams are not entropic 221

The 'Explicate Order' that David Bohm had referred to 222

Whom Einstein had once referred to as his successor 223

Finite resources subject to depletion 224

Without the utilization of cosmic energy 225

An interface to the cosmic power grid 226

Evident on the face of the Sun in UV light 227

Let's stop playing those silly and dangerous games 228

Soil temperatures at the Solar Terrestrial Institute in Irkutsk 229

NASA's Ulysses spacecraft 230

The phase shit to the next glaciation cycle 231

We are heading towards a new Ice Age in 30 years 232

Society likes to play fantasy games 233

The global warming hoopla 234

Give it more time to prepare 235

The 25th solar cycle 236

The Sun is already on the track of a large weakening trend 237

What our response should be is rather obvious 238

Part 6 - The UFO phenomenon 239

Rapid on-off conditions are natural occurrences 240

No UFO craft does actually exist 241

The UFO as an example 242

UFOs comparable to the sprites 243

Fragility of the sprites in the stratosphere, and UFO events on the lower atmosphere 244

UFO phenomenon fragile 245

Powered state of the Sun rare 246

UFO sightings from the 1950s on 247

The solar system is on a path to the vanishing point 248

The critical warning 249

The entire world is affected 250

Part 7 - Orbit dynamics of the planets 251

The Sun will 252

At a mere 71% of the escape velocity 253

Non-magnetic steel balls magnetically self-spacing 254

Large dust accumulations in the ice of Antarctica 255

The dust always stops 256

A wider field of phenomena stands as evidence for the Primer Fields 257

The near-geometrically expanding spacing 258

The geometrically expanding spacing of the orbits 259

The spin-rotation of the Earth itself is evidently electrically powered 260

Faster than the rotation of the Earth itself 261

Believed that the jet streams are powered by the Coriolis effect 262

Another phenomenon where rotational movements occur faster 263

Cold waters in the polar regions The Arctic pool has an outflow 264

Two currents 'spin' off from the rotating pool 265

The cold deep waters originating in the polar regions 266

Recycling system emits CO_2 dissolved 350 years ago 267

The recycling system incorporates three different transit times 268

The transit time to the Indian Ocean 269

Different transit times resurface CO_2 from three different eras 270

The large increase in CO_2 that has occurred 271

The Great Global Warming resulted from increase in solar activity 272

The Great Global Warming that has recovered the global climate 273

The measured 15% increase in CO_2 from the 1950s on 274

A quarter of the global atmospheric CO_2 gets recycled annually 275

It is further interesting to note 276

The sharp CO_2 increase that the ice core data indicates 277

Dr. Zbigniew Jaworowski, ... 278

With CO_2 being highly soluble in water 279

Dr. Jaworowski compared the CO_2 concentration in ice 280

Dr. Jaworowski notes that the hokey-stick phenomenon 281

In this sense the Primer fields even affect politics 282

The Great Global Warming that pulled us out of the Little Ice Age 283

The CO_2 subject is far bigger than a mere academic concern 284

The mandated mass-burning of food for biofuels production 285

The CO2 issue far out of mere academic concerns..286

Getting back to the phenomena of fluid movements that are faster 287

The Sun rotates significantly faster at the equator 288

The Sun's two high-activity bands spaced centered off the equator 289

The evidence illustrates the functioning of the Primer Fields 290

The ecliptic principle is also apparent..291

The resulting shape apparent in the ecliptic shape of galaxies 292

A solar system, that is in all major aspects electrically powered 293

This means that galaxies are not stable entities 294

For our galaxy, the Milky Way Galaxy...295

The two long intergalactic resonance cycles 296

The solar system and the Earth are not fundamentally isolated entities 297

This also means that the Sun is not its own master 298

The phase shift in our world, onto an inactive Sun 299

Part 8- The dawn of humanity, civilization, and God..300

Only during Ice Age environments...301

The development of the human species didn't even begin..302

The critical factor that dominates the timing 303

A type of nourishment for mental development...304

Cosmic-ray particles 305

The interaction between cosmic-ray flux and human development..306

Since it became possible with Carbon-14 measurements 307

Great progressive developments in civilization 308

The Maunder Minimum 309

In Palaeozoic history 310

Increases in cosmic-ray flux, ...311

The rise and fall of spiritual recognition 312

The warm periods stand out as periods of cultural destruction 313

With the higher-resolution Carbon-14 measurements 314

Beyond the insanity of the Nazi holocaust 315

The Sleep of Reason Produces Monsters 316

The future of humanity is not inherently bound to the cosmic default 317

Humanity has in its path become a highly developed species 318

This means that no huge feat is required for humanity to free itself 319

The biofuels holocaust can be stopped in a similar manner 320

Only the Ice Age Challenge cannot be so easily met 321

The critical breakthrough can be made in a short time nevertheless 322

The truth is dead 323

But will she rise again? 324

This is the truth 325

Part 9 - The Giza pyramids and Stone Henge 326

Grasping the fire 327

Some of the wonders stand as giant structures 328

In asking these questions 329

The builders of the pyramids 330

Built 12,800 years ago, and not as tombs 331

The complete absence of inscriptions 332

Built around astrophysical phenomena 333

Images of the Primer Fields in action 334

The type of image that we see today in the Red Square nebula 335

The famous three stars 336

Three stars determined the pyramids 337

Three pyramids as a single project 338

A silent testimony 339

Another significant story 340

Giza is located near Cairo 341

The Sahara was a lush region 342

Northern boundary of the ice age safe zone 343

The safe line 344

A perfect zone free of hurricanes 345

Floating agriculture is small stuff 346

The future of Canada, Russia, and Europe 347

The Stone Henge project 348

A ring of 56 chalk pits 349

A lab experiment, published in 2003 350

A ring of 56 distinct plasma filaments 351

A very-high power plasma stream 352

It is surprising to note 353

We no longer see these patterns 354

Part 10 - Politics versus science 355

Humanity urgently requires the ice age challenge 356

A spiritual task, more than a political task 357

We wield weapons out of weakness 358

We face a choice 359

Humanity still lives in the Roman age 360

We simply starve them to death 361

The ice age challenge as a new paradigm 362

Only we lose by our failing 363

To rekindle that flame in the heart 364

Not even the sky will be a limit 365

Will we dare to step up onto the wings and fly? 366

The greatest danger that we face 367

Brainwashed by the choruses of the professional scoundrels 368

People who have sold their soul for a song 369

Brainwashing to keep the internal-fusion sun theory alive 370

Children to become depopulated? 371

The Sun is an electrically powered star 372

Behind the scene of denial 373

The intelligence of humanity 374

Break through the fog of political games 375

Alert scientists and truth-seekers 376

Scientific progress the hallmark of humanity 377

Unlimited energy resources 378

Potentials stand before us right now 379

Maybe this is what the Universe recognized and put to our credit 380

To create a society without empire 381

The city on a hill 382

The greatest economic and scientific development 383

Warm climates after the Little Ice Age 384

The war against empire has not yet been won 385

Society's self-directed spiritual development 386

To meet the human need 387

Challenge of the dimming Sun 388

Something to celebrate 389

Harvest is Seedtime 390

Discovering Love 391

The melody of nature - what a song! 392

Lu Mountain 393

Listen to the song 394

Flight Without Limits 395

Brighter than the Sun 396

More Illustrated Science Books by Rolf A. F. Witzsche 397

Part 1 - Introduction to the dimmer Sun

Part 1

The Primer Fields

Introduction to the dimmer Sun

By popular request, I have updated and reproduced my video, 'Ice Age of the Dimmer Sun in 30 Years,' with a new voice, a new form, with leading-edge discoveries in science added, and with a new musical theme. And to make the content more accessible, I have divided the video's ten parts into a series of ten individual videos.

The science-focus of the series

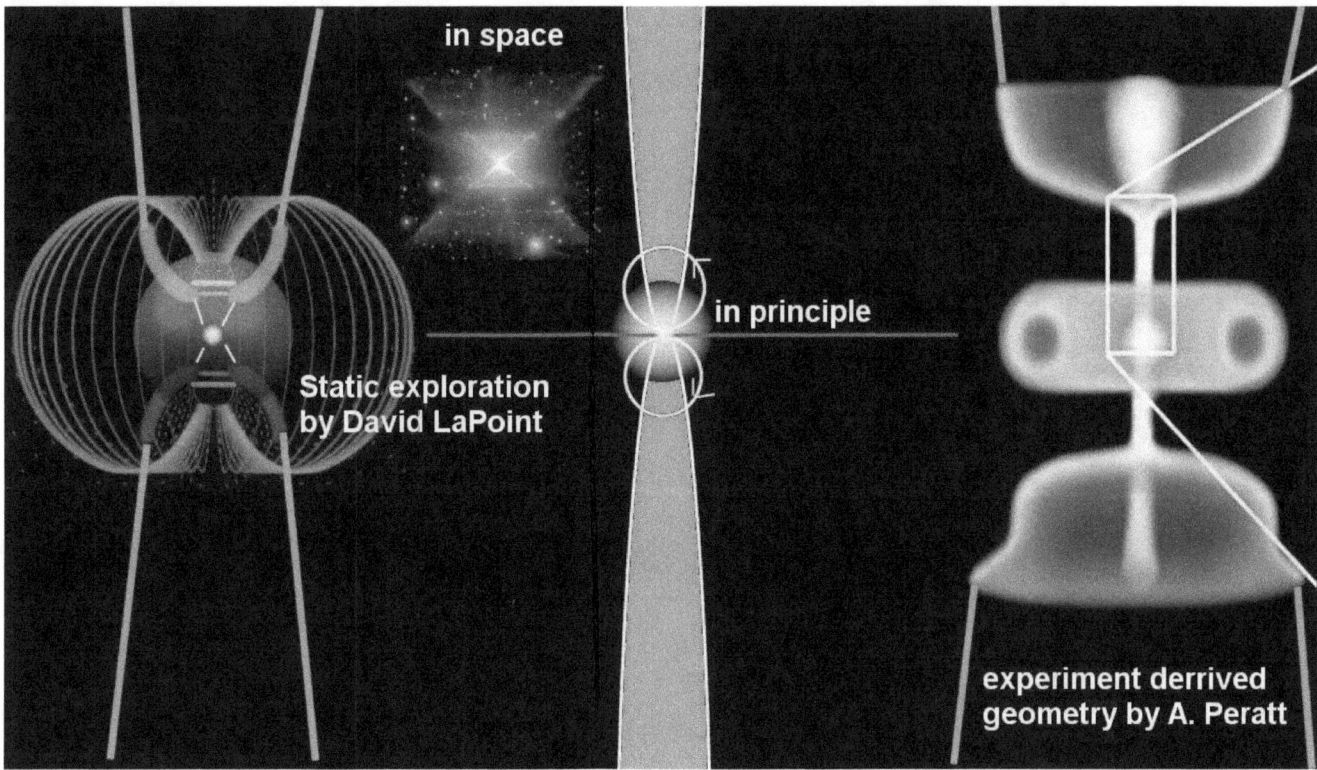

The science-focus of the series, of course remains the same as in the original video, being centered on the Primer Fields that are understood as one of the most fundamental aspects of the science of the universe, the galaxies, the solar system, the Sun, and the effects of the Sun on the Earth, ranging from the color of the sunlight to solar cosmic-ray flux affecting our climate that we enjoy and also fear.

So what are these all-pervading Primer Fields then, that seem to be affecting everything?

The Primer Fields

David LaPoint - The Primer Fields

In December 2012 the plasma physicist David LaPoint published a series of videos about a phenomenon that he called: 'The Primer Fields.' With his discovery he takes aim at long-held mistaken theories about the physical universe, including black holes, dark matter, and dark energy, which are not required to interpret the observed evidence of the universe.

The most basic electromagnetic fields

In his videos David LaPoint illustrates the existence, and the operating principles, of the most basic electromagnetic fields that shape the solar system, the galactic system, and ultimately the cosmic environment. He suggests that these basic primal fields, which he discusses, are fundamental to the order and operation of the universe, and are even reflected in the transmission of light and the structure of matter itself, all the way down to its smallest subatomic forms.

The Primer Fields and their effects on our Sun

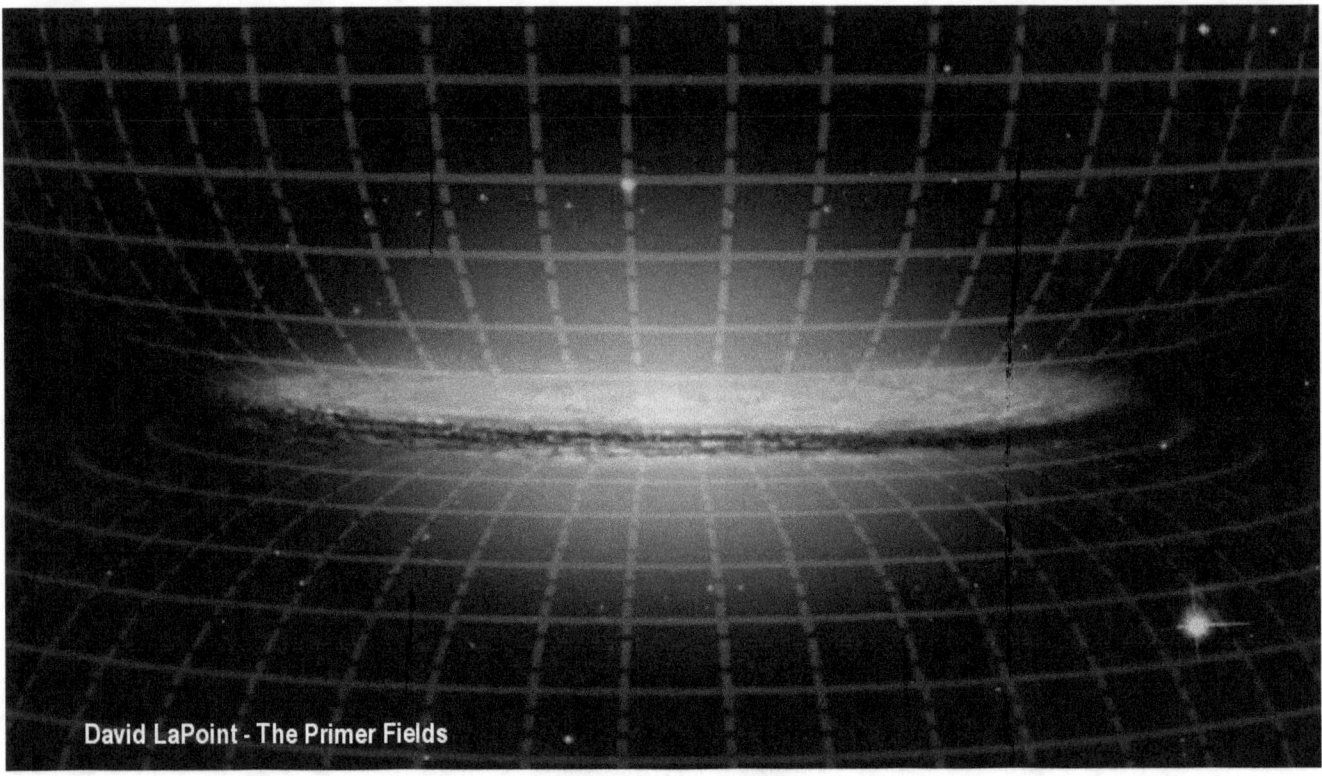

David LaPoint - The Primer Fields

On the gigantic scale the electromagnetic fields that he recognized to exist, and has explored in the laboratory, shape entire galaxies. They are so fundamental to almost everything, that he calls them "The Primer Fields." In fact, without the Primer Fields and their effects on our Sun and planets, even the galaxy itself, might not have been created, so that nothing would actually exist without them.

Intensely focused and compressed plasma

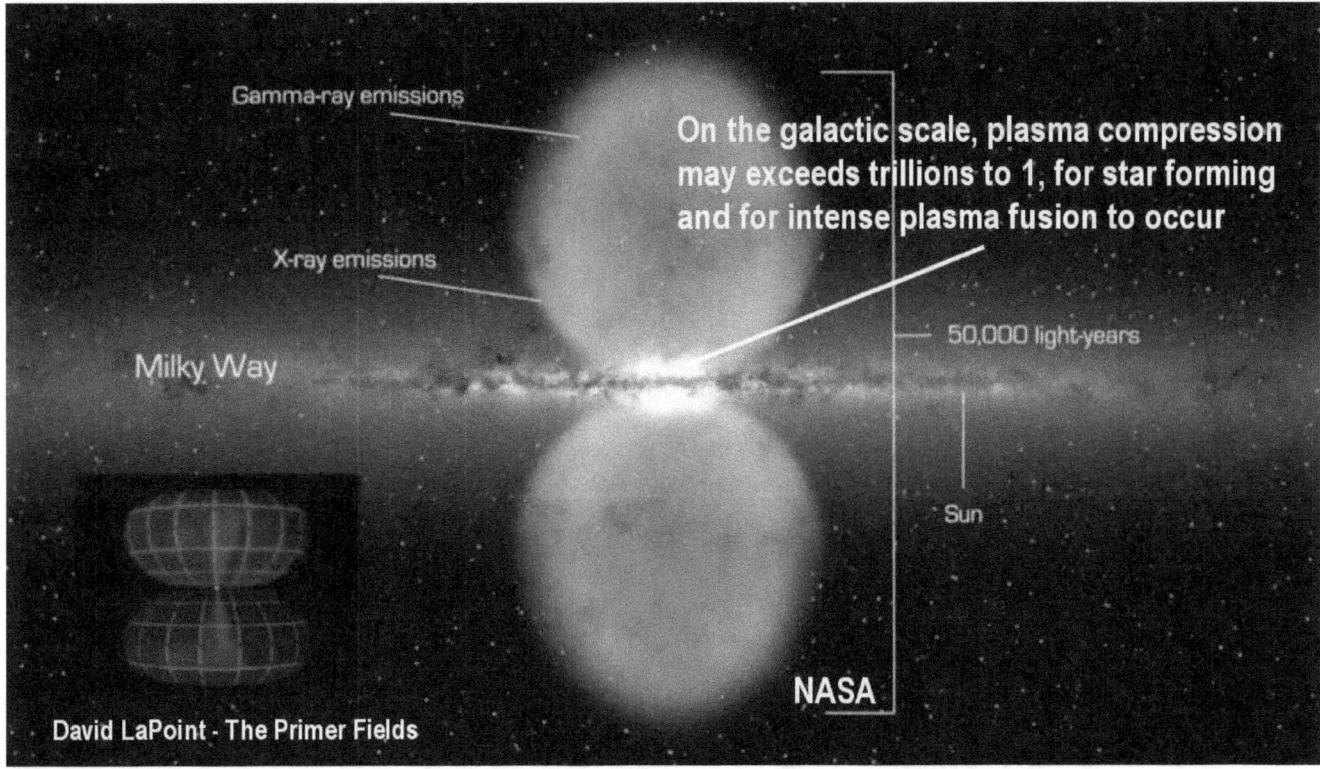

Star formation in a galaxy typically occurs in intensely focused and compressed plasma that the Primer Fields facilitate, to the point that nuclear fusion occurs, by which atoms, stars, planets, and entire galaxies are formed.

Our solar system as a single functional unit

David LaPoint - The Primer Fields

The term, Primer Fields, is also justified, because on the ' smaller' stage of our solar system, the operation of the Primer Fields, though they are correspondingly smaller in size than on the galactic scale seen here, is critical for the dynamic operation of our solar system as a single functional unit.

The dynamics of the ice ages

David LaPoint also discovered that the principles of the fields, that he calls the Primer Fields, is such that the galactic plasma streams that flow through the solar system become concentrated to a high degree of density around the Sun, which he replicated in principle in laboratory experiments. The result of his work brings a radically new dimension to the perception of the Sun and also of the dynamics of the ice ages.

The Sun as an electrically powered star

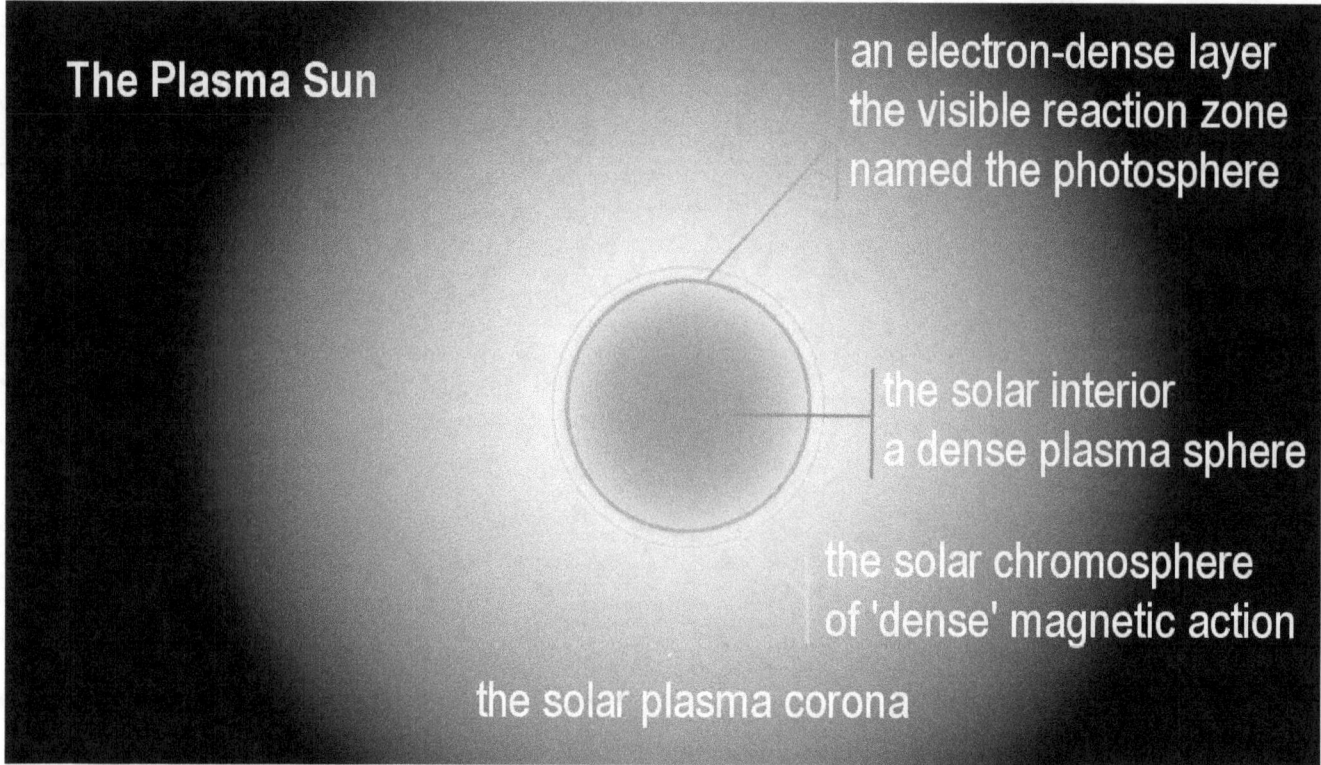

The perception of the Sun as an electrically powered star has existed for a long time already. A large body of evidence exists to support the electric Sun theory. But it had a weak point. For the theory to work, a high concentration of electric plasma is required to exist around the Sun, which put the theory in doubt. As an alternative, the theory of the nuclear fusion-powered Sun was created, except the observed evidence doesn't support this theory either.

Sunspots on the surface of the Sun

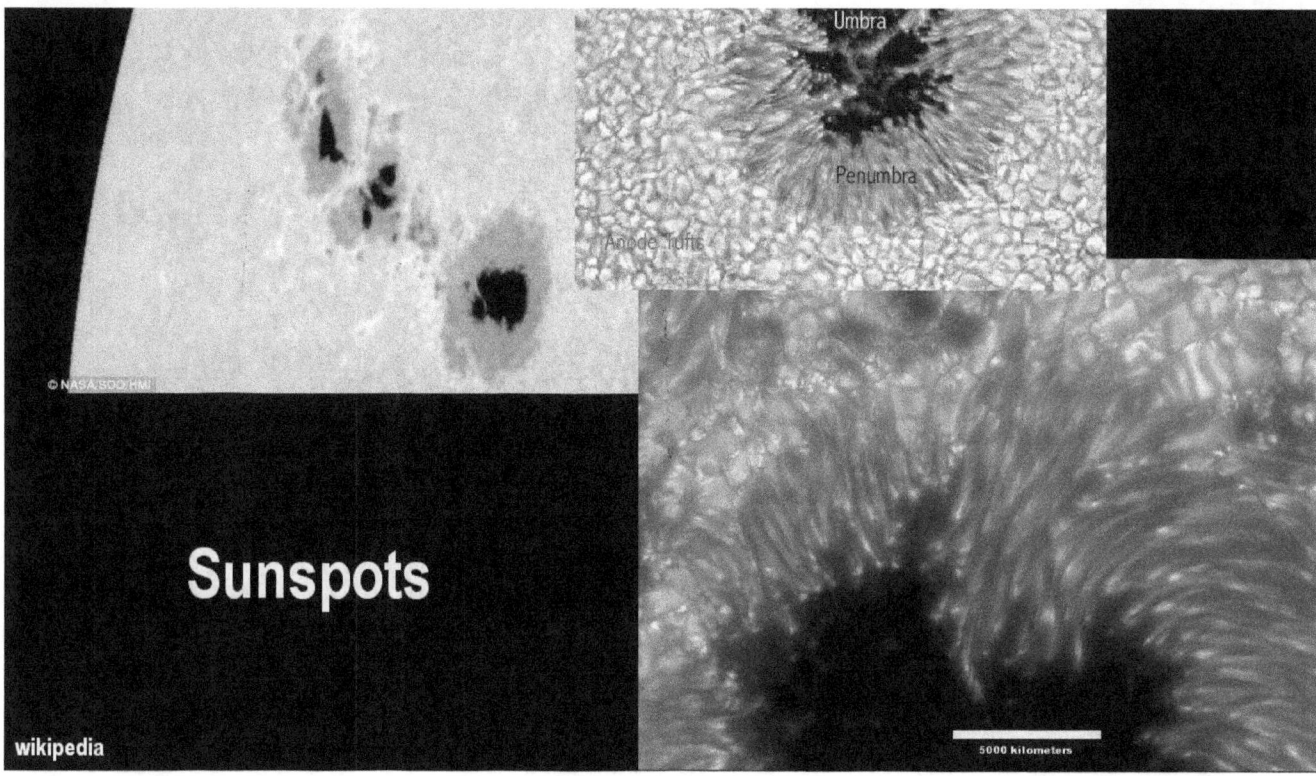

The sunspots on the surface of the Sun show the Sun to be darker below the surface. We should see the opposite if the Sun was internally powered, instead of it being electrically powered at the surface.

The fusion Sun theory

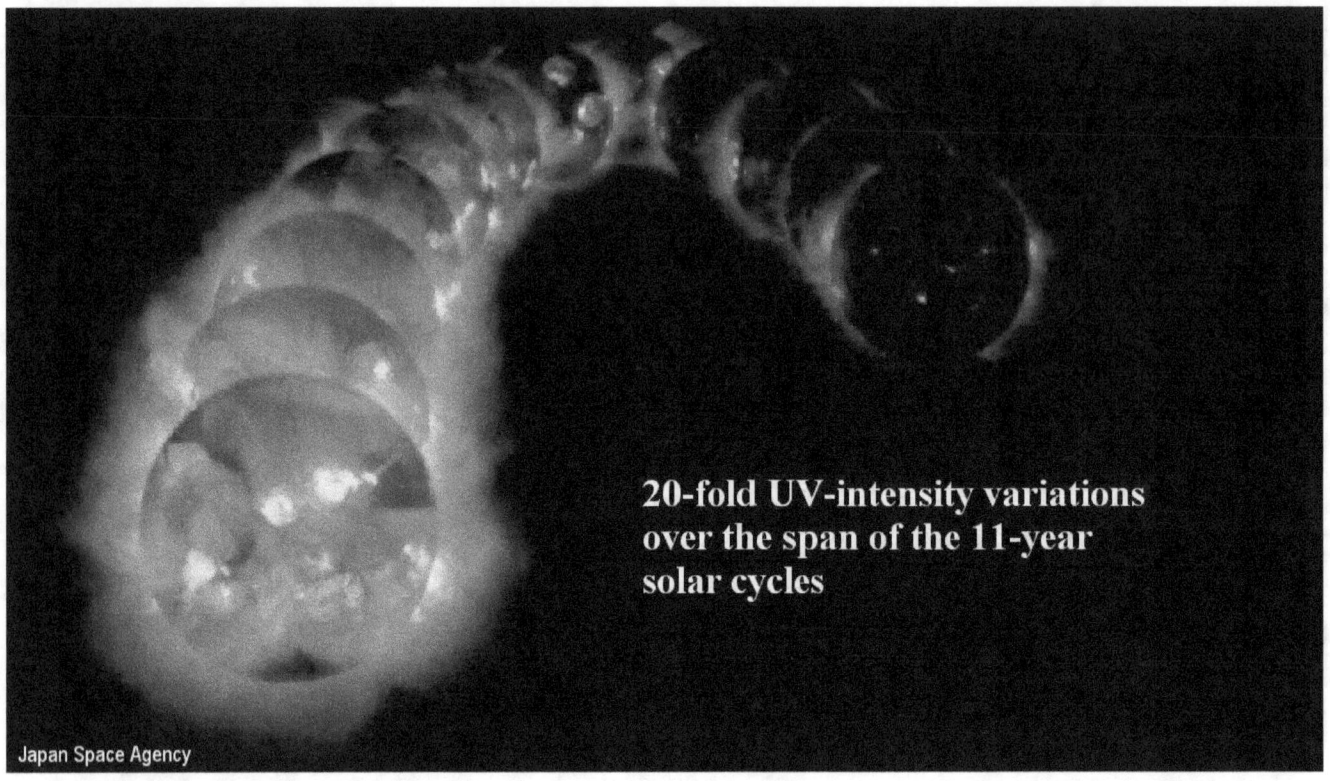

20-fold UV-intensity variations over the span of the 11-year solar cycles

Japan Space Agency

Also the fusion Sun theory cannot readily explain the 11-year solar activity cycles in which the brightness of the Sun varies by a factor of 20 when observed in the high ultraviolet band. The fusion Sun theory has it that the photon travel time from the Sun's deemed fusion center to the surface, is deemed to be in the range between 10,000 and 170,000 years, while the assumed energy-transfer time itself, from the center of the sun to the surface, is deemed to be on the order of 30 million years. This hardly supports the 11-year solar cycles, in which the Sun's magnetic field changes direction at every cycle.

A minimal threshold of conditions

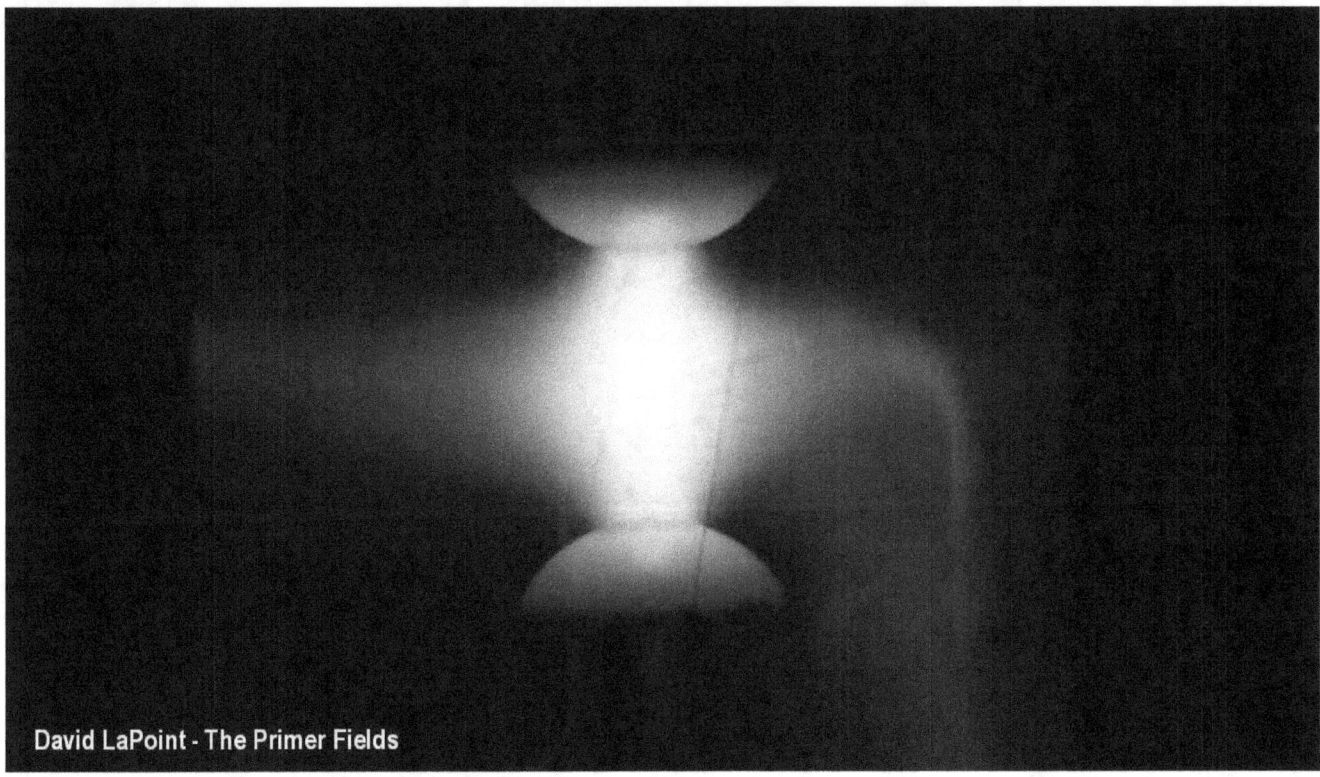

David LaPoint - The Primer Fields

The contribution that David LaPoint brings to the table in this debate weighs enormously in support of the electric-Sun theory by illustrating how the high density plasma field is created around the Sun that the theory depends on. The Primer Fields theory takes away the barrier against the electric-Sun theory, which thereby becomes totally plausible, together with a number of related theories that thereby likewise become plausible. However, the Primer Fields theory takes us one critical step beyond merely supporting the electric-Sun theory. It renders the functioning of the electric Sun subject to a minimal threshold of conditions that must be met for the Sun to be powered, which, when it is not met, causes the Sun to become inactive, dim, and 'cold.'

If the threshold is not reached

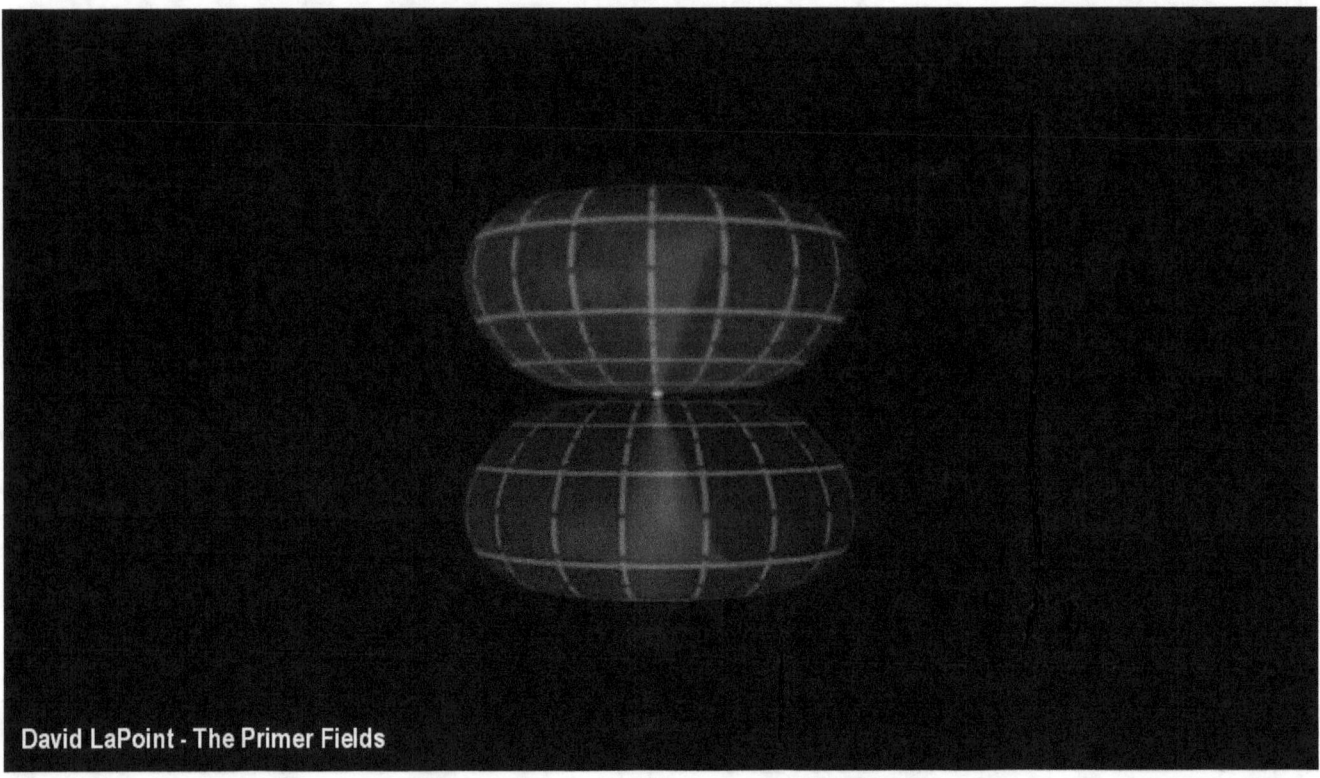
David LaPoint - The Primer Fields

While David LaPoint may not have intended to support the electric-sun theory, his work illustrates that the process that creates a dense plasma sphere around the Sun happens, because the Primer Fields have that effect. It is known in plasma physics that these fields themselves depend for their existence on a minimal density in the plasma that flows through the system. If the threshold is not reached, the fields do not form. If the fields do not form, or cannot be maintained, the Sun does not have the condition established for it to be powered.

The Sun simply turns off

The Sun simply turns off to an inactive state in which it glows dimly by its stored up energy and some nuclear decay processes within it.

When our brilliant sun is turned off

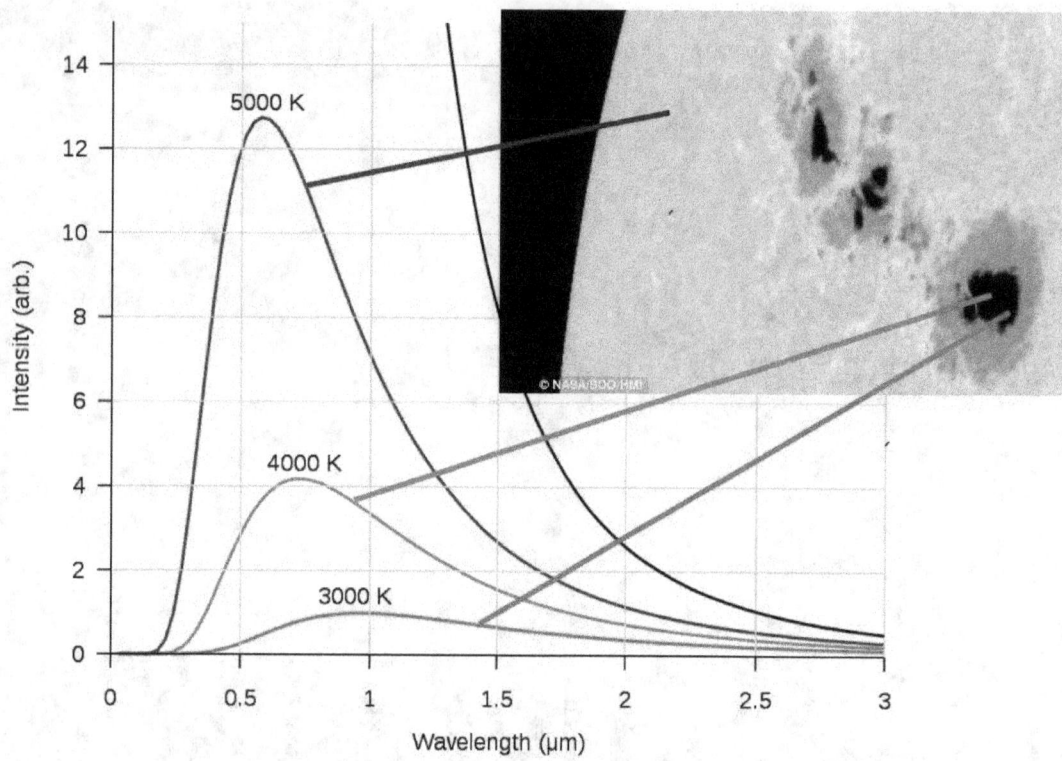

The required threshold, therefore, becomes an important factor in considering the dynamics of the ice ages. Just imagine the consequences on the Earth when our brilliant sun is suddenly turned off as the Primer Fields collapse, and looses more than two-thirds of its energy output in possibly a single day.

The Milankovitch theory

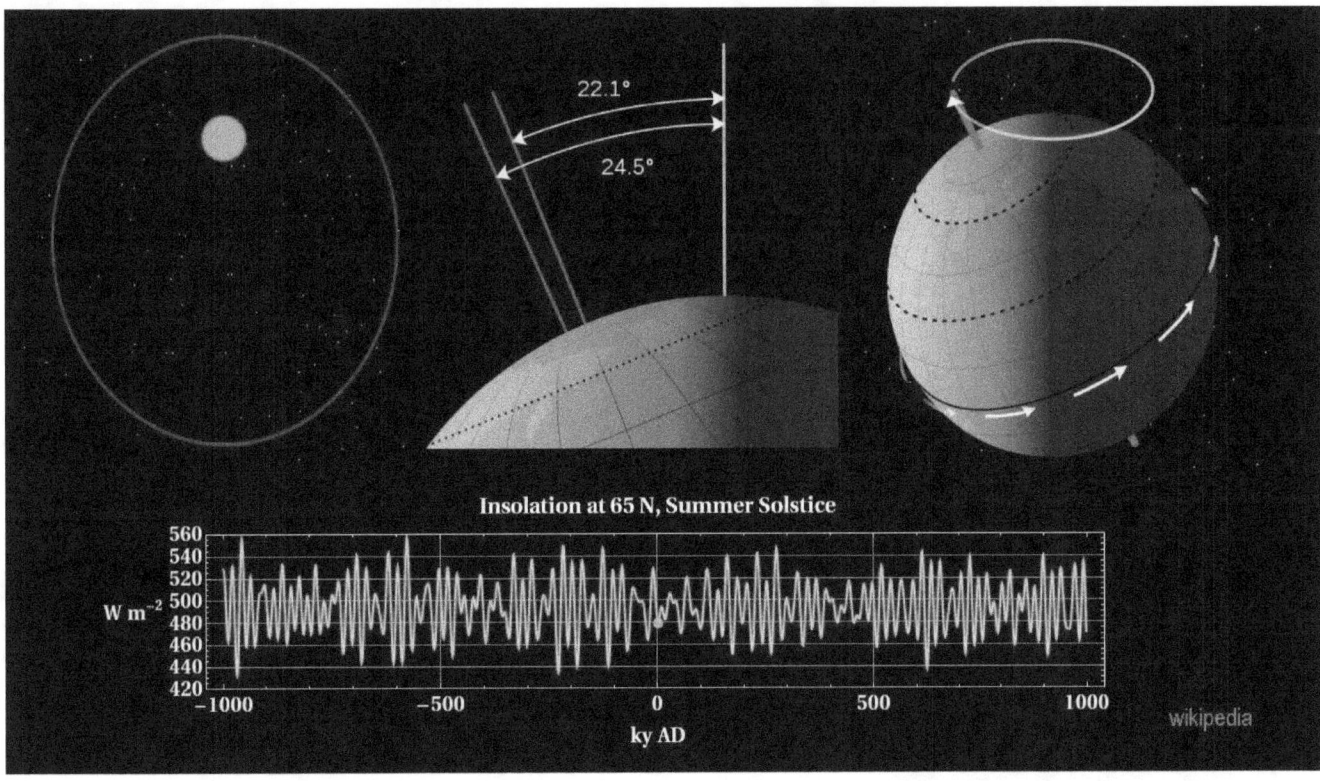

The conventional theories of the ice age dynamics all regard the ice age glaciation as the result of a gradual cooling of the Earth. One theory sees the cooling being caused by cyclical variations of the orbit of the Earth around the Sun, with the Sun remaining an invariable constant by this theory. In the fusion-Sun theory, the Sun is deemed to be an invariable constant, so that the ice ages are caused by anything else except the Sun.

The resulting orbital-variation theory, known as the Milankovitch theory, however has so many holes in its fabric that it is generally no longer seriously considered as a cause for the ice ages.

The large temperature fluctuations

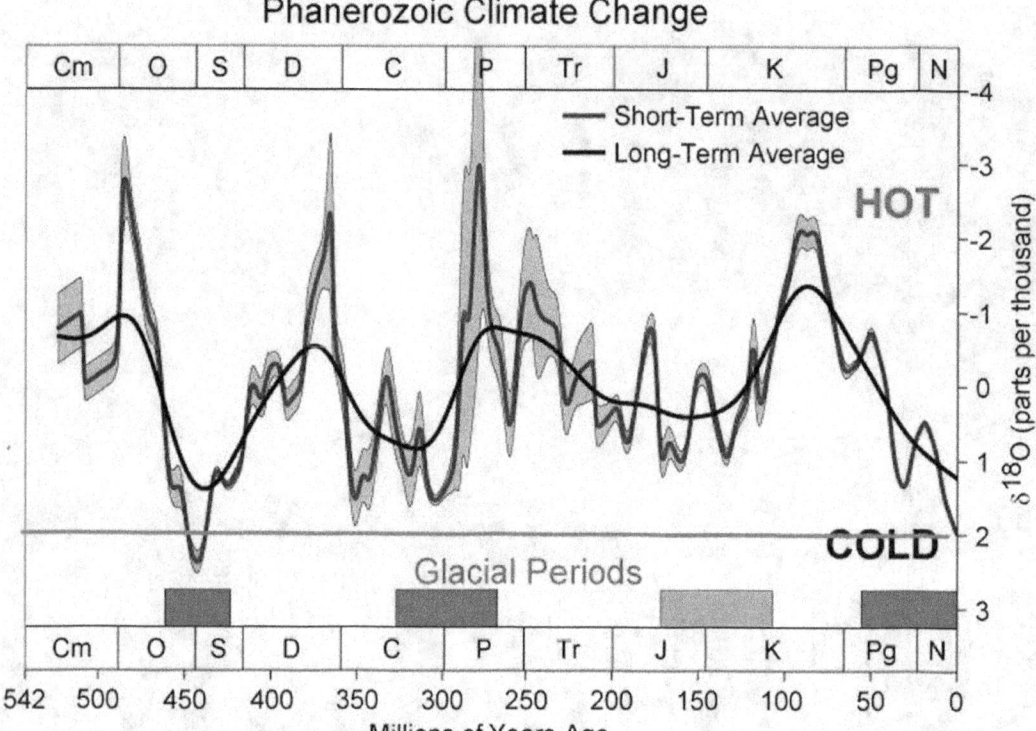

The standard electric-sun theory doesn't have this problem. Plenty of evidence exists that the power output of the Sun does vary with the density of the electric input. The large temperature fluctuations that are known to have occurred over the last half billion years are evidently the result of cyclical electric input fluctuations. The Primer Fields theory does not change that, but it does open the gate to the recognition that the entire process of the powered Sun can collapse when a minimal power-input threshold is not met, as we have seen happening during the ice ages.

Primer Field theory brings a huge difference

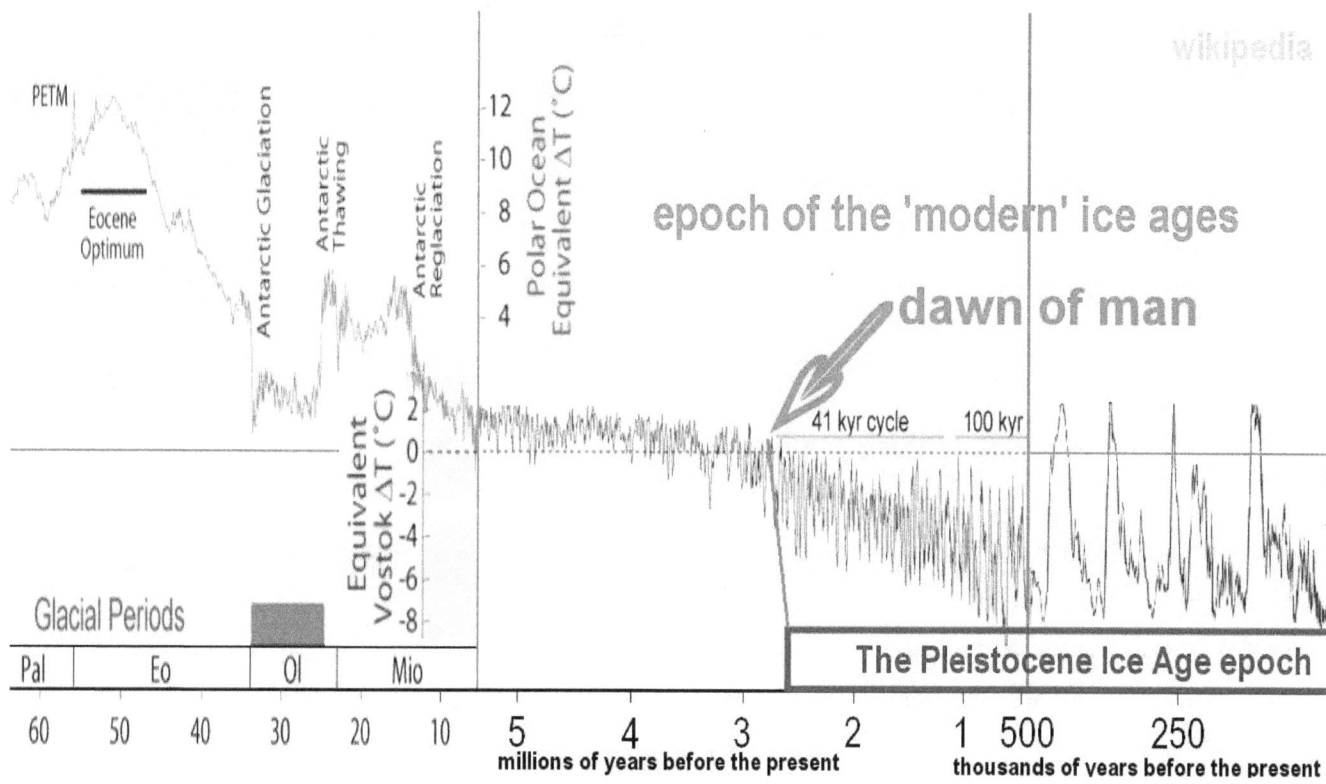

The Primer Field theory brings a huge difference to the table. In the past it was believed that ice age conditions develop slowly and gradually over extended periods, so that humanity can adjust itself to the changing conditions, even in cases when the Earth becomes five times colder than the Little Ice Age had been, over the span of 50 years. The Primer Fields theory jolts us out of this dream, or at least it should.

The Primer Fields theory presents us a great blessing

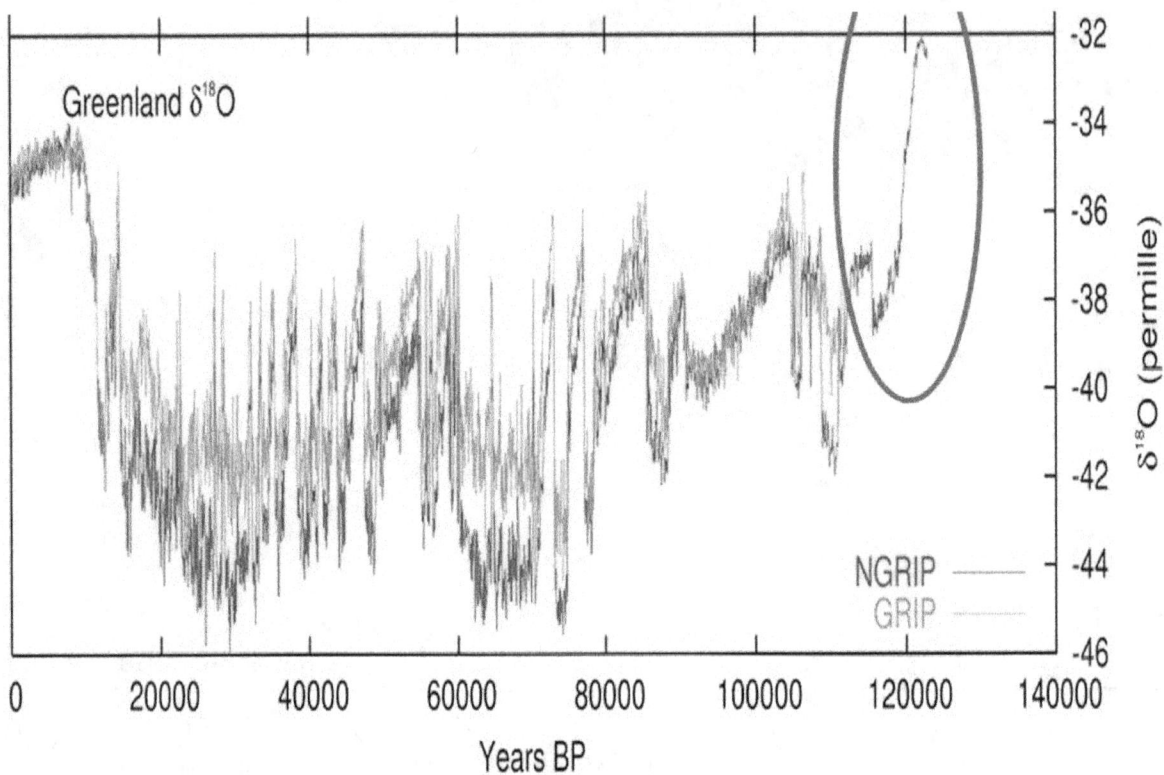

If the fields are not maintained sufficiently that they collapse and vanish, extremely radical changes occur in as short a time as possibly a single day. The Primer Fields theory takes away the general notion that we have a long time to prepare for the coming Ice Age and don't need to respond yet. It tells us that the opposite is true; that we need to get prepared as fast as possible, which almost no one is yet willing to even consider. In this context the Primer Fields theory presents us a great blessing in disguise.

To avoid the catastrophe of nuclear war

It forces us, if we are willing to respond, to re-develop our humanity as fast as possible, which we urgently need in the present to avoid the catastrophe of nuclear war that no one will likely survive, for which the preparations are evermore intensely pursued in numerous ways.

An incentive to get our act together

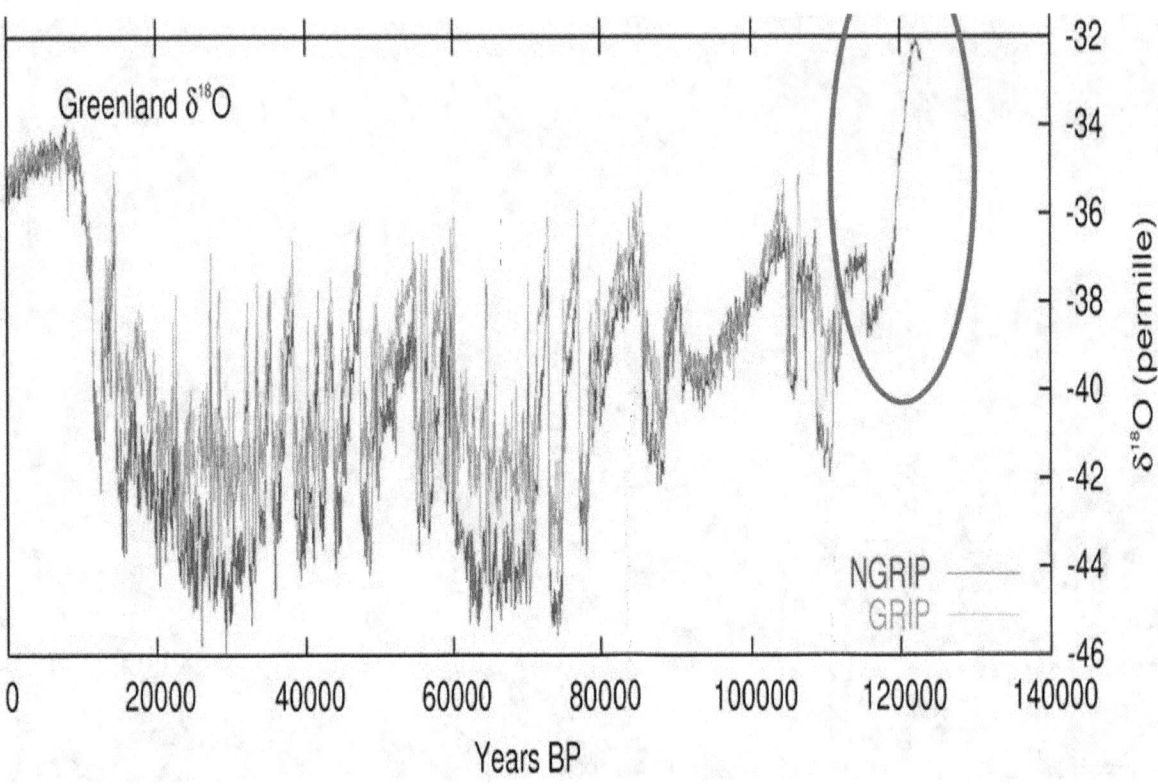

In this regard the potentially near, rapid ice age transition that the Primer Fields theory brings into view, offers us an incentive to get our act together, which would be the greatest salvation that we could experience on every front where civilization is presently fast collapsing.

A single basic principle

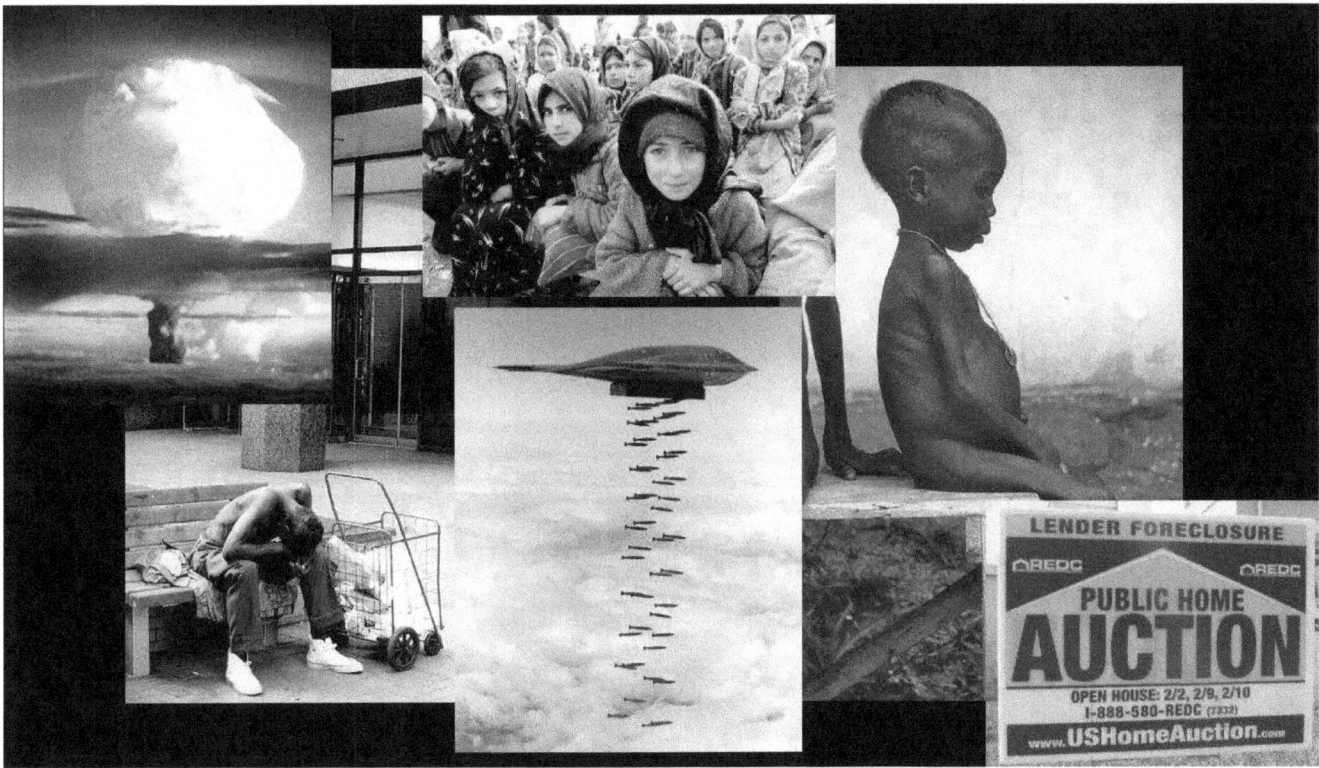

The Primer Fields theory illustrates a single basic principle that applies to all complex systems, including civilization. In economics, when the creative and productive power of a nation diminishes below a minimal threshold, the entire economic system disintegrates. It becomes inactive. The nations collapse. The world becomes cold. The people die.

Close to the minimal threshold

Castle Bravo - the first U.S. test of a dry fuel thermonuclear hydrogen bomb - March 1, 1954 at Bikini Atoll, Marshall Islands

This happens to cultures too when the recognized value of our humanity drops below a minimal threshold. Nuclear war happens then, when the humanist culture disintegrates.

We are close to the minimal threshold on a number of these vital fronts, especially in economics.

When the assumed value of money becomes uncertain

The High Five in Dallas, Texas, USA - Wikipedia

For example, when the assumed value of money drops below the point where the value of money becomes uncertain in the markets, it becomes meaningless at this point. At this point the financial markets collapse, banks close, stores close, gas stations close, transportation collapses, the food supply system stops, the entire platform of physical infrastructures becomes inactive.

The Glass Steagall law

1933 - the Glass-Steagall law to protect that value of the U.S. currency - now repealed

The world's money-value system can unravel in a single day. No one knows how close we are to this day. The Glass Steagall law that once protected the value of money in the USA, has been repealed in Congress in 1999 by a bunch of traitors who were bribed to do so with a $350 million slush fund. To date all efforts have failed to restore even this minimal protection of the value of currency that the repealed law had afforded for the 66 years in which it was active. Against the background of the now fast collapsing value of money the entire economic system can collapse in a single day, just as the presently powered Sun can go inactive in a day when the fields collapse that prime its operational environment.

During the interglacial warm period

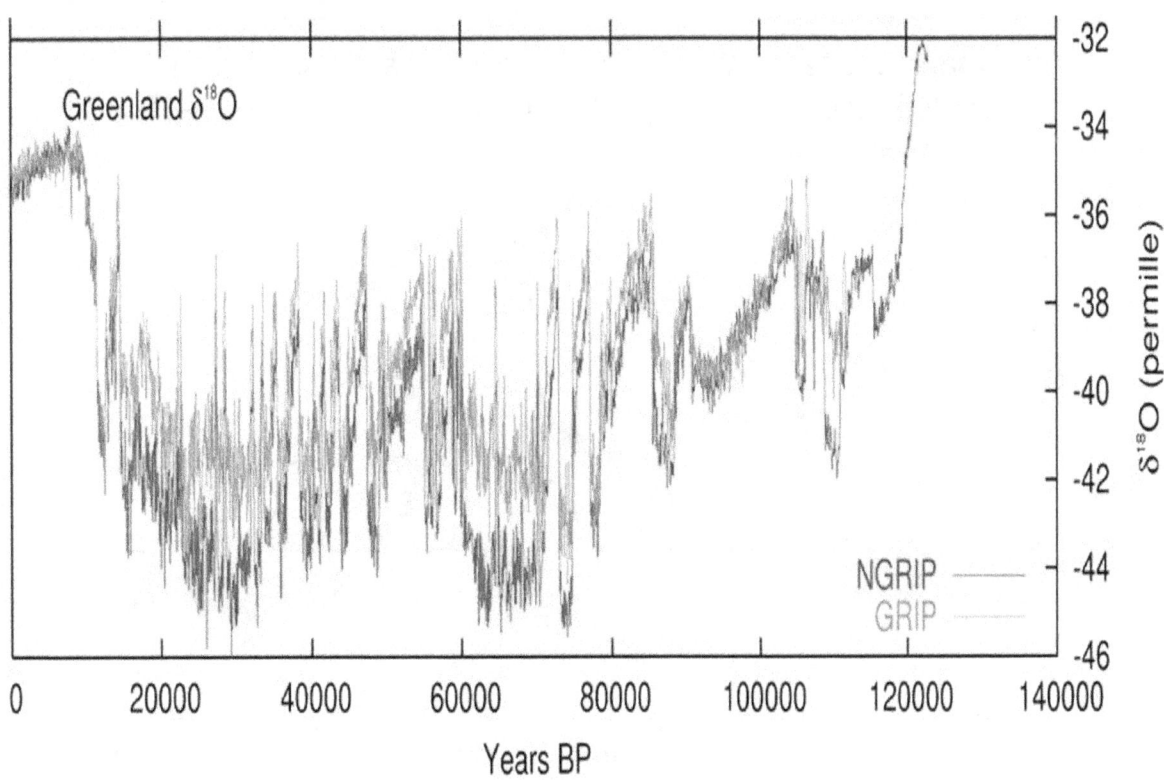

David LaPoint's work illustrates that the electric Sun, which varies only slightly during the interglacial warm period that we are presently enjoying, will with great certainty go completely inactive when the plasma density of the electric streams feeding into the solar system falls below the threshold level where the primer fields collapse.

In astrophysics, that's the point where the ice age transition begins. That's what happened 120,000 years ago.

We will cross the cut-off threshold

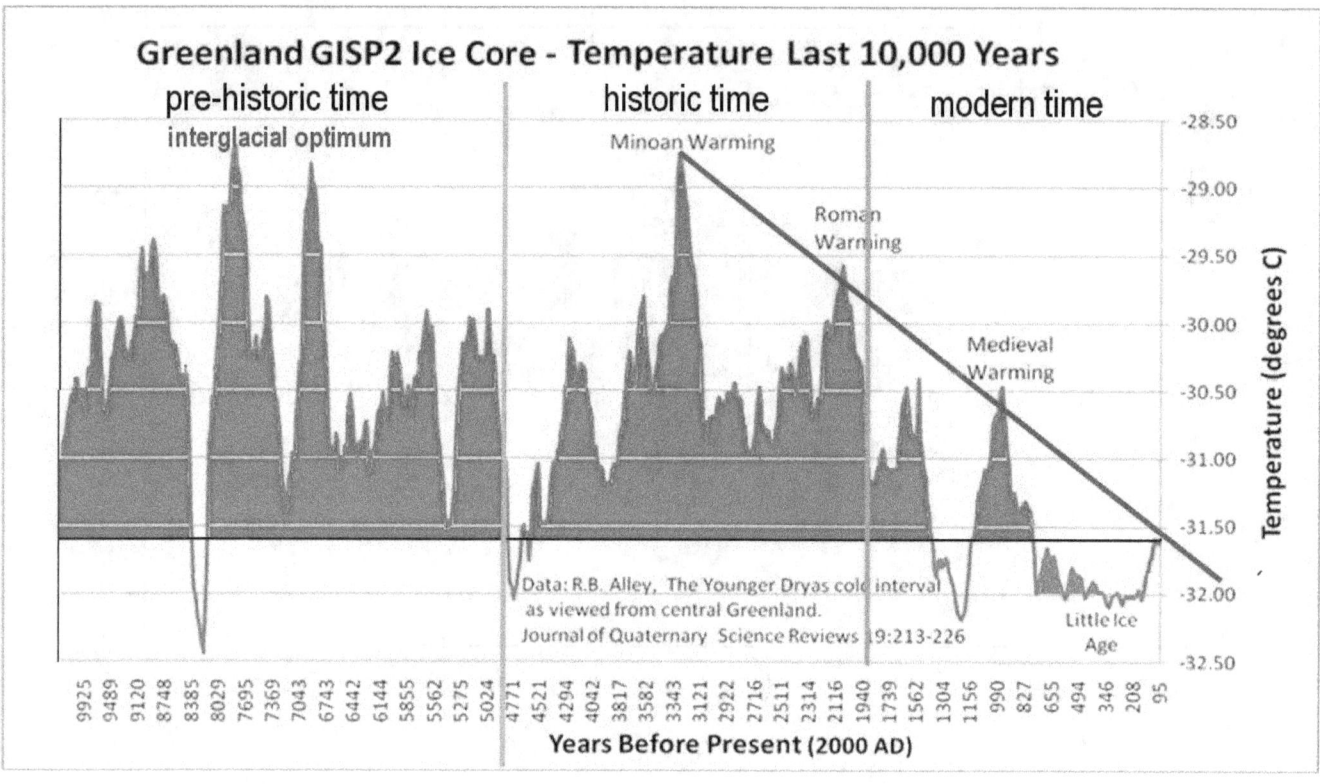

The current trend suggests that we will cross the cut-off threshold in the near future. An ice age begins when the minimal conditions no longer exist for the system to function, that powers the Sun.

The Sun reverts to its default state then - its inactive state. It turns dim in a single short step, and the climate on earth that looses two thirds of its energy input, turns cold. At this point rapid deep cooling begins all across the Earth.

Before this point is upon us, all the preparations for the dramatic changes in the Earth's environment will have to be completed. To fail is not an option if we want to continue to live. For this reason it is not an option either that we raise the value of our humanity far above the present level and protect it on all fronts. And this needs to begin now.

The Sun became inactive 120,000 years ago

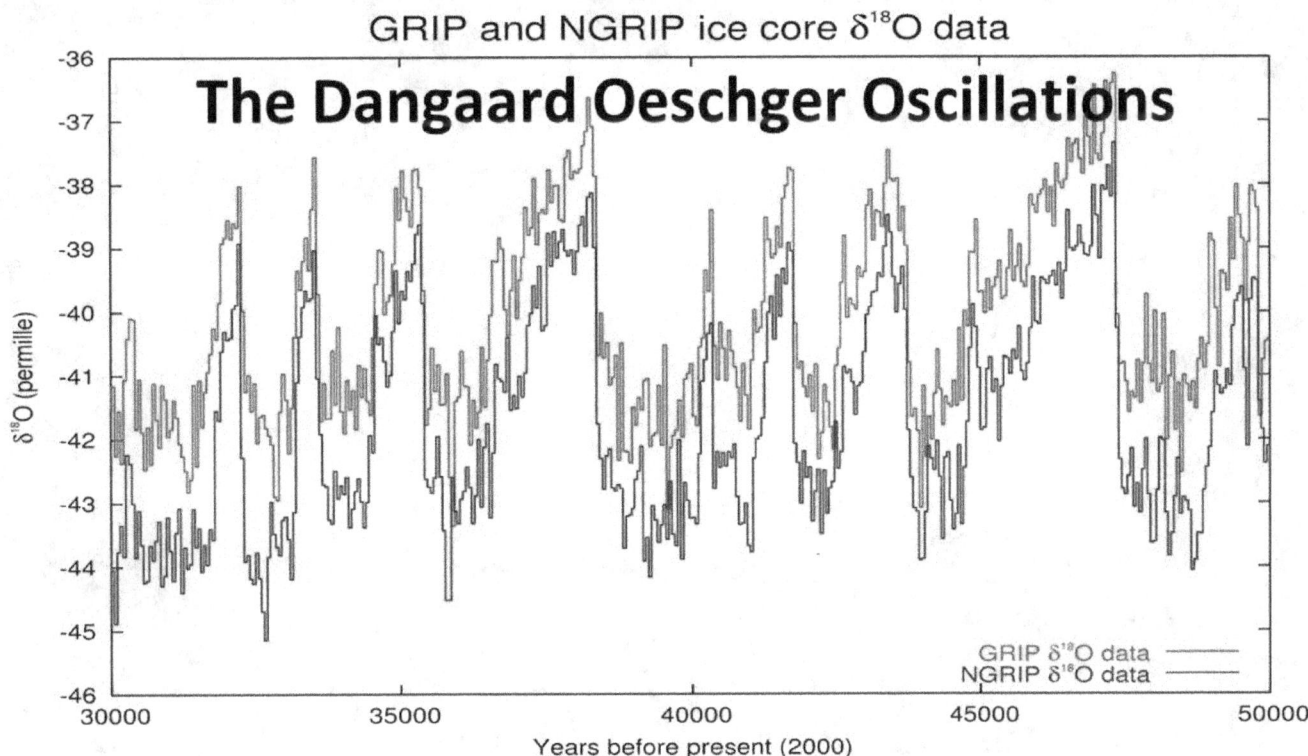

The analysis of ice cores drilled from the ice sheets of Greenland indicates that the Sun became inactive 120,000 years ago. That's when the last Ice Age started. The ice core records also tell us that the Sun became activated for short periods, every 1470 years, throughout the entire glaciation period. The periodic re-warming may have prevented the Earth from freezing up completely.

Ice sheets more than 10,000 feet deep

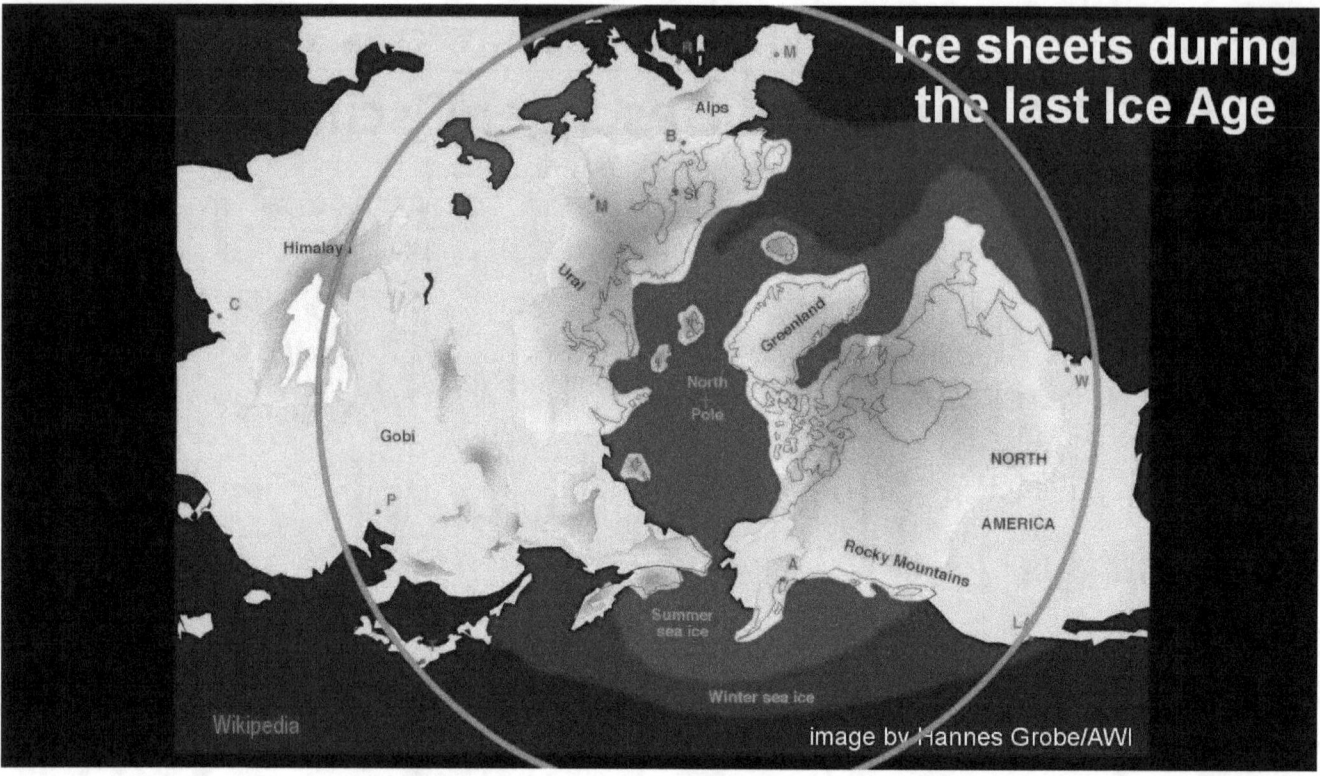

The inactive periods of the Sun were nevertheless long enough to create the huge cooling on earth that lays up ice sheets across the northern hemisphere more than 10,000 feet deep, which are known to have existed.

The Sun inactive in 30 to 50 years

It is here, that the theoretical becomes important for us all in a critical practical manner. The present trends suggest that the turn-off transition that makes the Sun inactive, might be upon us in 30 to 50 years time. When this happens the entire world will have to live with a dimmer Sun on a colder earth. Of course this poses some challenges for maintaining our food supply.

The chlorophyll in plants

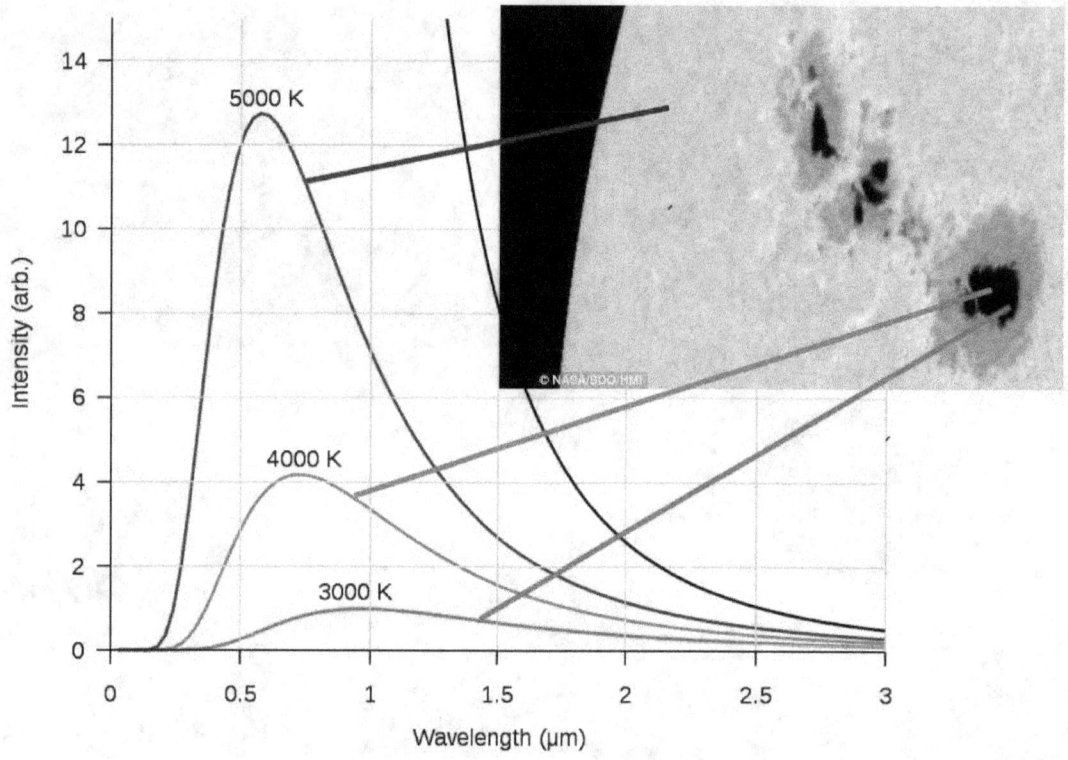

The chlorophyll in plants will have to function at reduced energy levels and with a shifted radiation spectrum.

Here a great challenge arises for humanity as a whole, to work together to create infrastructures that are needed to maintain our food supply in the coming dimmer and colder world.

Yes, we will do this

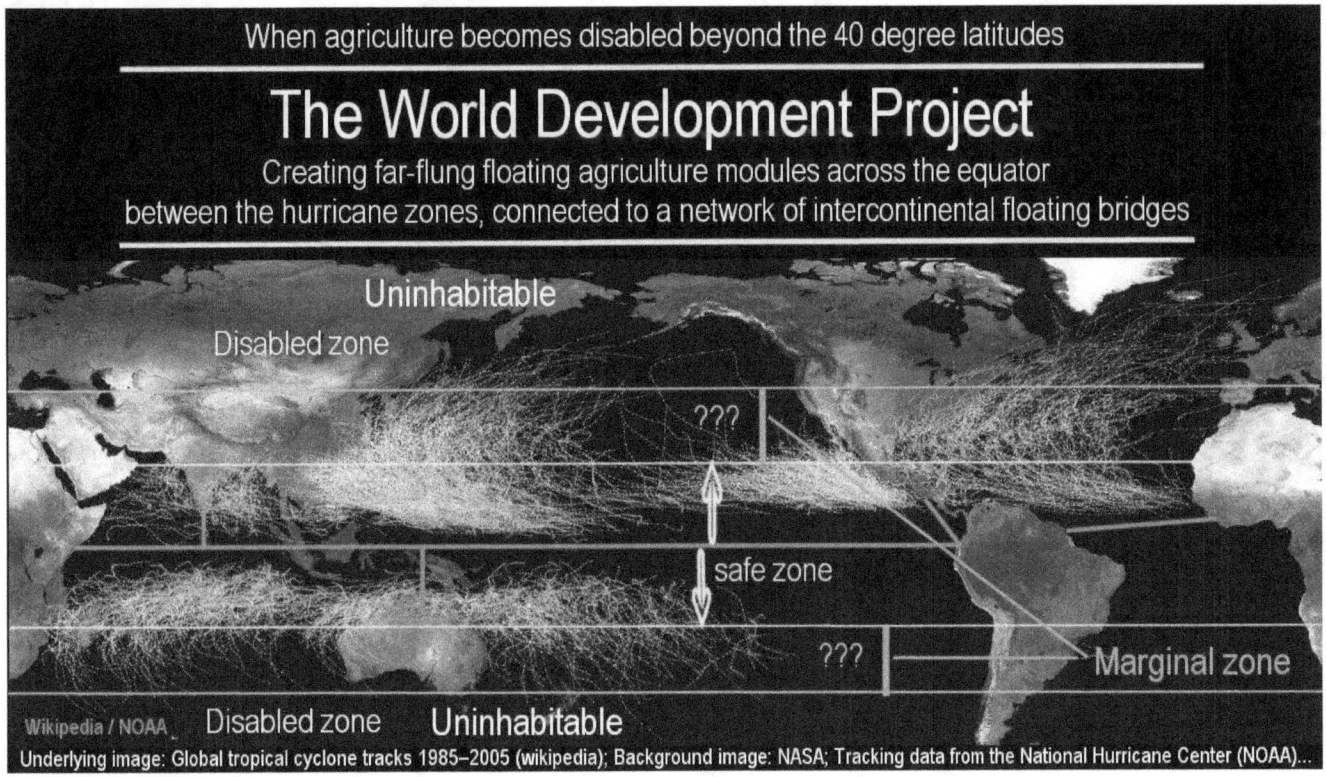

While it is technologically and economically possible to meet the challenge to protect our world, our civilization, and our future, the question remains to be answered whether we will do what is needed to maintain our existence?

The answer must be: "Yes, we will do this. Yes, we will do what is critically necessary for us to survive."

Truthful science supplies the potential

We will even do it for far lesser reasons, such as for becoming free of the terrible strangulation of our world that constricted scientific perception has brought upon it. Truthful science supplies the potential to open the door for the liberation of humanity that false science inflicts massively, and globally.

But what is love?

Our physical civilization is largely built on science, first and foremost, just as our social civilization is built on love.

But what is love? We cannot weigh it, measure it, quantify it, but we know that it is so immensely substantial that civilization would collapse without it.

Much of the same can be said of science.

Without science standing at the heart of it

Without science standing at the heart of it, civilization with all its numerous freedoms would not exist.

False science is destructive

Bertrand Russell

He foresaw the doom of empire resulting from human development by cultural, scientific, and technological progress, enabling larger populations to exist. This he fought to prevent by all means possible.

Lamenting the world population increase, enabled by scientific and technological progress, one of the masters of empire, Bertrand Russell, wrote:

"But bad times, you may say, are exceptional, and can be dealt with by exceptional methods....

War, so far, has had no very great effect on this increase, which continued through each of the world wars. ... War ... has hitherto been disappointing in this respect ... but perhaps bacteriological war may prove more effective. If a Black Death could spread throughout the world once in every generation, survivors could procreate freely without making the world too full. ... The state of affairs might be somewhat unpleasant, but what of it? Really high-minded people are indifferent to happiness, especially other people's."

Bertrand Russell, The Impact of Science on Society (New York: Simon and Schuster, 1953), pp. 102-104

That false science is destructive is well-recognized by those who would destroy civilization with it. False science has a destructive effect.

Society raising itself out of the trap of false science

Rotor of a steam turbine by Siemens

wikipedia CC BY-SA 3.0 Photo w. permission of Siemens Germany by Christian Kuhna

With society raising itself out of the trap of false science, a path opens before it to universal liberty, by scientific and technological progress, towards the brightest cultural renaissance of all times. That's not being too optimistic. The potential exists. The principle is applied. It reflects the power of our humanity as human beings.

False science to forcefully strangle civilization

Poster of the Climate Conference. Licensed under Fair use via Wikipedia

Our present world is full of examples of false science being massively applied to forcefully strangle civilization, and to a large degree voluntarily.

One false science concept is that the amazing recovery of the climate of the Earth from the devastating Little Ice Age in the 1600s, until 1998, was a manmade phenomenon caused by human economy and energy applied for living that must be eliminated. But false science remains false, regardless of how deeply it has penetrated society and controls its responses.

When honest science displaces constricted science

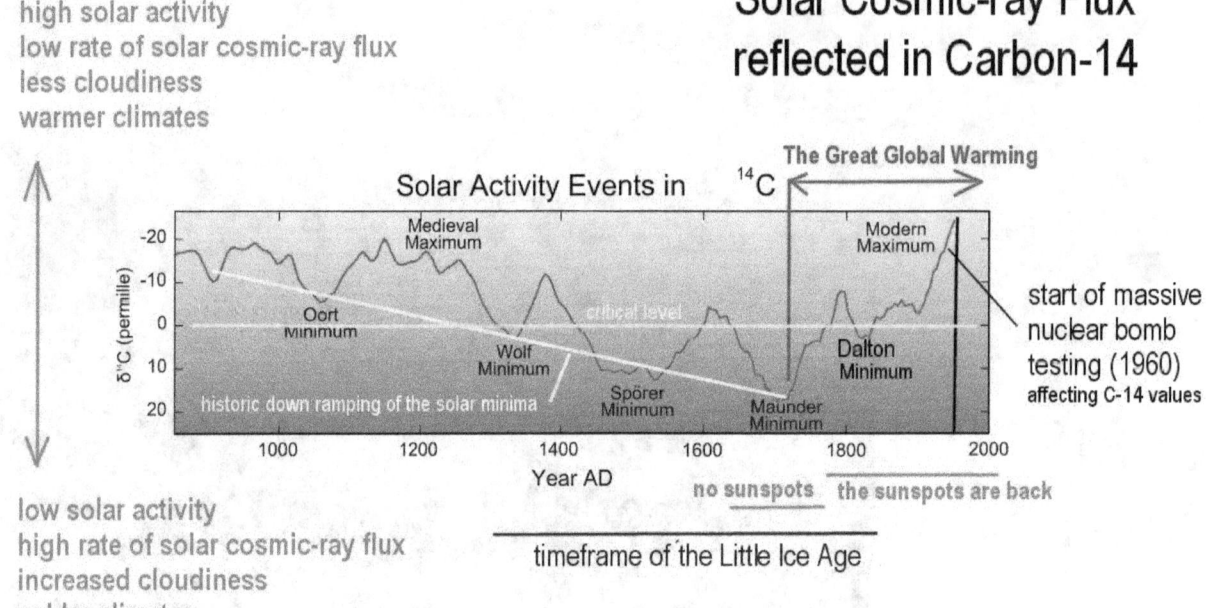

"Carbon14 with activity labels" by Leland McInnes at the English language Wikipedia. Licensed under CC BY-SA 3.0 via Commons

Unrestricted science demonstrates with recognizable certainty that the re-warming of the Earth from its ice house condition in the 1600s was not a human achievement, was not a consequence of mankind's industrial development and its burning of carbon fuels, but was absolutely, and measurably, and scientifically demonstrably, a solar-forced climate effect that has staged the Great Global Warming that enabled modern society and its efficient agriculture to become possible. When honest science displaces constricted science, a great liberation becomes potentially possible.

The solar-forced Great Global Warming

Work of the United States Federal Government (2007)

Melting pots on the Greenland Ice Sheets

In the carbon-14 documented example of the solar-forced Great Global Warming, which comes to light as a matter of basic fact in un-constricted science, the liberating effect of truth lifts the burden from humanity of being a climate villain on the Earth that false science has brought like an indictment against humanity, for which the nations are presently induced to stop the burning of carbon fuels and thereby commit economic suicide.

The Greenland Ice Sheet experienced some melting

Yes, the Greenland Ice Sheet experienced some melting of its ice masses that were built up in the deep cold of the Little Ice Age, with the Arctic's sea ice likewise experiencing some thinning out, as it is shown here. However, one doesn't need to have a Master of Science degree to recognize that such effects are in line with the measured increase in solar activity that supplies our climate in the first place.

The Primer Fields stand in the background to the unfolding scientific recognition of the natural dynamics of the forever ongoing climate changes on Earth, slow as the recognition process may be.

Humanity never was a climate factor

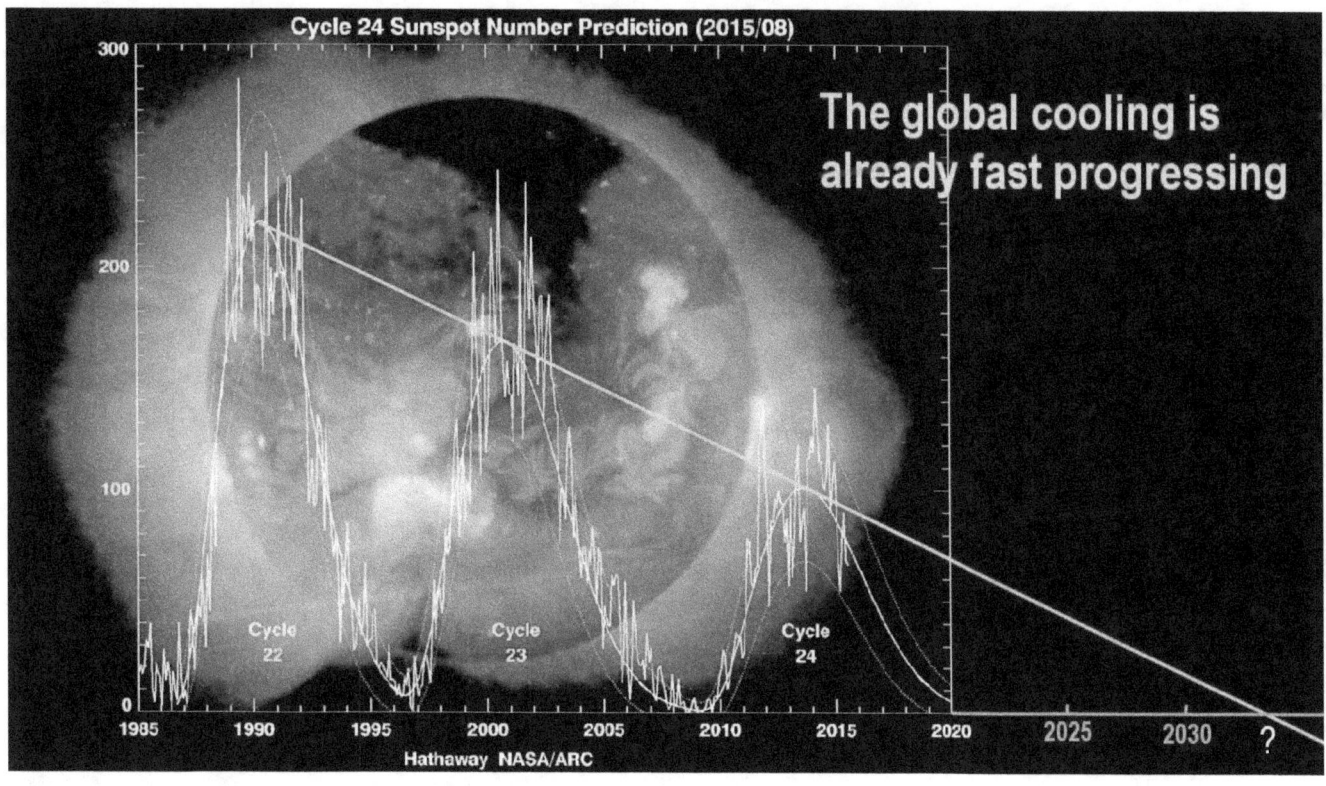

Humanity certainly does not have the capacity, by any means within its grasp, to affect the climate of the Earth. The Earth's climate changes with the effects forced on it by the dynamics of the Sun. Humanity never was a climate factor.

The carbon gases, such as CO_2, that humanity generates by its living, are not a climate factor either - never have been, or ever will be.

The most horrific holocaust

With its simple, wide open scientific recognition, honest science presents itself with the potential to liberate humanity from many constricting dogmas, including that for the most horrific holocaust in the history of the world, that is presently ongoing.

In the name of reducing CO2 emissions by automobiles, vast amounts of agricultural resources are being diverted from the nourishment of people to be burned in the form of biofuels. The burned food resources would nourish upwards to 400 million people, if they were not burned. In a world that has a billion people living in chronic starvation, the food burning unfolds as a holocaust in which 100 million people are forced to die by starvation. The result exceeds the Nazi holocaust 100-fold, and Napoleon's total-war holocaust 1000-fold. Honest science can bring an end to this tragedy that false science has invoked. And here too, the Primer Fields stand in the background to the liberation that simple, truthful, science recognition can inspire.

The electric motivator

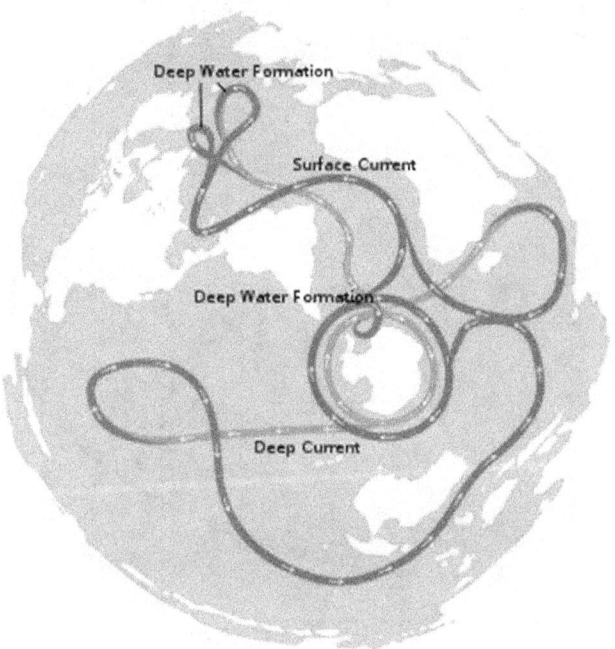

The ocean currents conveyor belt centered on the deep cold waters encircling Antarctica

"Conveyor belt" by Avsa - under CC BY-SA 3.0 via Commons -

The Primer Fields stand in this scene as the electric motivator that in part drives the gigantic global seawater recycling system that transports cold, CO2-rich sea water, from the ice-cold polar oceans that are able to dissolve CO2 more readily, into the tropical oceans where the cold waters warm up and release the high concentration of CO2 that rides along in the recycling streams.

The cold deep currents flow slowly

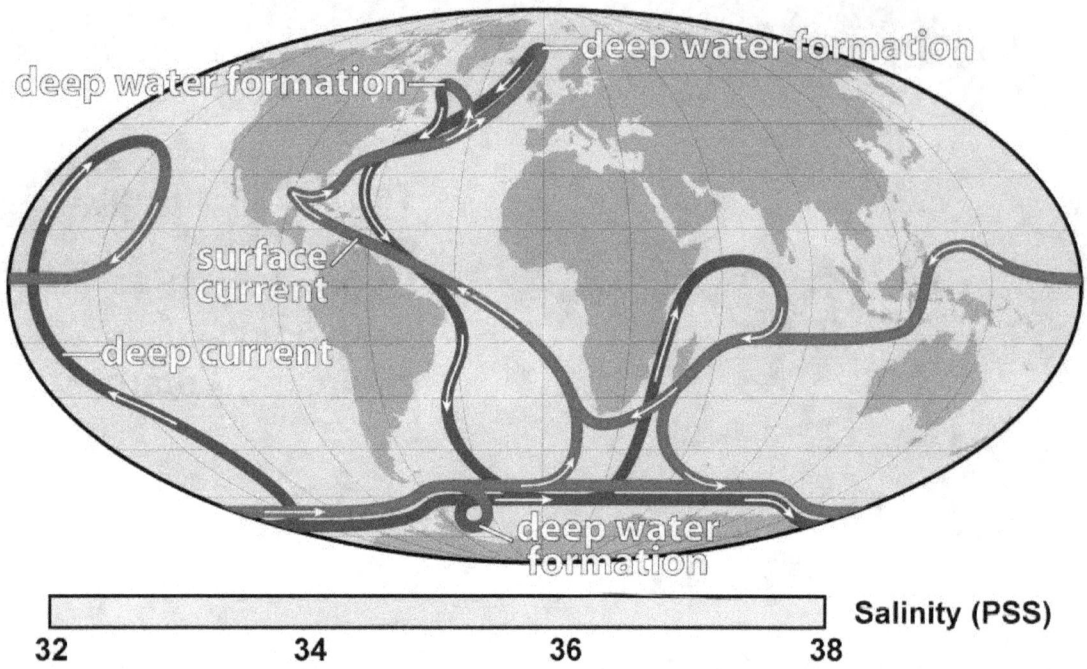

The cold deep currents flow slowly. They flow so slowly that the dissolved CO2 from Antarctica will remain in transit for roughly 350 years untill it becomes released back into the air in the tropical oceans near Africa. For the waters from the Arctic, the recycle time likely exceeds a thousand years.

CO2 is coming back to us

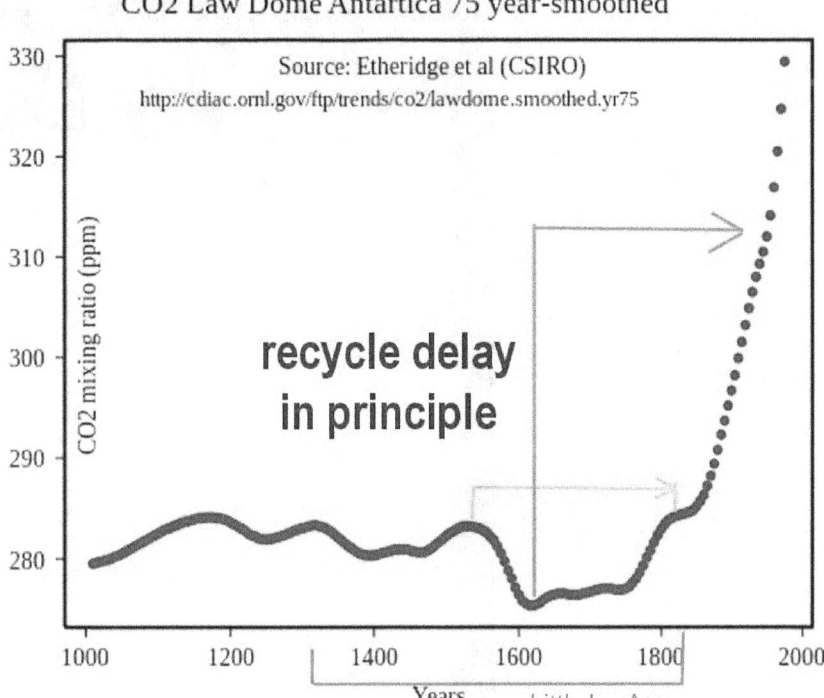

This means that the high rate of CO2 being dissolved into the oceans during the Little Ice Age and earlier cold periods, is coming back to us, into the atmosphere in our time, 350 years later. With this simple scientific recognition in un-constricted science, the noted long recycle delay completely exonerates humanity from the charge of having caused the large increase of atmospheric CO2 that is measured in the modern world.

Here. simple science puts the cause where it belongs, into the court of the natural system, and not into the court of human activity and human energy production. And as I said before, the Primer Fields stand in the background as an element of the electric cosmic process that powers the recycling system.

In the order of a millionth part of it

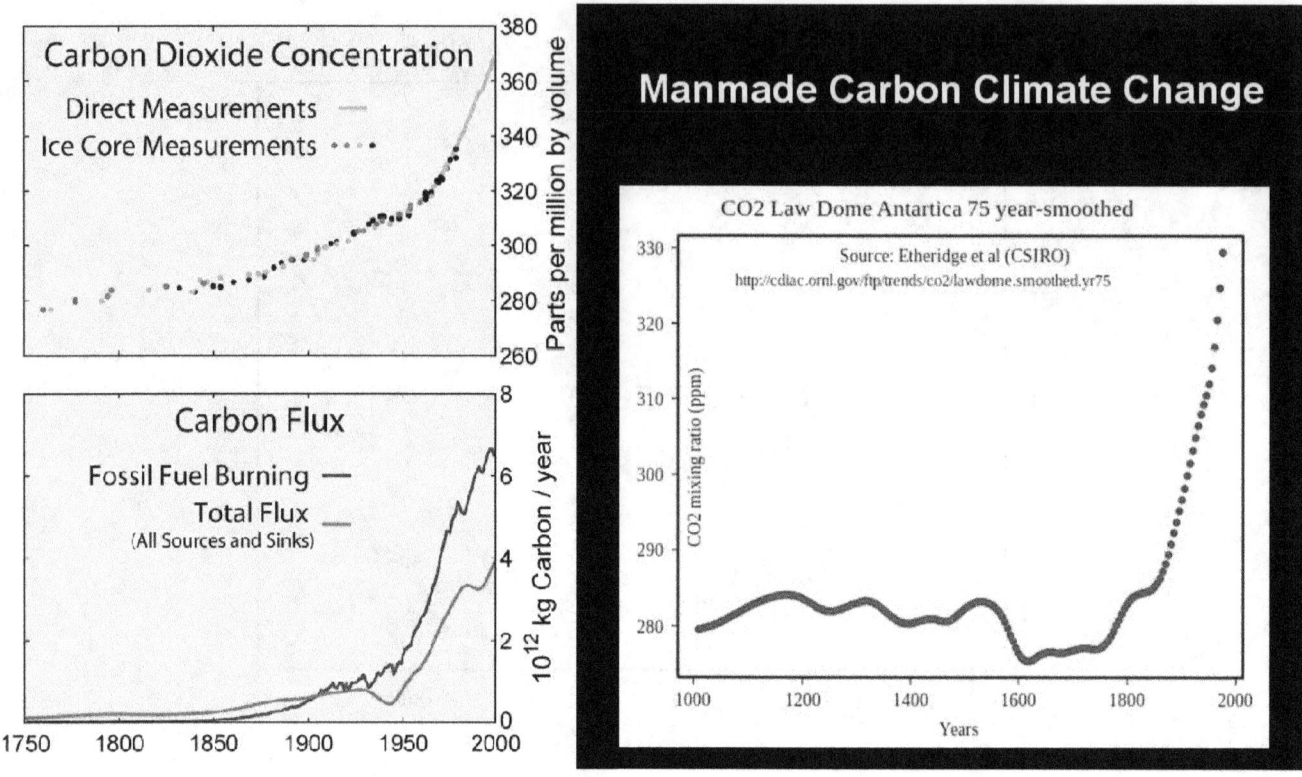

Nor is the CO2 in the air an actual climate factor anyway. It is not physically possible for it to be that. While CO2 is a greenhouse gas that absorbs radiated light energy and reflects it back in a scattered fashion, it is also a scientific fact that its action in the overall context of the greenhouse climate system is extremely minute, in the order of a millionth part of it.

CO2 has no climate-effect

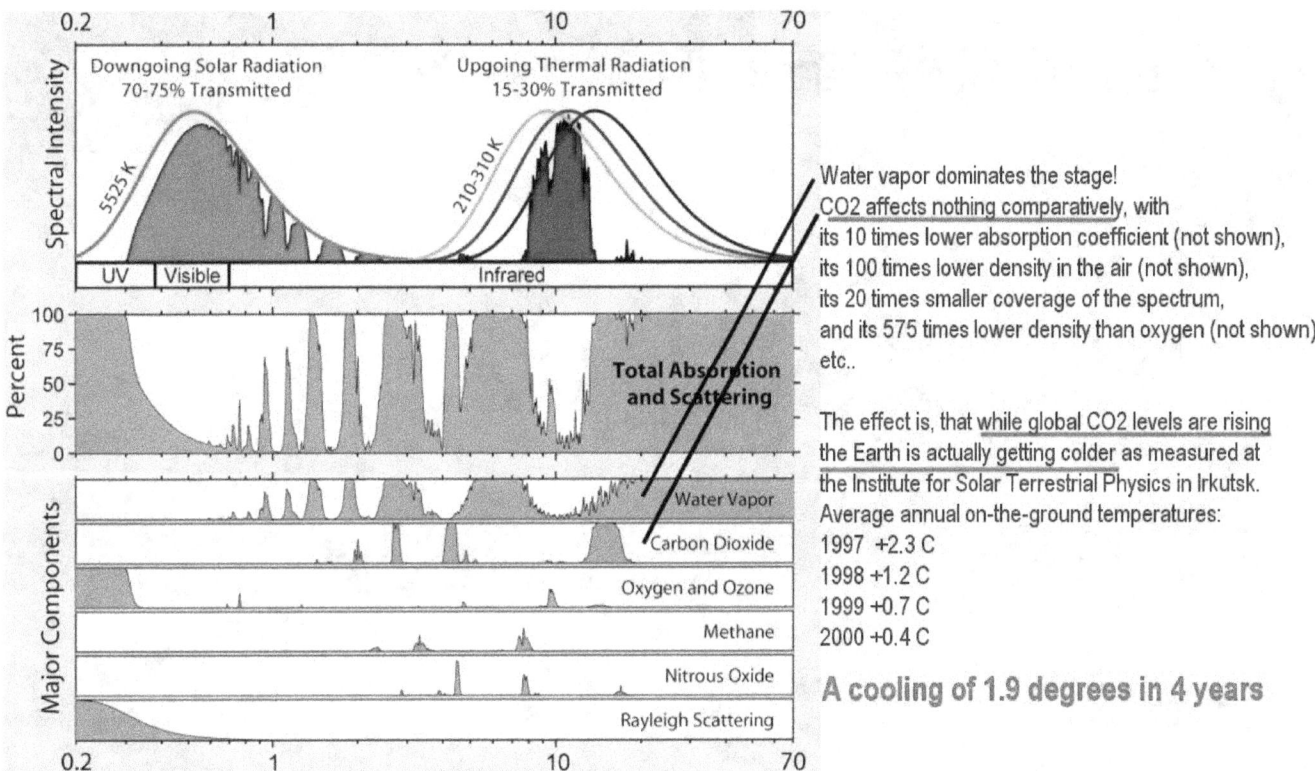

For all practical purposes CO2 has no climate-effect, and certainly not one that is actually measurable. Its absorption coefficient is 10 times weaker than that of water vapor, which is 100 times more densely present in the atmosphere and is 20 times more widely responsive across the spectrum. And all this is vastly overshadowed by oxygen with an absorption coefficient as high as water vapor, but with a 5-fold greater density in the air. And not less significant is the Raleigh Scattering effects of oxygen and nitrogen in the air that adds to the greenhouse radiation. CO2 adds up to nothing in comparison.

Constricted science sometimes claims that CO2 captures outgoing energy radiated back from the Earth. It is not acknowledged however, that outgoing radiation contributes only about 9% of the atmosphere's total heat budget. But there too, CO2 is vastly overshadowed by water vapor.

Most of the outgoing energy

Most of the outgoing energy has actually nothing to do with greenhouse gases at all, not even with water vapor. The largest climate factor is the effect of cloudiness. The white top of clouds directly reflect a portion of the incoming solar energy back into space, which is thereby lost to us. The amount of cloudiness has a huge effect on the climate.

Latent heat that is released

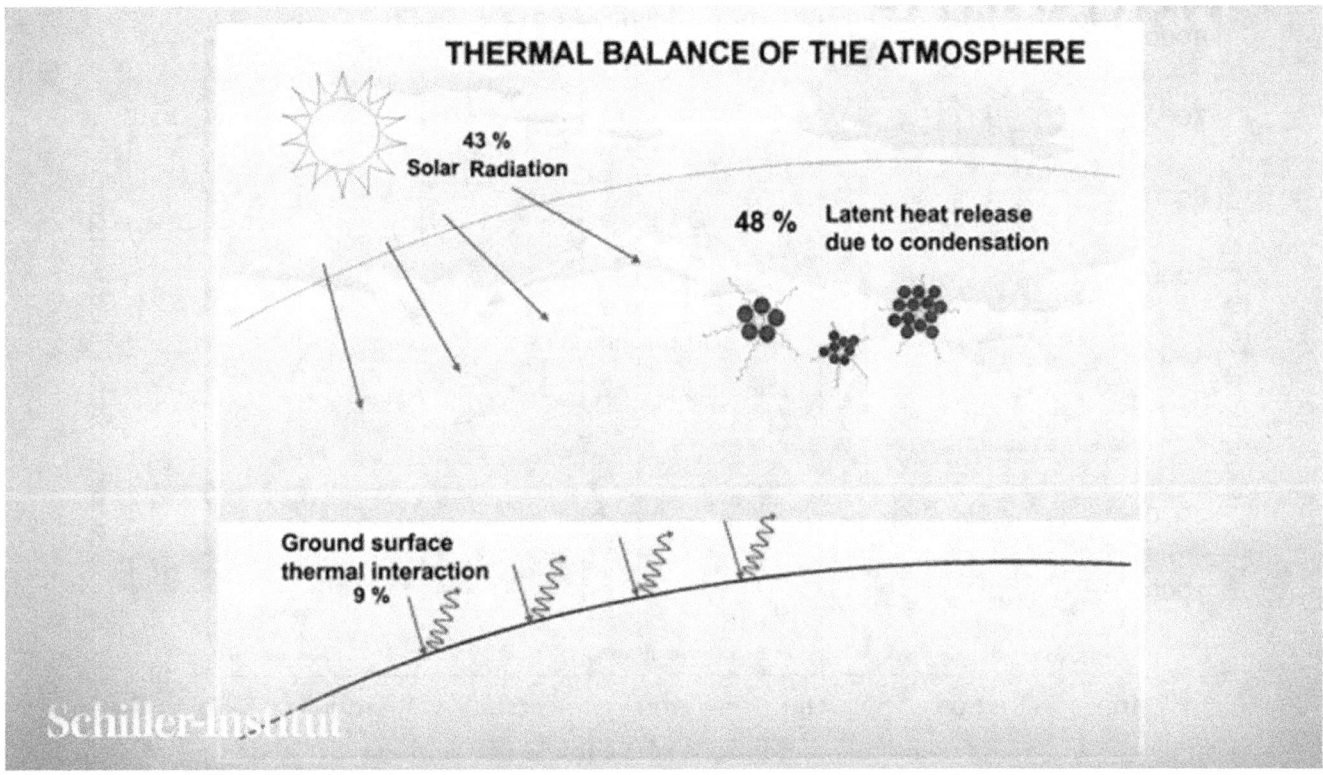

Cloudiness also has a large effect on the atmospheric heat budget. Slightly less than half of the atmospheric heat budget is supplied by latent heat that is released when water vapor condenses back into liquid droplets. This heat is generated in the clouds, high in the atmosphere, and is to a large degree cooled off into the colder region above the clouds.

The Primer Fields stand in the background of the process that largely affects and controls cloudiness on Earth, as the biggest climate factor. CO2, in comparison, adds up to nothing. Its effect is that minuscule. It doesn't even enter the scene where effects are measurable.

Even if the CO2 density in the air was increased 2000%, the increase would have no measurable effect on the climate. A 2000% increase of a millionth part still doesn't add up to anything significant. The increase would still remain below what is measurable in the context of the total climate effects.

That's where CO2 is a factor

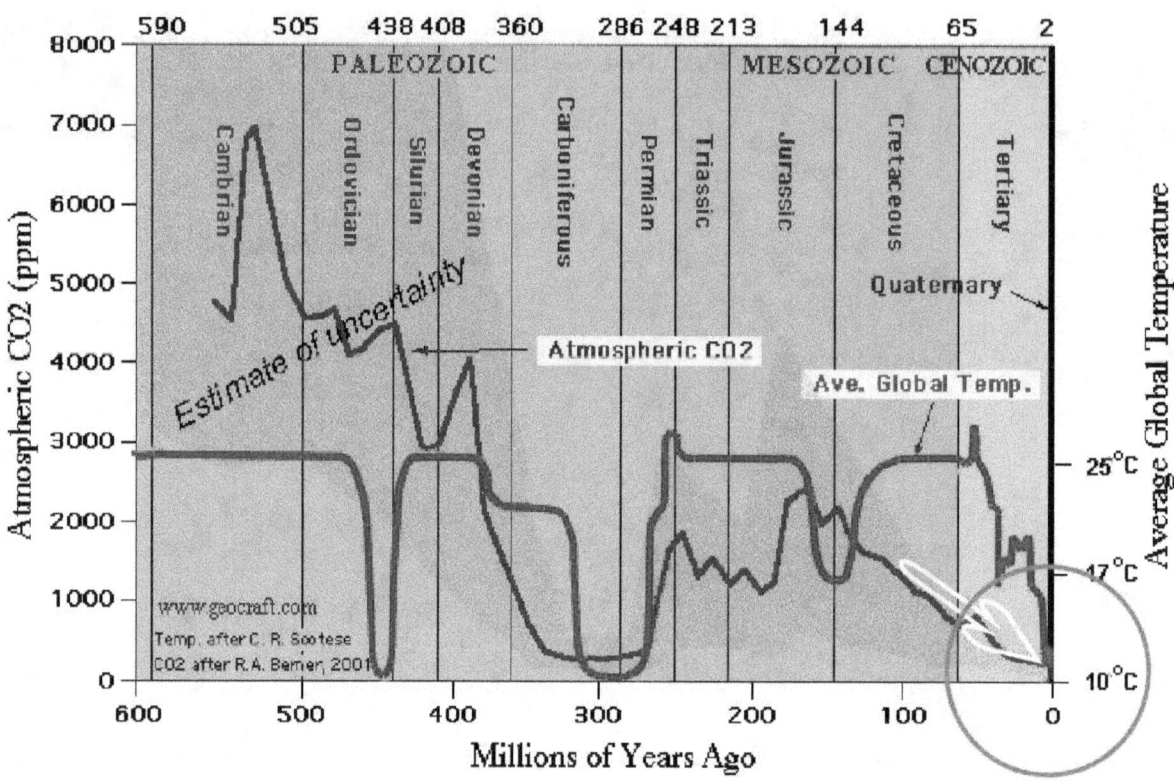

The 2000% increase, which adds up to a 20-fold increase of the global atmospheric CO2, would of course become supremely measurable in the arena where CO2 does have a measurable effect. We presently live in a severely CO2-starved world, with the lowest atmospheric concentration in hundredth of millions of years. The world is presently so severely CO2 starved, that when greenhouse operators increase the CO2 density in their facilities 2-fold, a 50% increase in plant-growth results. Imagine what a 20-fold increase would accomplish, of the type that we had in historic times, or even just 10-fold, as we had it in the Jurassic age, the world would become wonderfully green again, with richer harvests for human living. That's where CO2 is a factor. It's not a climate factor. It simply isn't.

The solar-forced Great Global Cooling

Cloudiness is the big, and hugely variable climate factor.

The solar-forced Great Global Cooling has begun.

Part of the system that affects cloudiness on Earth

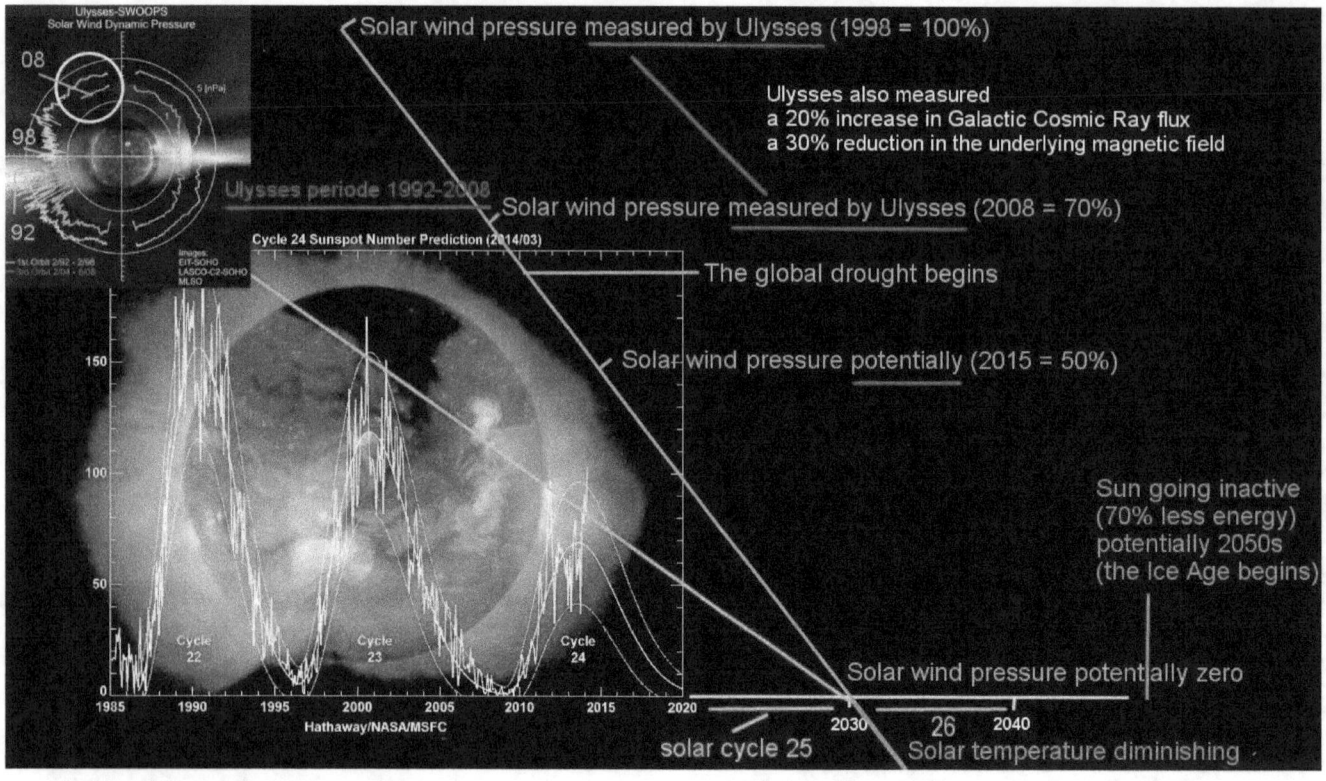

Since the Primer Fields are a part of the system that affects cloudiness on Earth, the subject of the Primer Fields is an important subject in science to be considered, even on the smaller scene that is not directly related to Ice Age dynamics, which, however are ultimately linked to it.

The wide range of considerations that come into focus here, in the context of the Primer Fields, of which the dimming Sun is of course the ultimate concern, evidently determines the wide range of topics that this video series must focus on, and does so.

A scientist is an economist

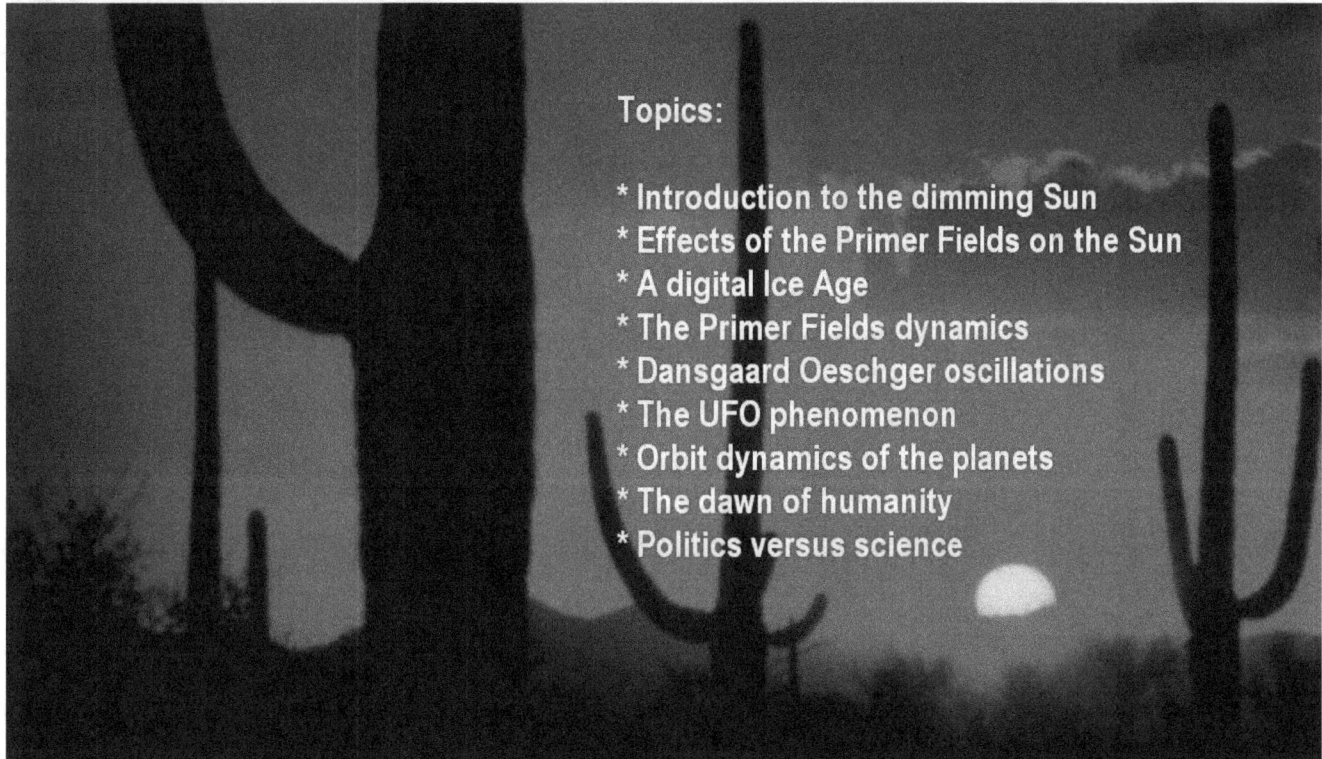

The exploration series focuses on the theories and the evidences, and how they will impact us all when the changing conditions unfold as they are scientifically understood to unfold by the nature of the principles involved. In the final video of the series, I will also focus in retrospect on the dimension of our response to the known conditions that affect our future, which are known by their principles. The human response factor to what is fully known, remains presently the biggest open question of them all. Will we become true to ourselves as human beings - true to what we know - and direct our future with what we know? the answer is critical for the very existence of humanity in the near future. It also has the potential to be an open door to the brightest, scientific, technological, economic, and cultural renaissance of all times.

As I said in one of my novels, in essence: a scientist is an economist. We utilize science to protect and enrich our world for the wider benefit of humanity as a whole, which, invariably is to our own benefit.

This means that economists should also be scientists, which may be required in the future, even for politicians as an entrance criterion.

Part 2 - Effects of the Primer Fields on the Sun

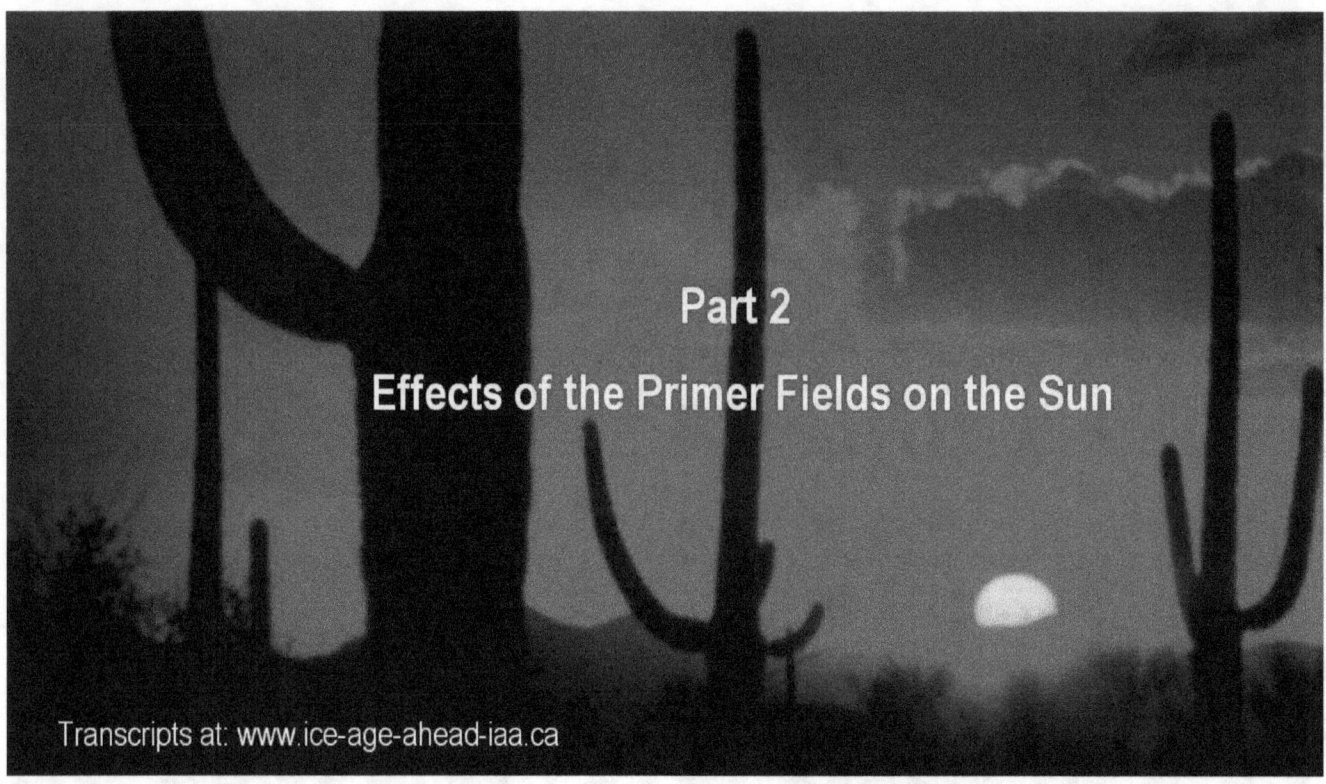

The Effects of the Primer Fields on the Sun

The plasma environment in our solar system

The shape of the galaxy shown here illustrates to some degree, in principle, the shape of the plasma environment in our solar system, though the density in the solar system is presently too low for the organizing plasma streams to be seen.

In the laboratory environment

A plasma sun born in the laboratory

David LaPoint - The Primer Fields

However, by replicating the shaping process in the very small, in the laboratory environment, we can explore the forces that shape these types of phenomena.

David LaPoint discovered in laboratory experiments

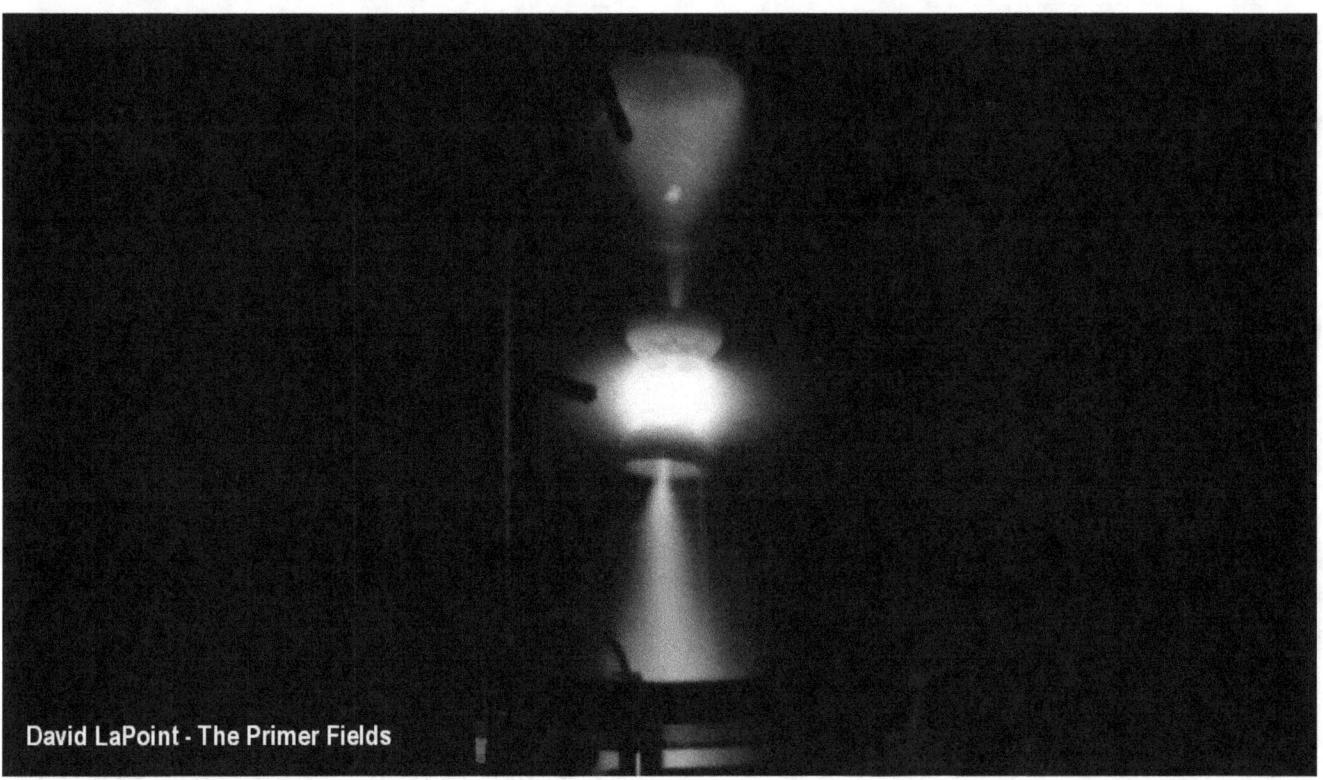

David LaPoint discovered in laboratory experiments that the Primer Fields, when they exist and are functioning, physically prime the environment of the solar system with a densely compressed sphere of plasma centered on our Sun.

The principle is illustrated here.

The effects are amazingly critical

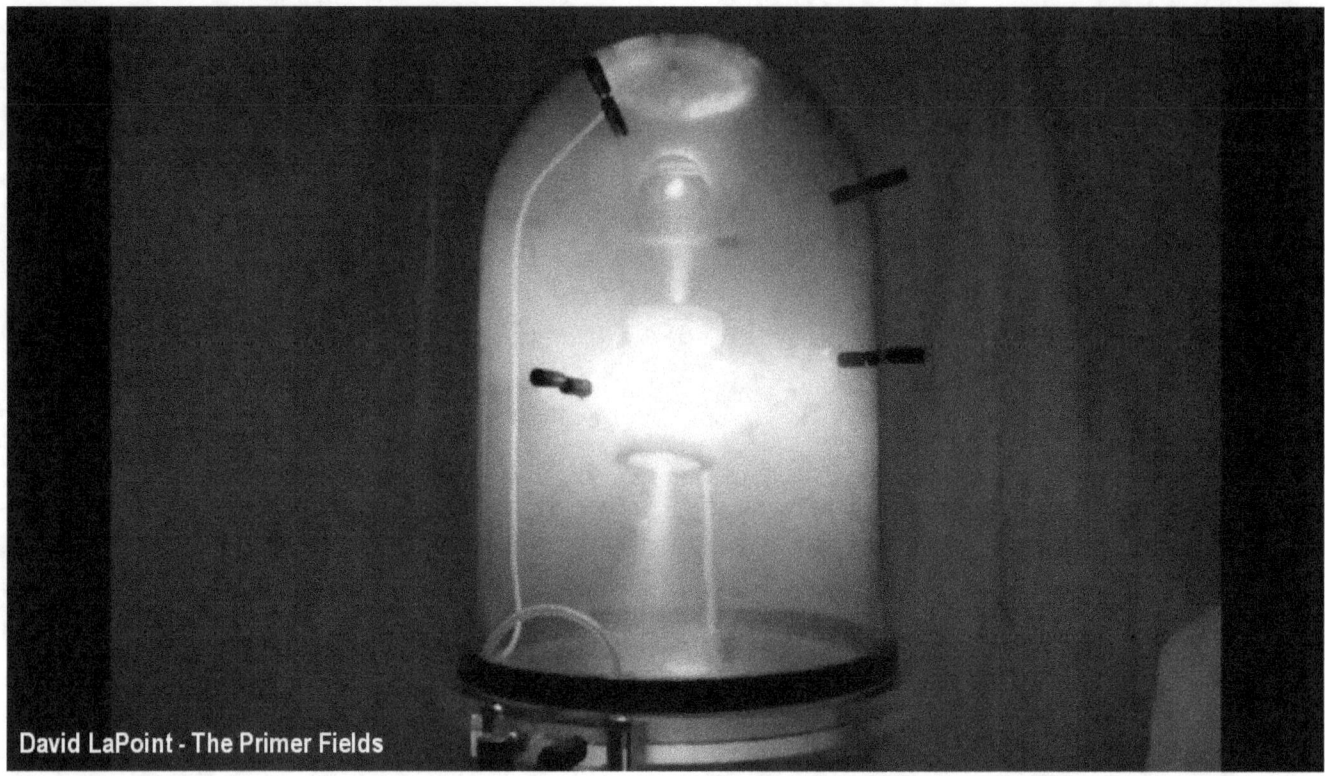
David LaPoint - The Primer Fields

The effects, however, that David LaPoint discovered, are so amazingly critical that great problems arise for humanity when their functioning is impeded or is collapsing.

Condensed plasma interacts with the Sun

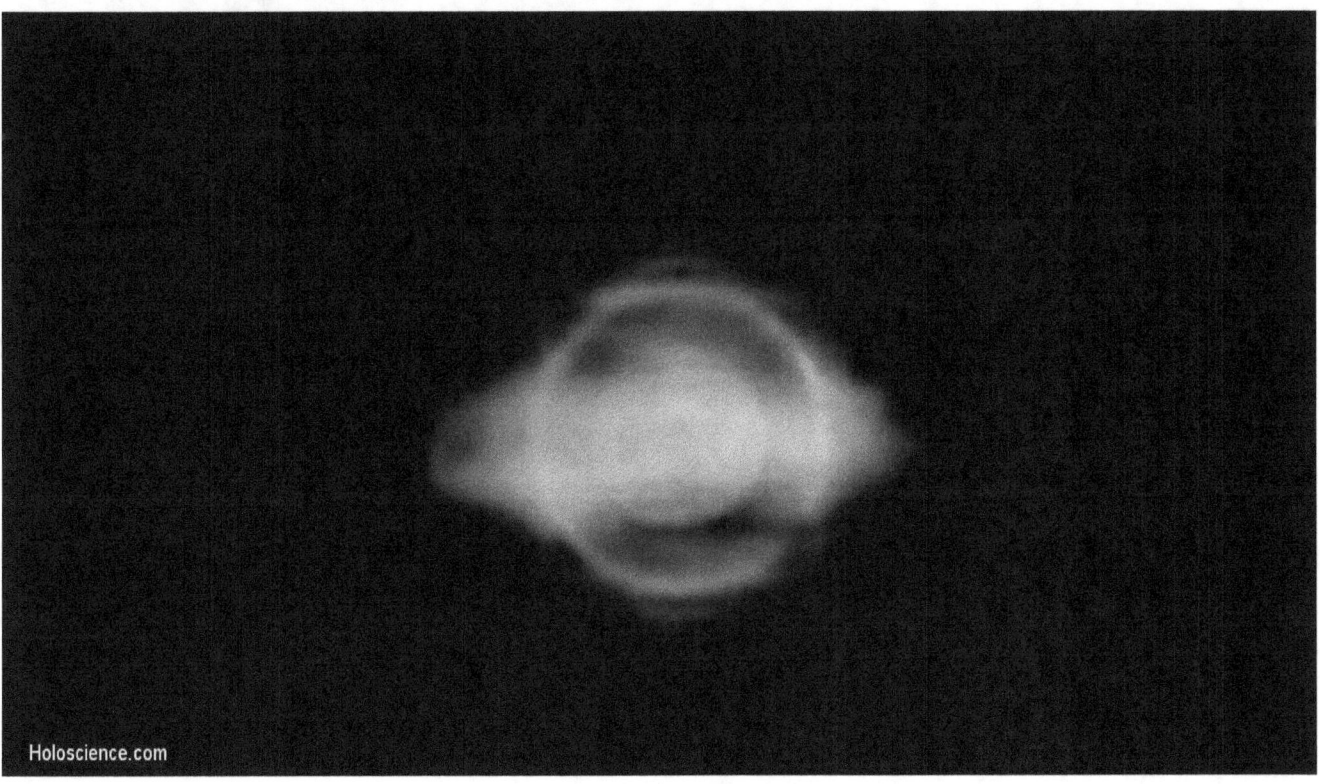

The evidence tells us that the dense plasma environment that is focused on the center of the solar system enables our Sun to be electrically powered from the outside by electric arc reactions. In these reactions the condensed plasma interacts with the Sun's photosphere.

Illustrated in the Red Square nebula

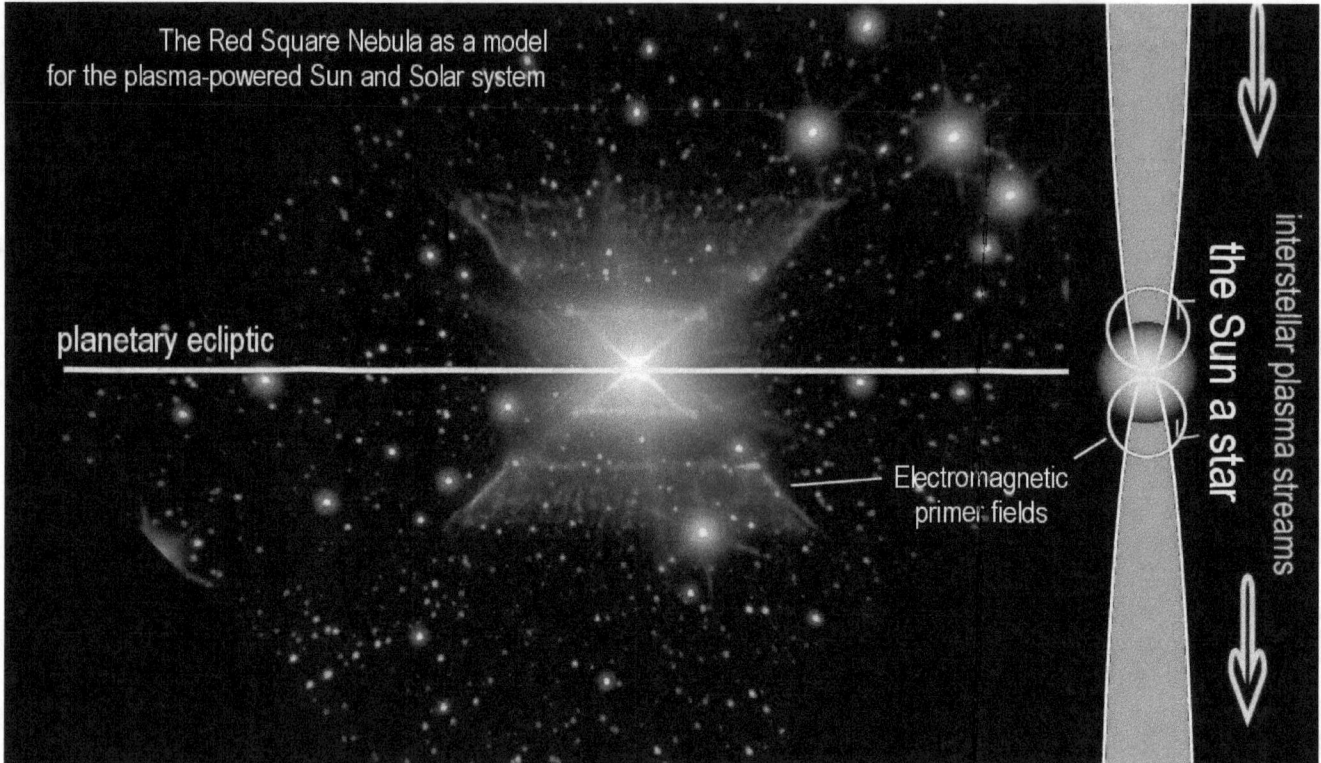

The condensing process is a multi-stage process that is illustrated to some degree in the Red Square nebula. The nebula is not the remnant of an exploding star as nebulas are often regarded. Instead it is an example of the typical features of the Primer Fields in operation. The fields form by the principles inherent in the flow of electric plasma in space. The details will be discussed later, though the overall effect is noteworthy here. The effect of the fields is that plasma streams that exist in galactic space in wide channels, becomes drawn together, like by a wide funnel, and becomes magnetically focused from the funnel onto a central sun, or a system of multiples suns that become intensely powered by this process. The planets exist outside of this densely powered sphere, on an ecliptic centerd on the Sun, between the two Primer Field structures. The plasma concentration process that we see in operation here, which is basically the same for every sun, renders our Sun as an intensely powered catalytic energy converter, that is powered not from within, but is powered externally at its surface.

The Sun's great brilliance generated at its very surface

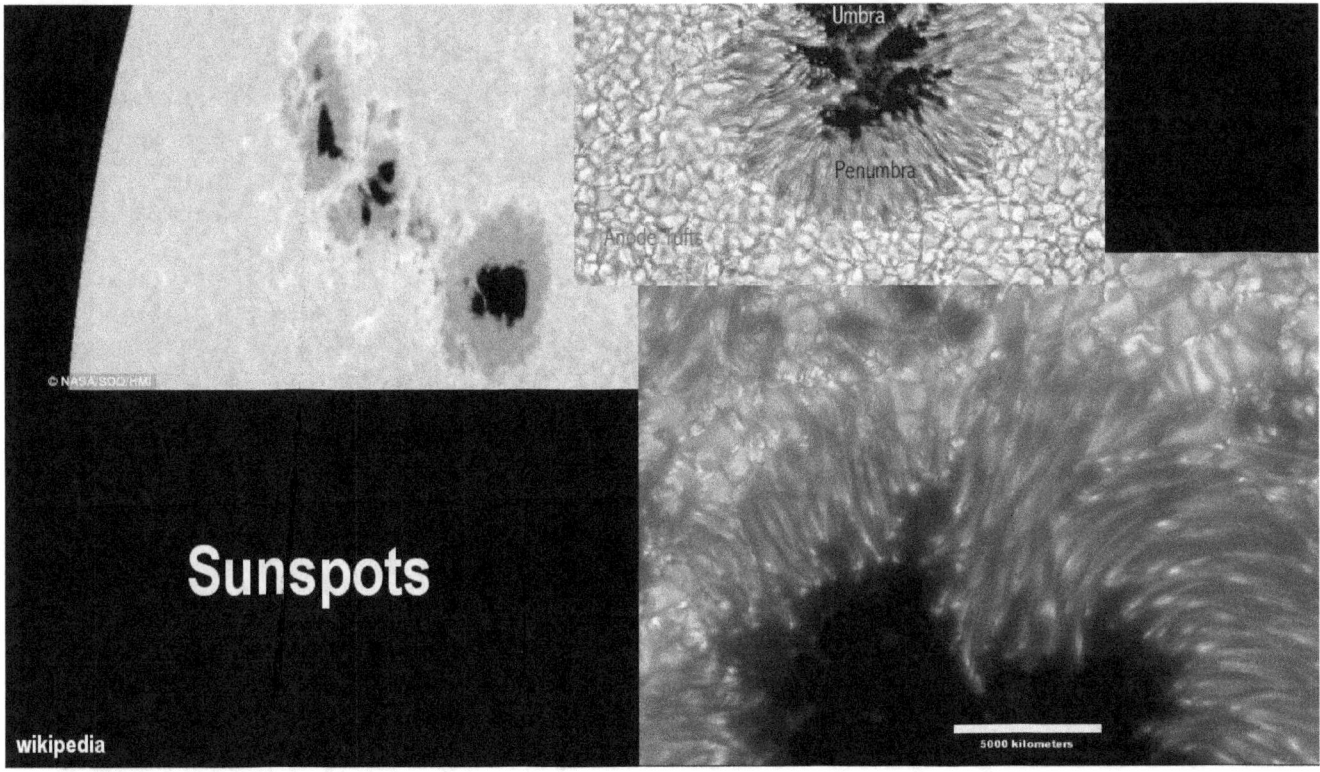

The evidence that the Sun is externally powered is fairly obvious. When we look beneath the high-powered photosphere, through the open space at the umbra of the sunspots, the Sun reveals itself as being much darker, and therefore cooler inside. The photosphere has been measured at a whopping 5,870 degrees Kelvin, and the surface below at a mere 3,000 degrees. This means that the Sun's great brilliance is electrically generated at its very surface and does not emanate from within from nuclear fusion reactions as it is generally believed.

The plasma concentration process

The plasma concentration process that enables a sun to be electrically powered, has been replicated in lab experiments.

A highly compressed plasma sphere was formed

There a highly compressed plasma sphere was formed in a thinly filled chamber of gas between two bowl-type permanent magnets that replicate the functional elements of the Primer Fields.

Magnetic fields operating environment

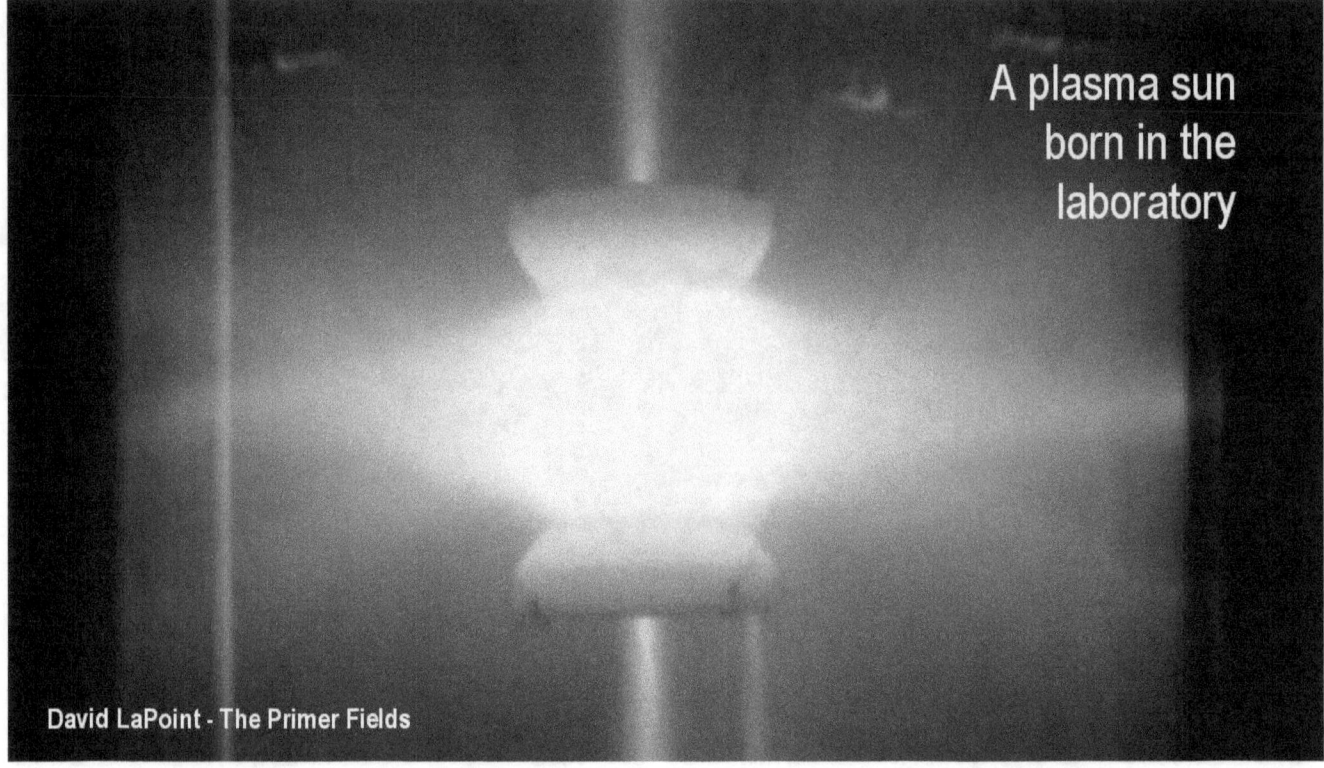

The result, which is shown here, was develop by the functions of the magnetic fields acting on the plasma flow to concentrate it.

It is interesting to note that in the lab experiment the plasma sphere at the center did not form instantly. The operating environment needed to be established first.

In the real world

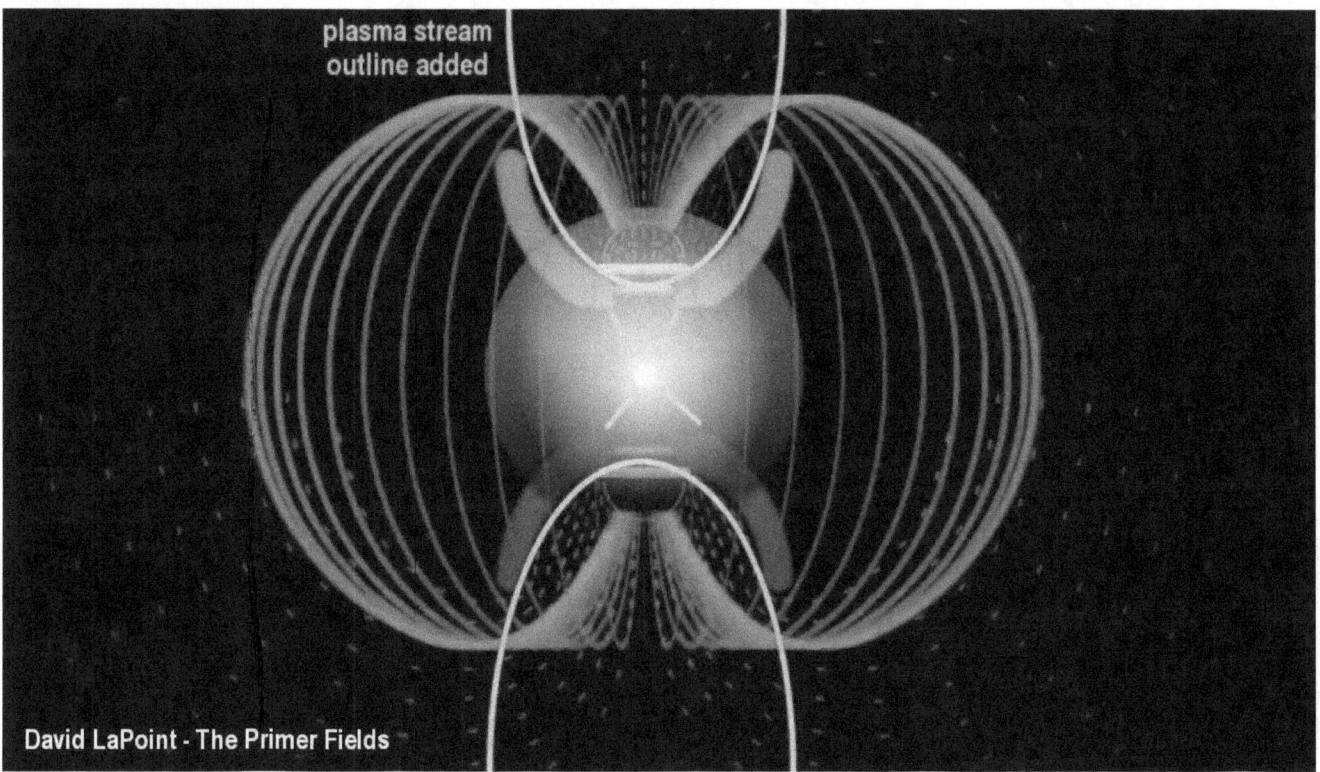

However, in the real world this establishing-function also includes the forming of the bowl-shaped magnetic fields themselves by which the plasma becomes concentrated. In the lab, the magnetic fields were provided with manufactured magnets. In the real world the bowl-shaped magnetic fields are electrically created structures that are formed by moving electricity in plasma streams.

The plasma sphere in which our sun is located

This means that the plasma sphere in which our sun is located, is the composite result of complex processes working together, for which a certain level of plasma input density is required for the forming of the magnetic bowl structures to happen.

I will illustrate later how the bowl-shaped magnetic structures are formed, that furnish the Primer Fields, which in turn create the conditions that focus plasma onto our Sun that lights it up to great brilliance.

Different types of atoms emit light in different bands

While different types of atoms emit light in different bands, the numerous types of atoms that are created in the photosphere, in combination, create photons of all possible shapes and sizes that together cover the entire light spectrum and beyond, in a seamless field of colors.

The color-rich world that we cherish

The white sunlight that we see reflected in the color-rich world that we cherish, is the direct result of the plasma-fusion process in the photosphere of the Sun that is facilitated by the actions of the Primer Fields that focus interstellar plasma unto the Sun. The white sunlight with its colorful spectrum is not possible on any other basis.

The Sun cannot be as a sphere of hydrogen gas

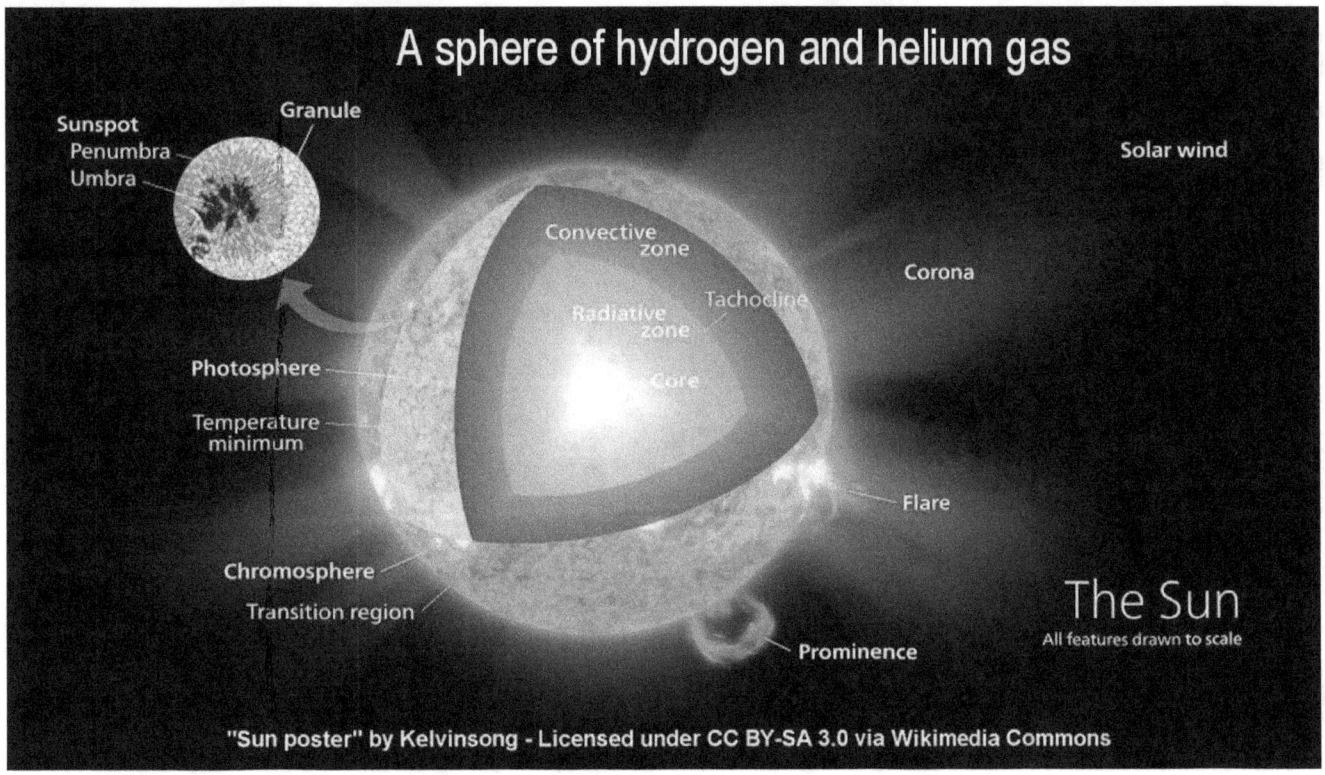

The historically theorized concept of the Sun as a sphere of hydrogen gas with a hydrogen-fusion process at its core, is obviously false, because it cannot produce the white sunlight spectrum that we see.

Hydrogen atoms emit light in only a few narrow bands

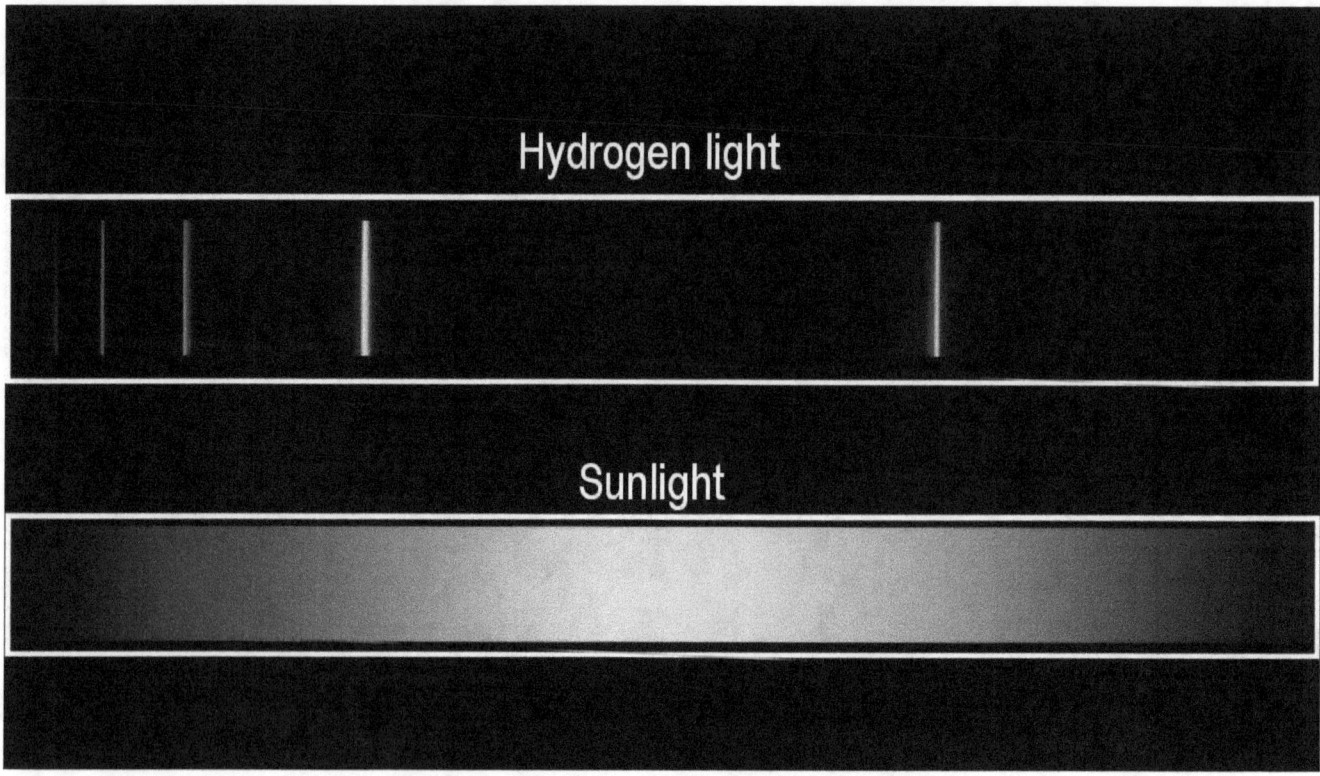

Hydrogen atoms emit light in only a few narrow bands.

The white sunlight spectrum is only possible by solar surface plasma fusion where all known types of atomic elements are created.

The color-rich white sunlight is clear tangible evidence

The color-rich white sunlight is clear tangible evidence that we live in a solar system powered by plasma streams that the Primer Fields are an active element of, centered on a plasma Sun.

The surface plasma fusion also emits highly energetic solar cosmic-ray flux

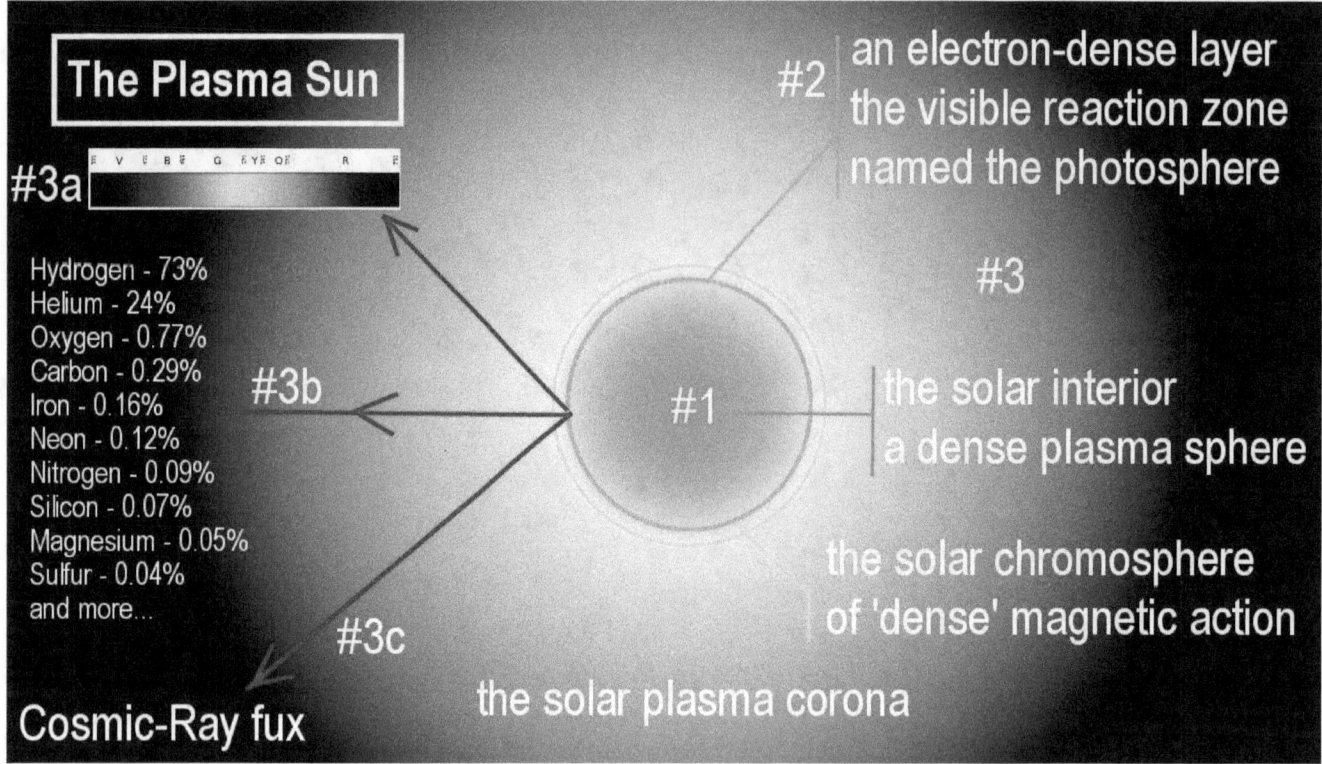

However, the surface plasma fusion does not only create atomic elements and streams of light, but also emits highly energetic solar cosmic-ray flux, marked 3C. Cosmic rays are not streams of light, but are events of highly energized individual electrons and protons escaping from the magnetic confinement of the primer fields of the plasma fusion cells in the photosphere. Most of the solar cosmic rays are trapped in the corona, marked #3, but when the corona weakens, more of them penetrate the barrier and reach the Earth.

Solar cosmic rays have an ionizing effect

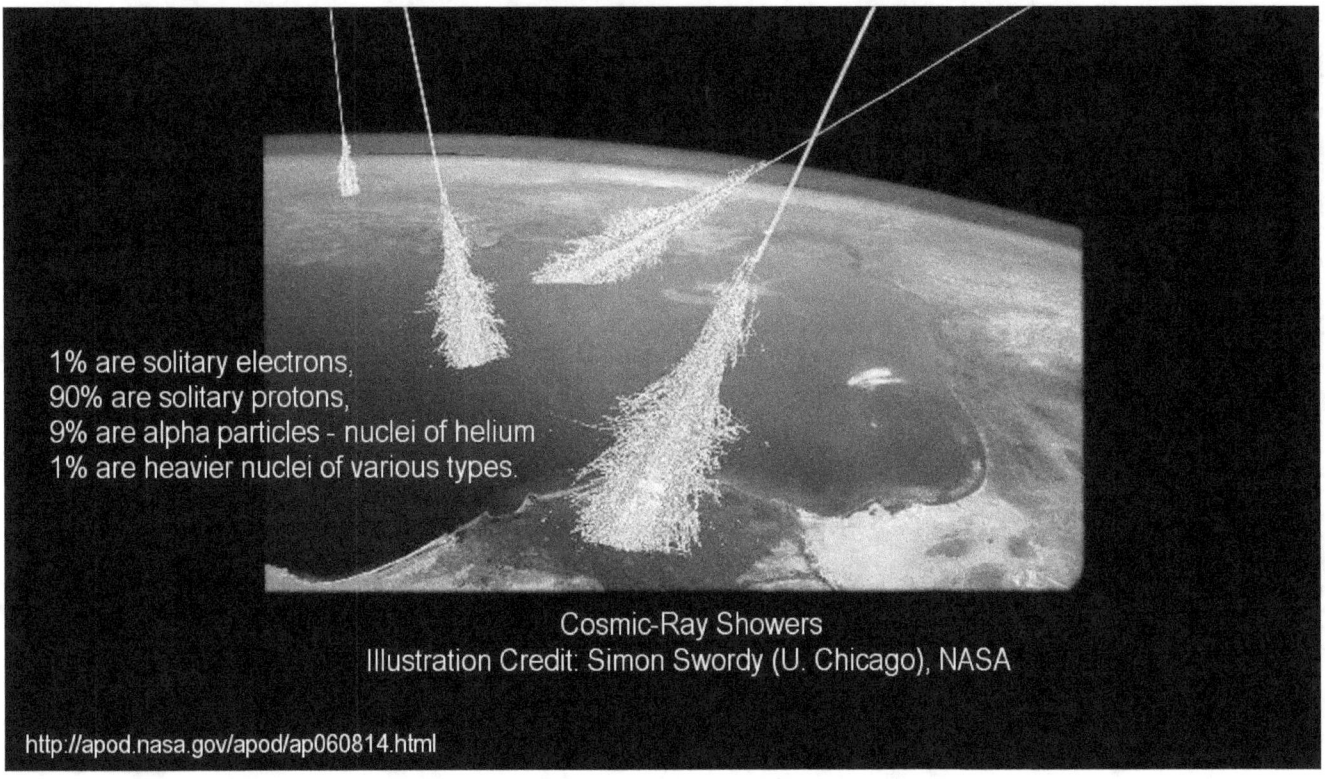

Solar cosmic rays have an ionizing effect in the atmosphere that enhances the cloud forming process, which affects the climate on Earth.

Increased cloudiness results in colder climates

Increased cloudiness results in colder climates. The white top of the clouds reflect a portion of the incoming solar energy back into space, which thereby becomes lost to us. The top of clouds also radiate latent energy from the cloud-forming process into space. Latent energy is released when water vapor is condensed into liquid droplets.

When a pot of water is boiled into steam

When a pot of water is put on a stove and is boiled into steam, the energy that is invested in the process is released as latent energy when the steam condenses back into water.

Clouds cooling latent energy into space

In the atmosphere, 45% of the thermal budget is derived from latent energy. Much of this energy is released in the clouds, at the edge of the atmosphere, where much of it is cooled into space.

By the combined effect of clouds reflecting solar energy directly back into space, and the clouds cooling latent energy into space, the rate of cloud forming is the major climate-determining factor on Earth, and the rate of this process is absolutely determined by the Sun, by the volume of cosmic-ray flux that escapes from it.

When the solar corona is weak

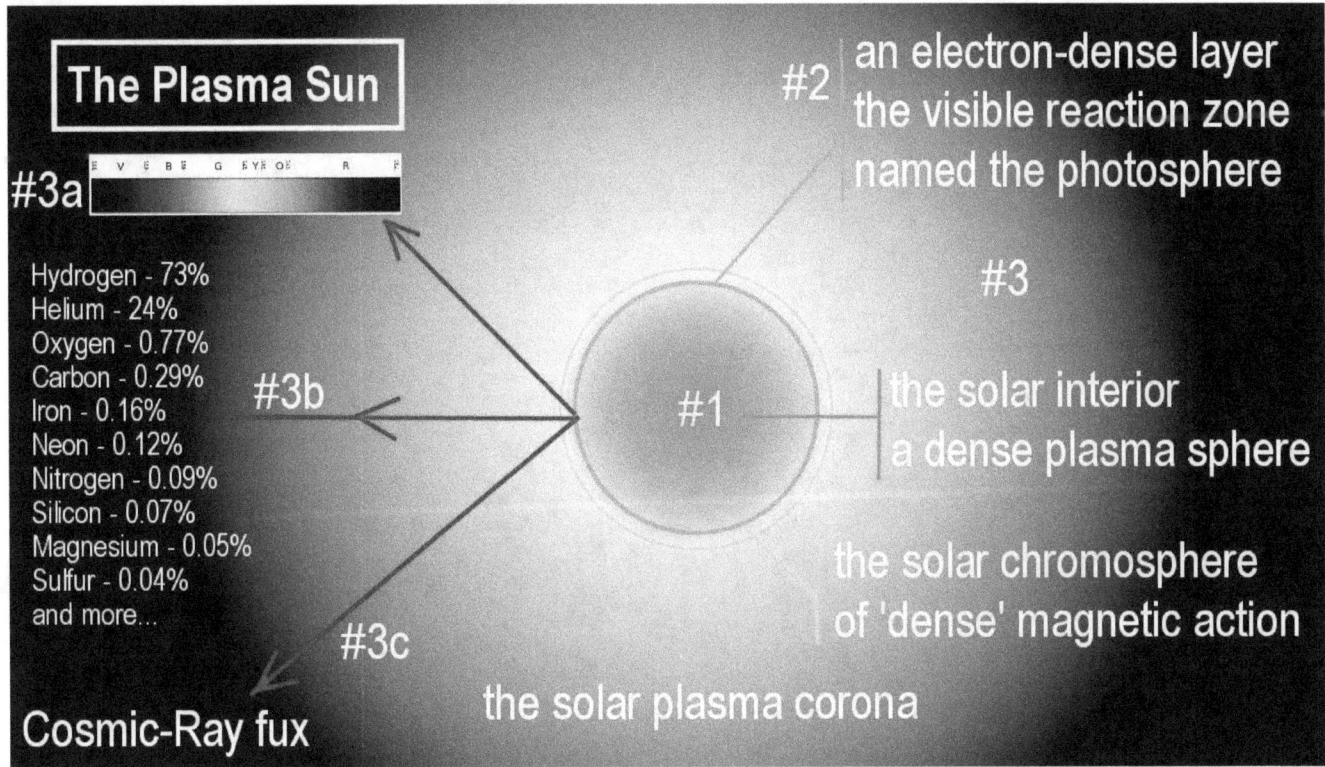

When the solar corona is weak, which also results in weak solar activity, the solar cosmic-ray flux is increased, cloudiness is thereby increased, and the Earth gets colder. When this happens in a big way, a Little Ice Age results.

That's what we saw in the 1600s

That's what we saw in the 1600s, at the time of the Maunder Minimum of the solar activity. The global cooling was so massive that rivers became skating rinks in the winter.

Indicated in Carbon-14 measurements

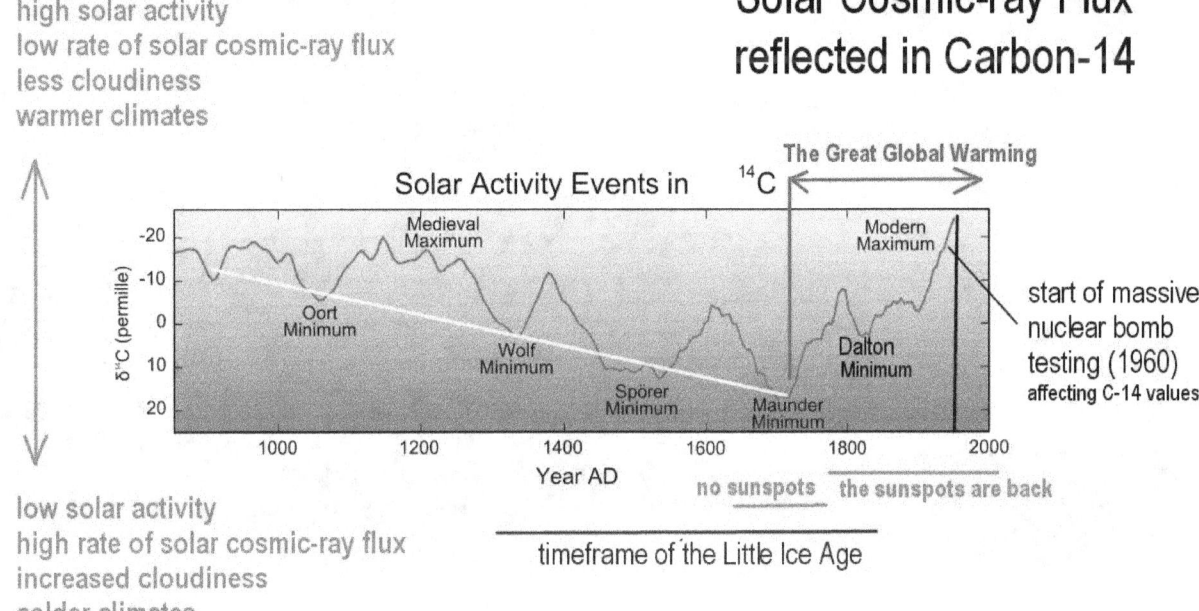

That this time of the Little Ice Age was a time of large volumes of solar cosmic-ray flux, is indicated in Carbon-14 measurements. Carbon-14 results from Solar cosmic rays affecting the atmosphere. All the cold periods that we have measurements for were periods of weak solar conditions that result in extremely high rates of solar cosmic-ray flux, and by implication also high rates of cloudiness.

The ever-changing climate on Earth

This means that the ever-changing climate on Earth, is directly caused by changing density in the interstellar plasma stream, which, through the Primer Fields system, is focused onto our Sun. It also proves that the Sun is surface-powered by the Primer Fields, because no other platform than surface plasma fusion is able to generate solar cosmic-ray flux, and this so massively that it affects the climate on Earth.

Global warming after the Little Ice Age, was not manmade

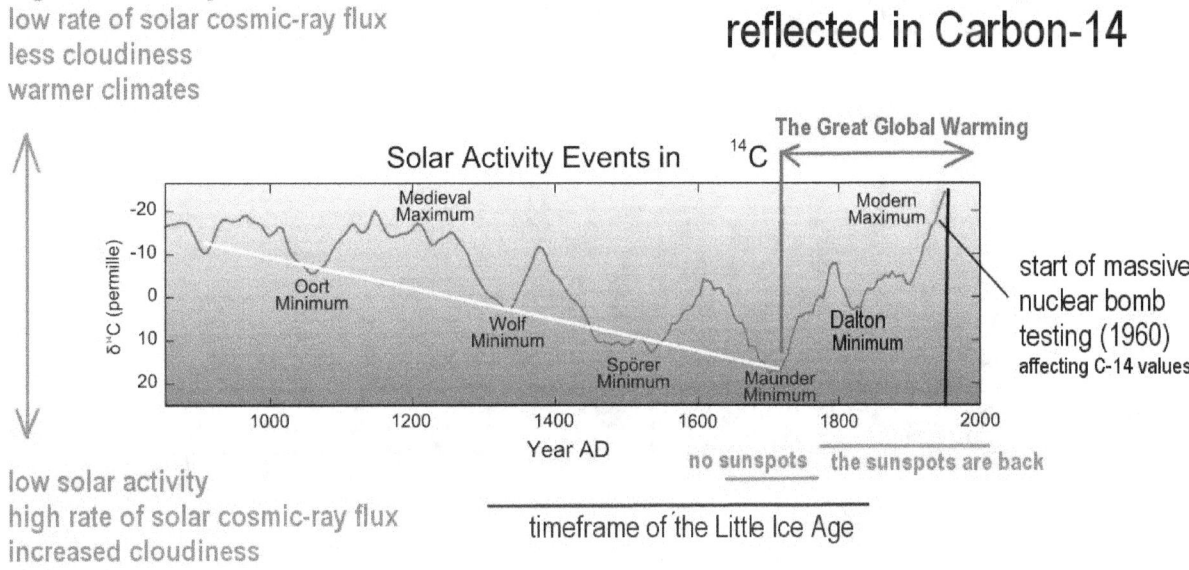

It also means that the great global warming after the Little Ice Age, was not manmade by industrial activity and fuel burning, but was the direct result in changing cosmic conditions that the Primer Fields focused onto the Sun where it affected the Sun's delicate operating dynamics.

The recognition of these changing cosmic conditions is critical for our time, because we see in them a long-term down-ramping in progress towards another Little Ice Age coming up fast, from which the solar system may not recover.

The solar system barely recovered from the Little Ice Age

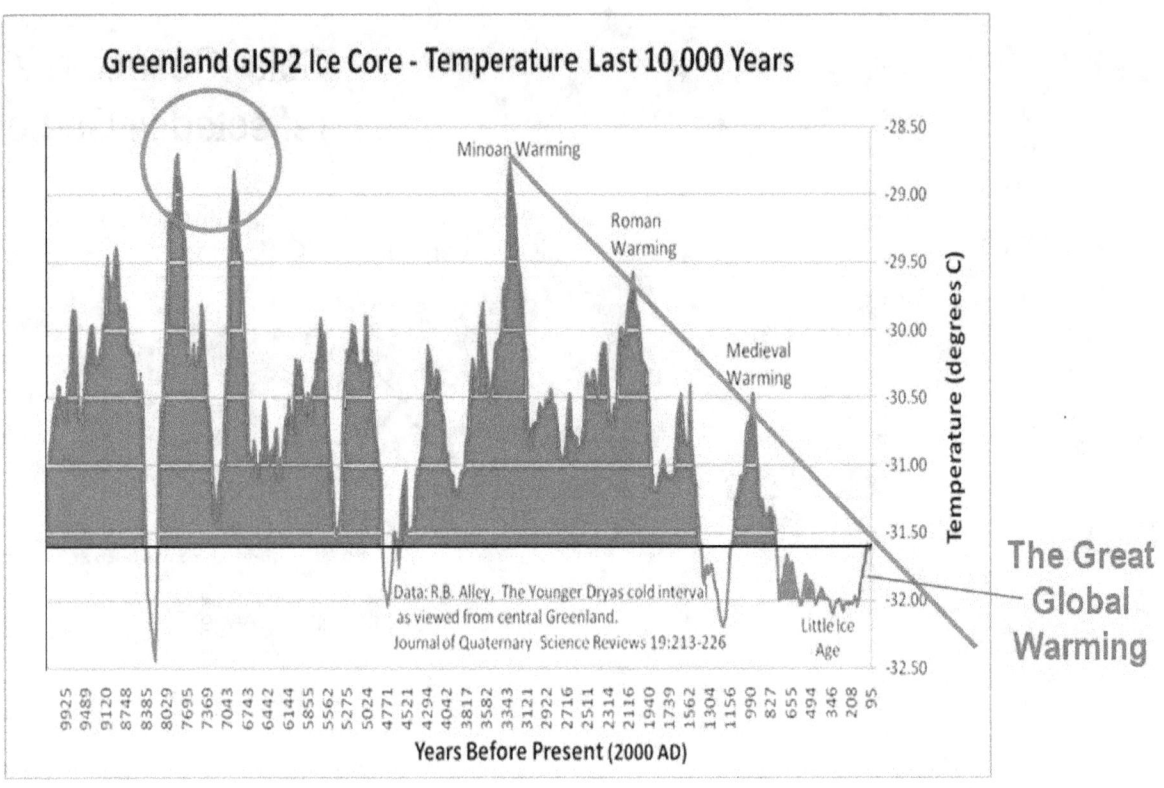

The solar system barely recovered from the Little Ice Age. The down-ramping that we see in suggests that the next minimum in the solar system may take us below the minimal density that is needed for the Primer Fields to be maintained.

When the Primer Fields collapse, a phase shift occurs

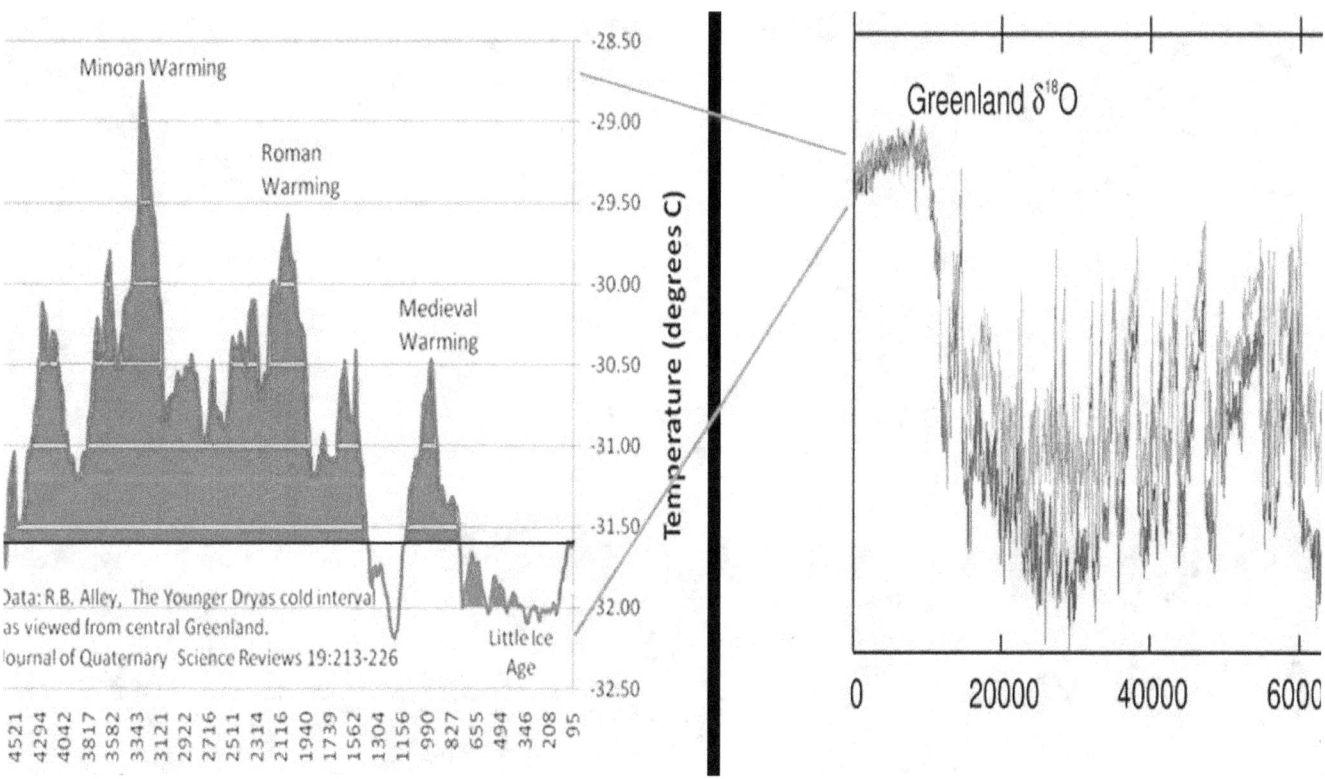

When the Primer Fields collapse, a phase shift occurs. Then all the climate changes that have occurred in the last 10,000 years will appear as nothing in comparison. With the Sun going largely inactive, the glacial climate begins that ice core measurements tell us, will be 40 times colder than the worst of the Little Ice Age had been.

It has become evident in lab experiments

It has become evident in lab experiments and cosmic observations that when the input density of the plasma streams that feed into the Primer Fields system drops below a certain minimum threshold, the critical fields don't form, whereby the entire process stops.

The solar system is not as robust as is generally believed

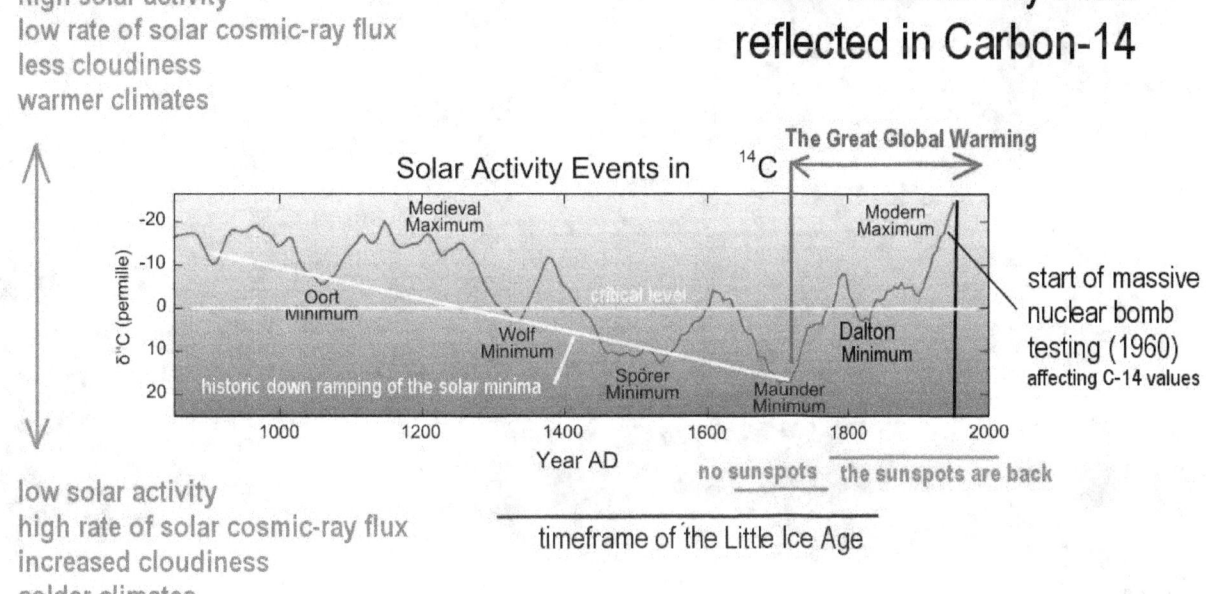

"Carbon14 with activity labels" by Leland McInnes at the English language Wikipedia. Licensed under CC BY-SA 3.0 via Commons

The climate fluctuations that have been experienced, that we have records of, tell us that the solar system is not as robust as is generally believed, and is down-ramping.

A colder, darker, yellow sun

When the phase-shift happens in the real world, the brilliant photosphere of our sun is no longer being powered. It becomes in effect, turned off. It fades into a thin haze, leaving in the wake a colder, darker, yellow sun that glows dimly by its internally stored-up energy like ambers of a fire gone out, which too, then slowly diminish.

This means that the transition to the next ice age will not be a slow process that is drawn out over thousands of years, but will be a radical turn-off transition in which the Sun goes dim in possibly a single day.

A deactivated Sun

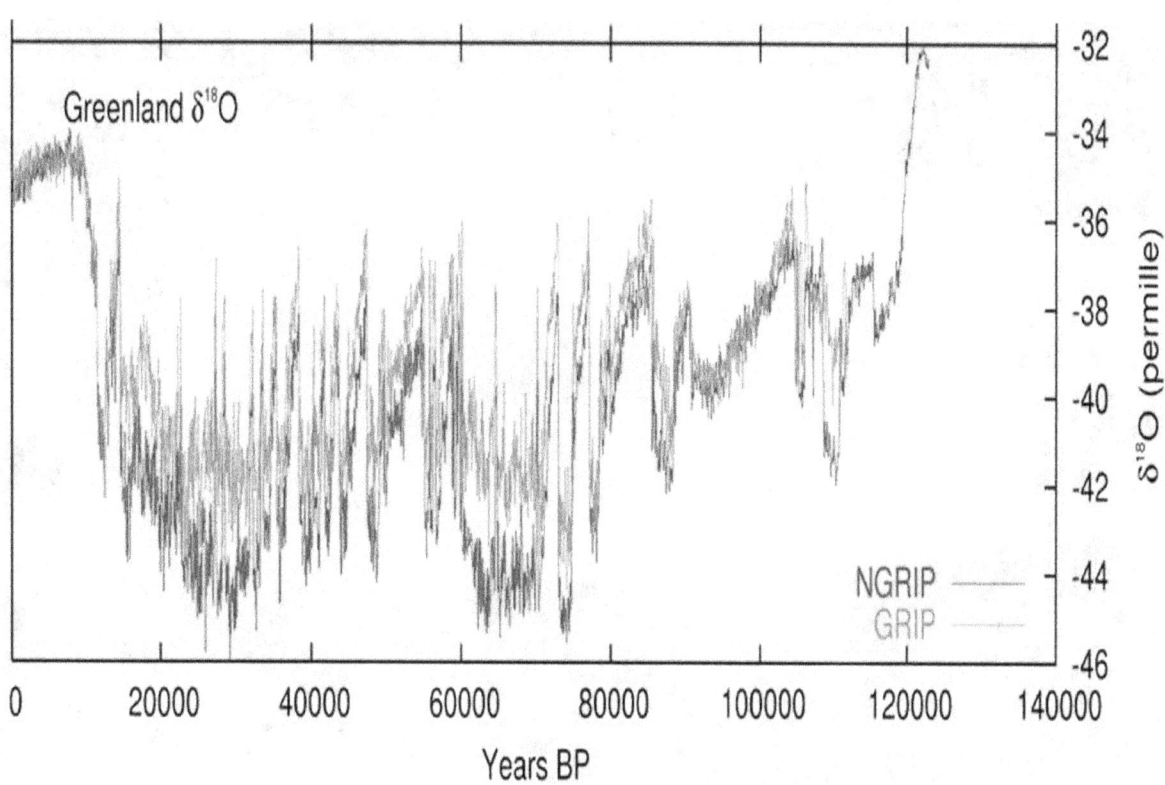

There exists plenty of evidence in ice core samples that the kind of rapid transformation of the climate on earth has happened that reflects an on-off transition, from an active Sun, to a deactivated Sun.

Nothing short of such a radical transition can explain the massive, rapid cooling that occurred when the last Ice Age began 120,000 years ago.

A massive reduction in solar energy

It takes a massive reduction in solar energy input into the Earth's climate, to cause the formation of the enormous ice sheets that become spread across much of the northern hemisphere as it is shown here, which piled up 10,000 feet thick, or more in some places.

A radically different world unfolds

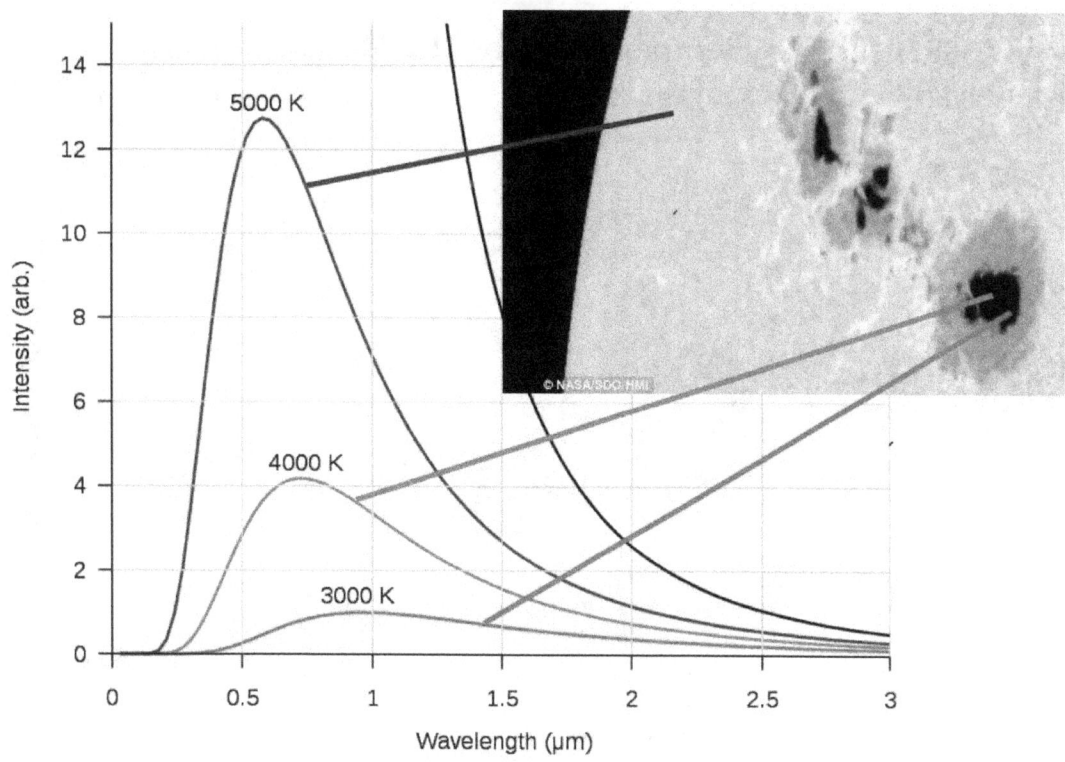

When the phase-shift happens a radically different world unfolds. The actual timeframe in which the phase-shift unfolds may be extremely short, probably spanning less than a single year for all practical purposes.

We are presently near the phase-shift point

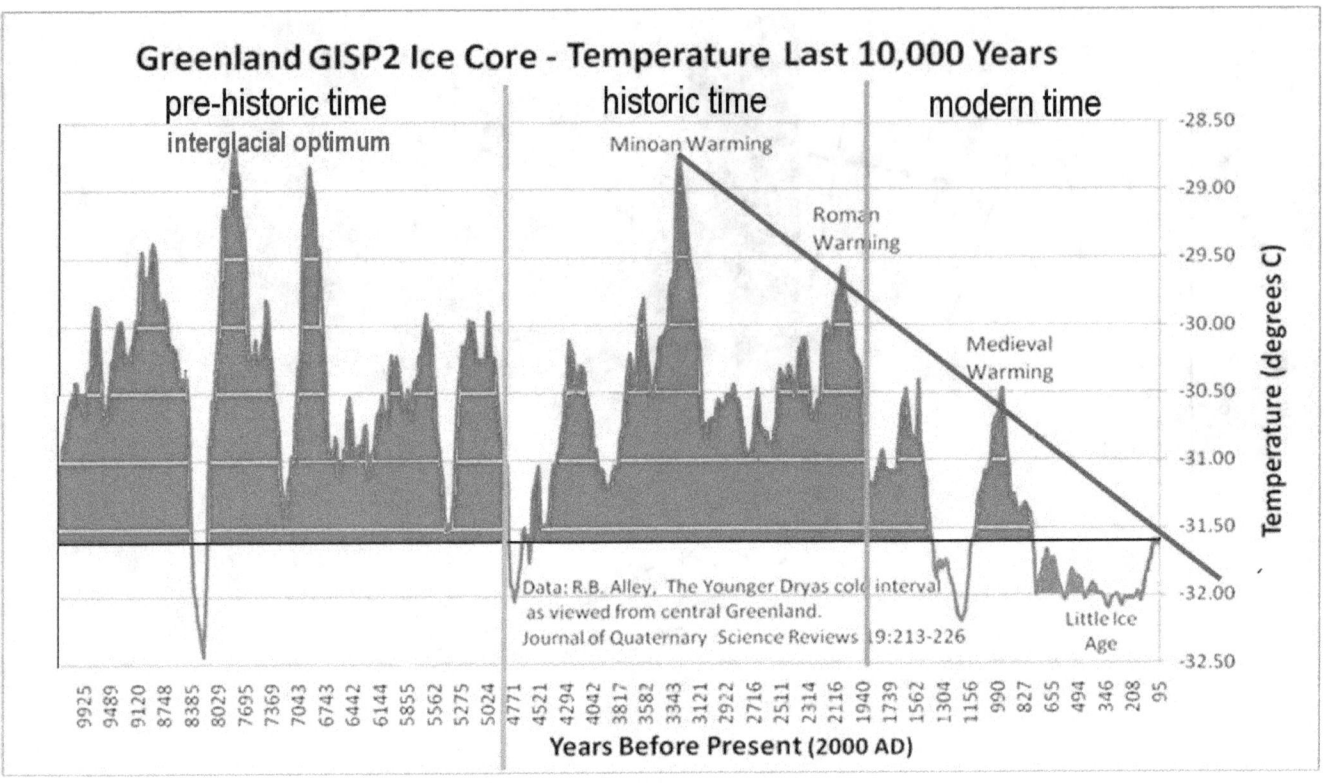

We are presently near the phase-shift point. We have come to the end phase of the actively powered period of the sun, and the beginning of the next 90,000-year glaciation period.

Of course we don't know the precise day or year at which the Primer Fields will collapse and the plasma sphere around the Sun will vanish, but do we really need to know this?

We know already enough to be inspired by it to take the appropriate actions to build new infrastructures for our food supply and for our living. We know with measured evidence on hand that the Earth has been in a diminishing trend for 3,000 years already, that appears to be now accelerating towards the critical cut-off point.

NASA's Ulysses spacecraft

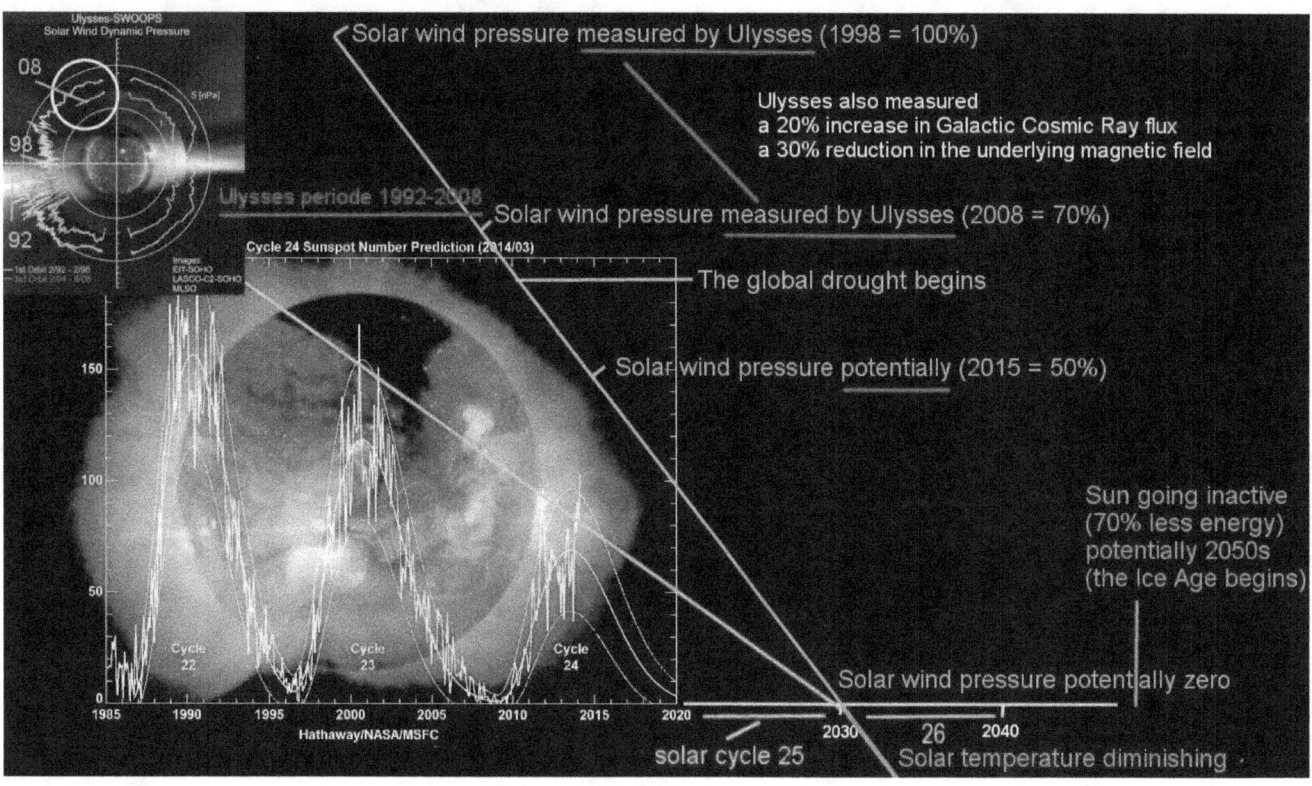

We see the solar system getting weaker. The tell-tale sunspot cycles are diminishing. Also, over the timeframe of a single decade, NASA's Ulysses spacecraft has measured an amazing weakening of the solar wind pressure by a whopping 30%. This is an enormous drop-off for such a short period.

When the Sun's powered state ends

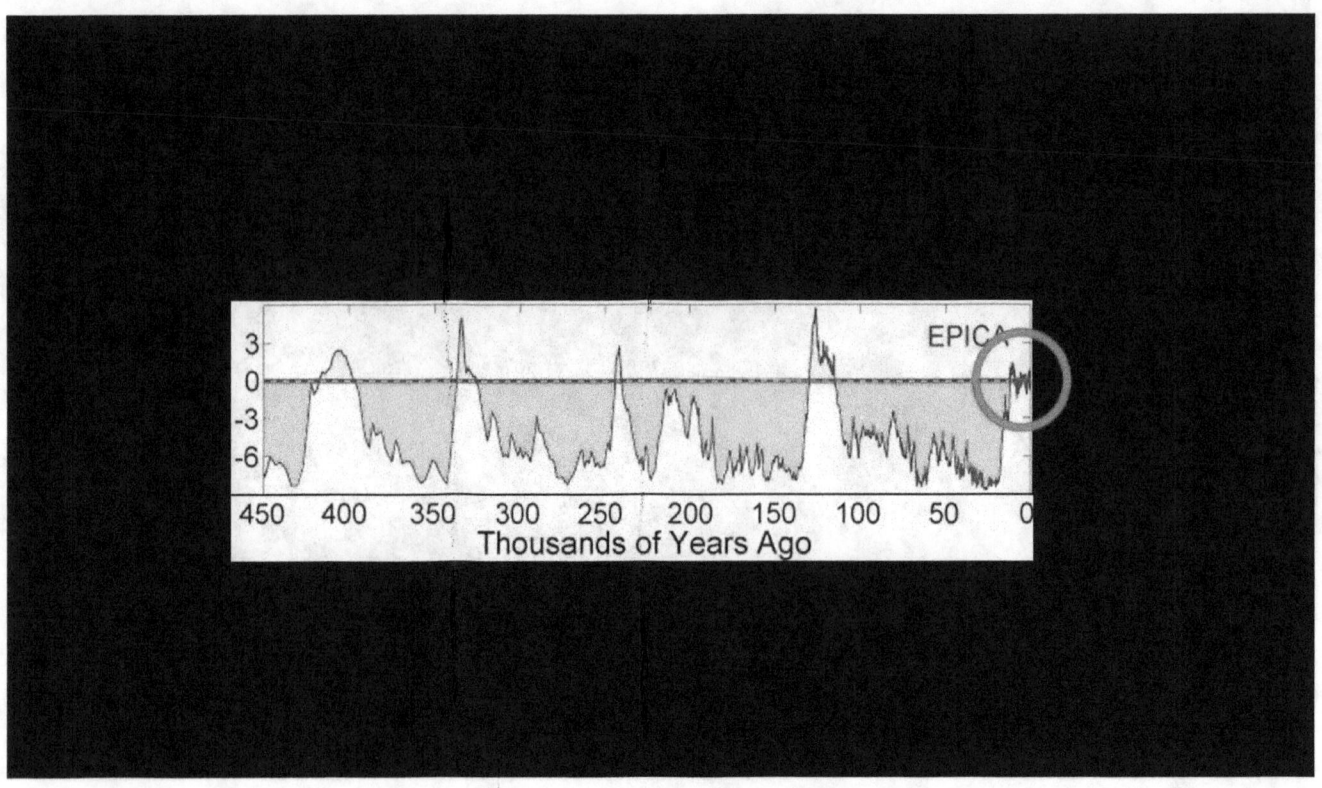

We also know that when the Sun's powered state ends, that is when it gets turned off to a lower intensity state, the big ice sheets will form. It takes a big change in the incoming energy that warms our planet, for big effects to happen.

No one is prepared for the consequences

photo by Andrew Mandemaker

Let's hope that the cut-off point that we are moving towards is still more distant in time than it appears to be, because at the present time no one of humanity is prepared for the consequences.

The Ice Age consequences promise to be far bigger

Antarctica, photo by Vincent van Zeijst - Wikipedia

Not the least preparations are even considered to be made, much less are made, even while the Ice Age consequences, when they begin, promise to be far bigger than most people dare to imagine.

Part 3 - The digital Ice Age

The digital Ice Age

The two long climate cycles that overlap

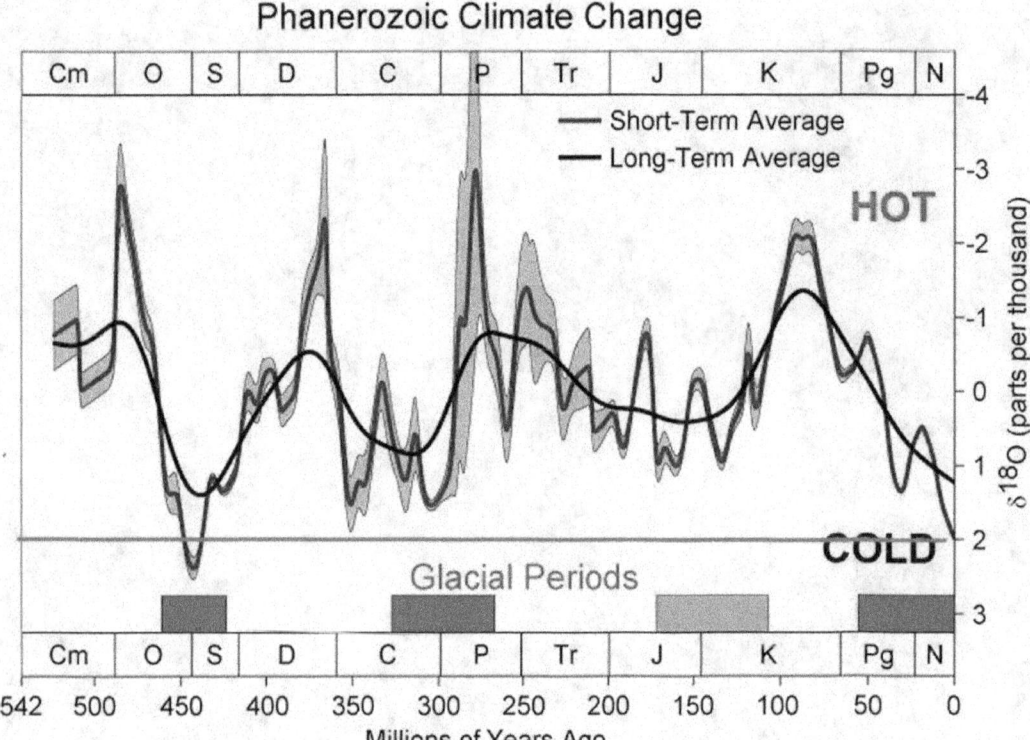

Major glaciation periods have occurred 4 times in the last half-billion years that we have records of. Some were so severe that major extinctions occurred as a consequences. We are presently in such a period that is extremely severe. The two long climate cycles that overlap and determine the strength of our solar system, are both approaching their minimal point together.

In the last 500,000 years of the resulting glaciation epoch

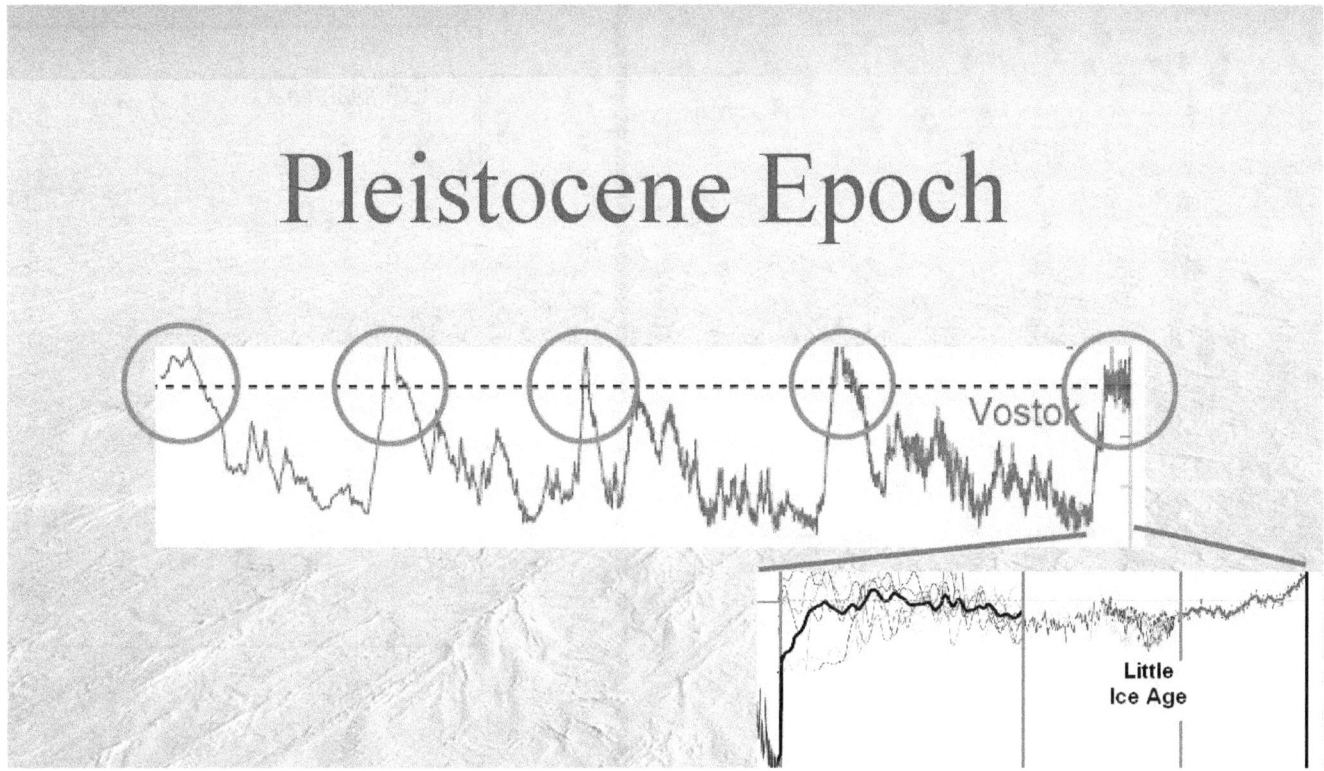

In the last 500,000 years of the resulting glaciation epoch, glaciation conditions occurred for 85% of the time, interspersed with brief interglacial warm periods, like the one we presently enjoy, which we erroneously regard as normal, but which has run its course and is now ending.

This means that the current warm period is a climate anomaly that is actually rather fragile. The Ice Age conditions are the normal state on Earth.

The difference between the two climate states

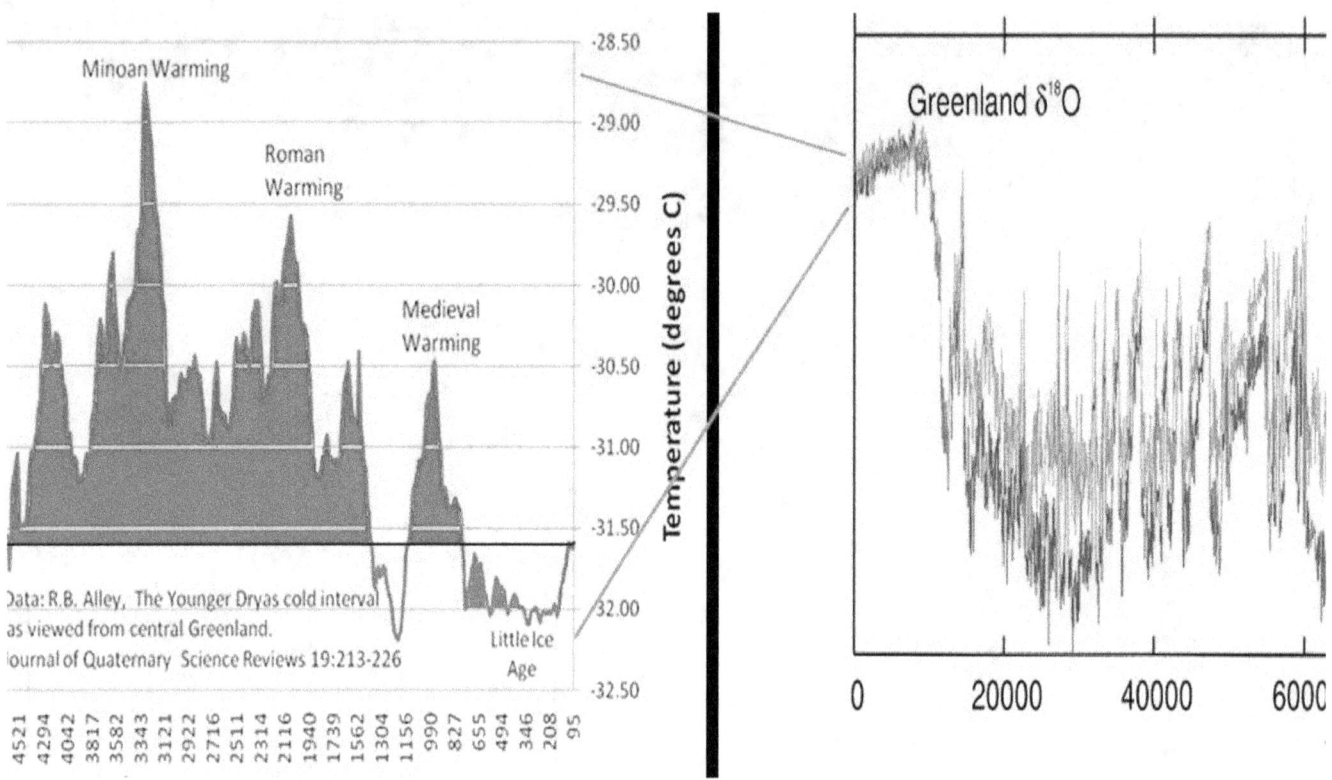

The difference between the two climate states is so enormously large, and occurs with such a swift transition between them, that it can only be rationally understood as the result of solar on-off transition. That's what the ice core records indicate.

Produced from two different drilling sites

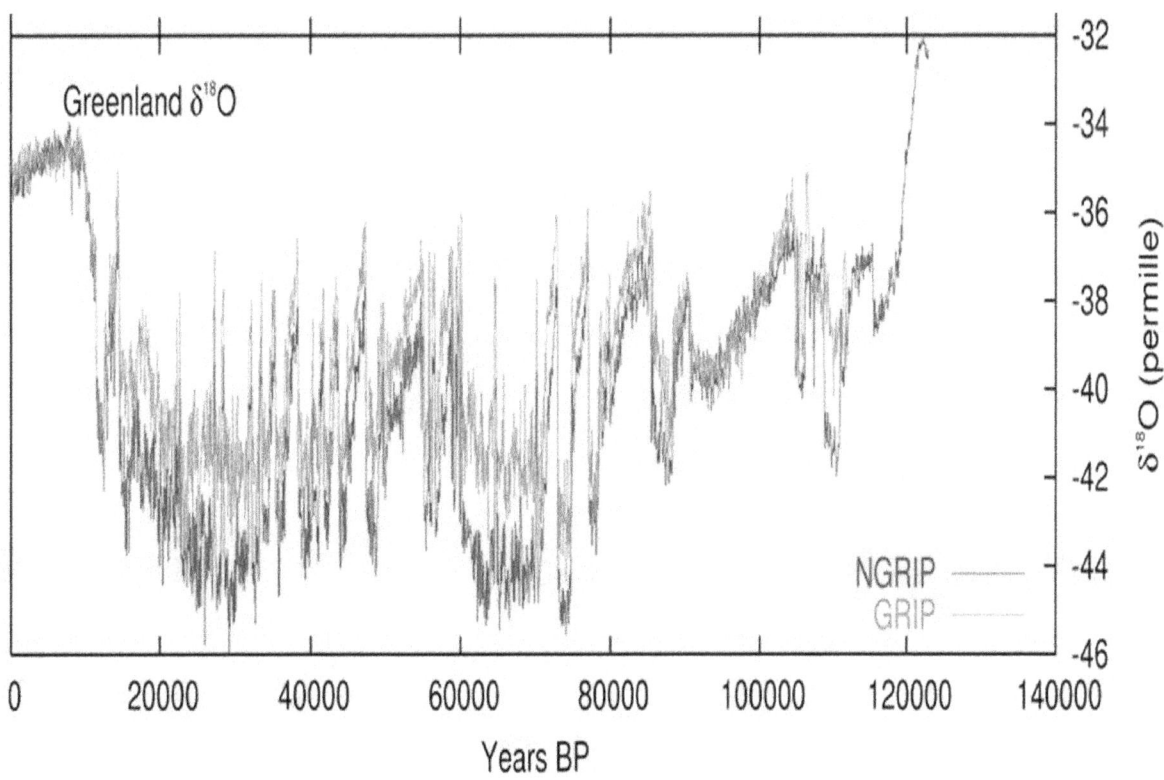

The ice core records from Greenland, that were produced from two different drilling sites, tell us both, that even during the deep glaciation period numerous events occurred that have rapidly warmed the Earth up from deep interglacial cold climates to near interglacial conditions.

These extremely large and rapid oscillations make perfect sense in the eclectic world of the Primer Fields where that come to light as miniature interglacial events that are caused by the Sun becoming periodically active again for short intervals, with the Primer Fields becoming re-established for as long as the conditions hold, which is totally possible in a resonating electric system.

At the beginning of the last Ice Age

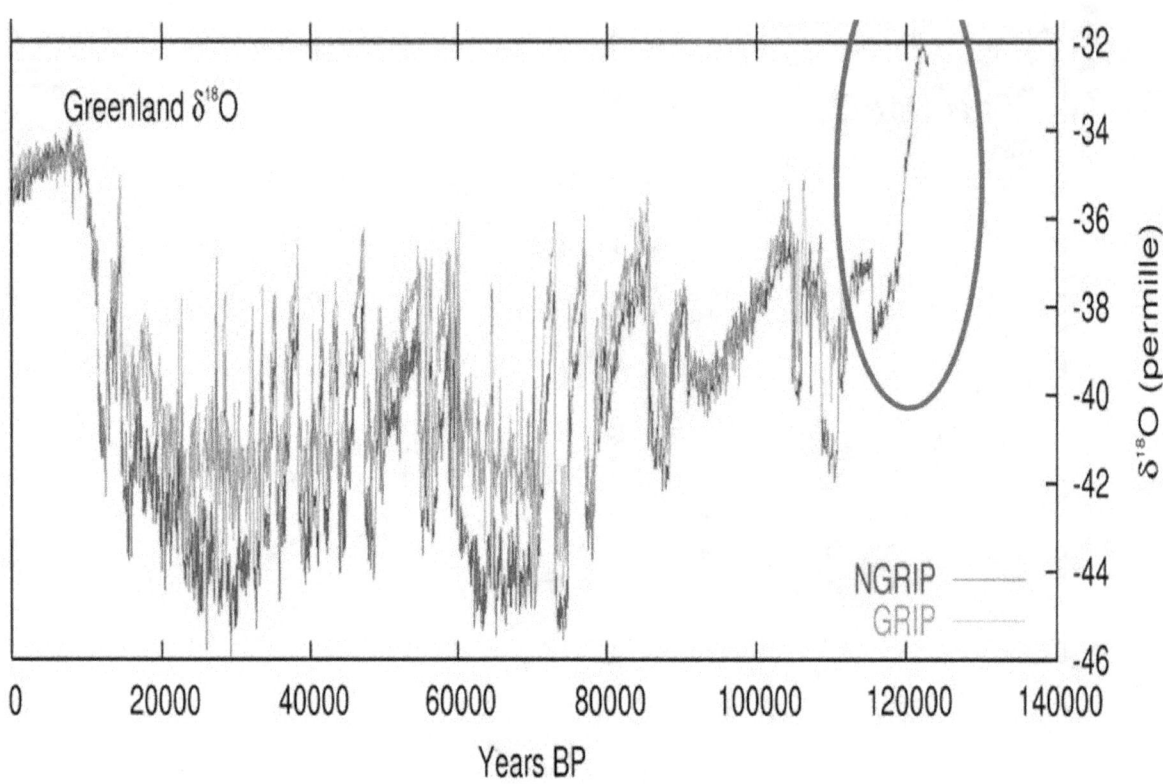

At the beginning of the last Ice Age the temperature derived from ice core samples in North Greenland dropped off steeply to about the mid-point of the deep glaciation level. This made the Earth about 20 times colder than the Little Ice Age had been.

Rapid oscillations in Greenland ice

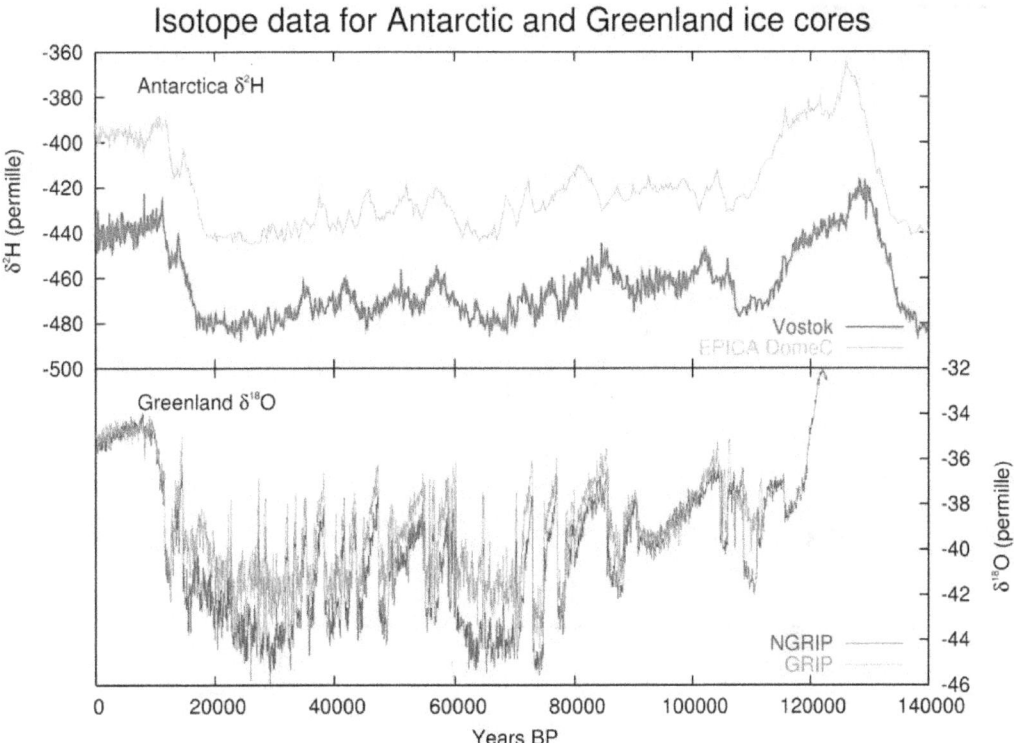

However, the sharp drop-off doesn't show up in ice core samples drilled from the ice in Antarctica, nor do the rapid oscillations show up that span the entire glaciation period, which are clearly evident in the Greenland ice core samples.

The Sun can alternate on and off states

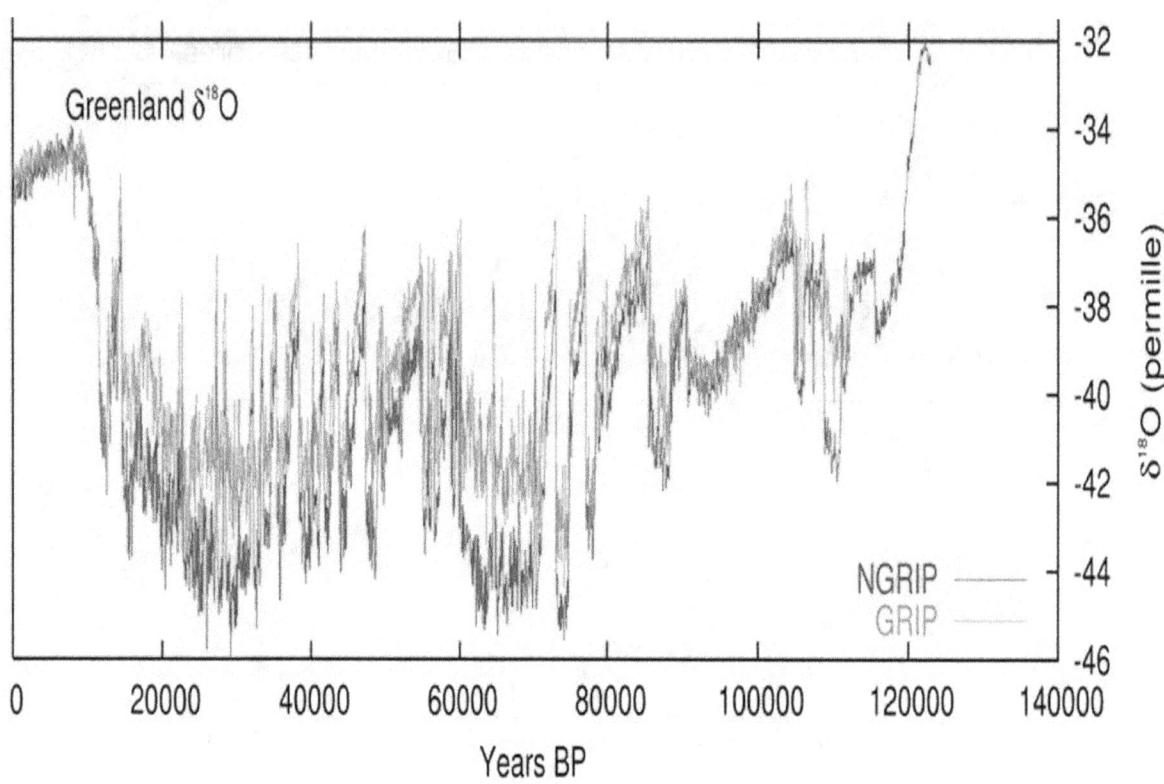

Does this mean that the rapid oscillations did not occur, or might be local occurrences? It is tempting to assume this, because without the Primer Fields theory that makes it rather plain that the Sun can rapidly alternate between on and off states, it is almost impossible to explain the large and fast climate oscillations that the Greenland ice core samples tell us of.

Greenland ice is much more sensitive

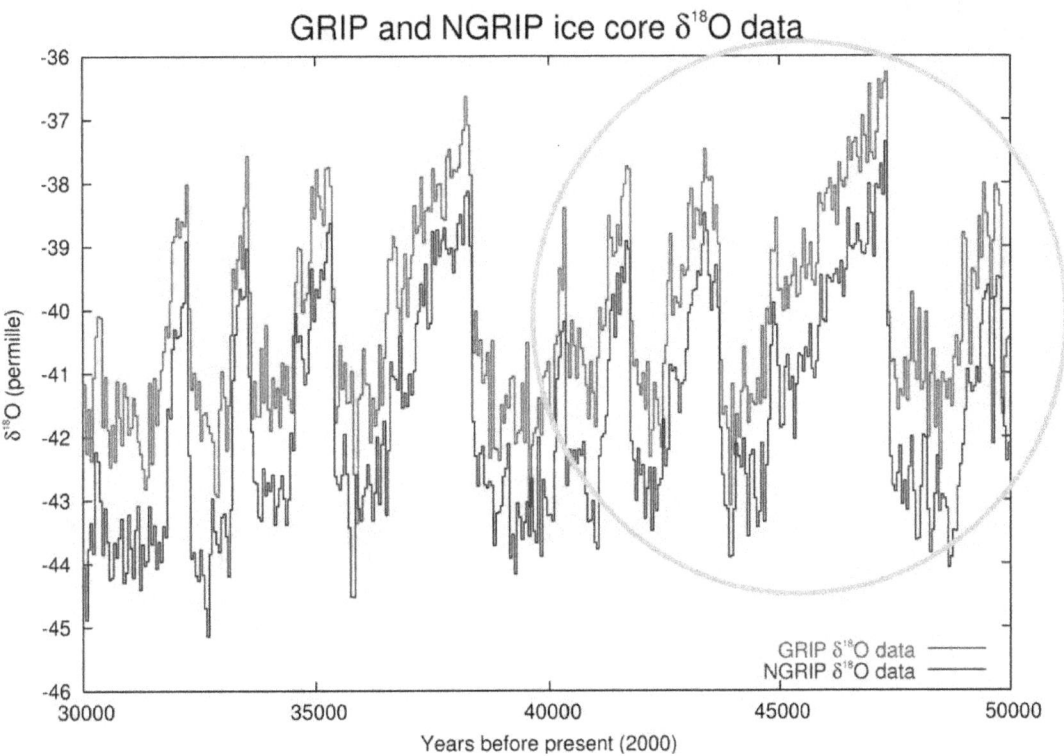

But with the Primer Fields theory considered, the enigma resolves into simply a series of events in which the Sun is actively powered for a brief span, followed by a longer span of the Sun becoming inactive again.

The Greenland ice is much more sensitive, and able to preserve these fast fluctuations, than the ice in Antarctica.

Antarctica, being an ice desert

Antarctica, being an ice desert, one of the driest spots on earth, it would be naturally less responsive to short-term variations of the type that we would expect to see when the Sun switches from its powered state to its non-powered state, where it shines only 'dimly' with its internally stored up energy.

Oxygen isotope O-18 ratio is temperature sensitive

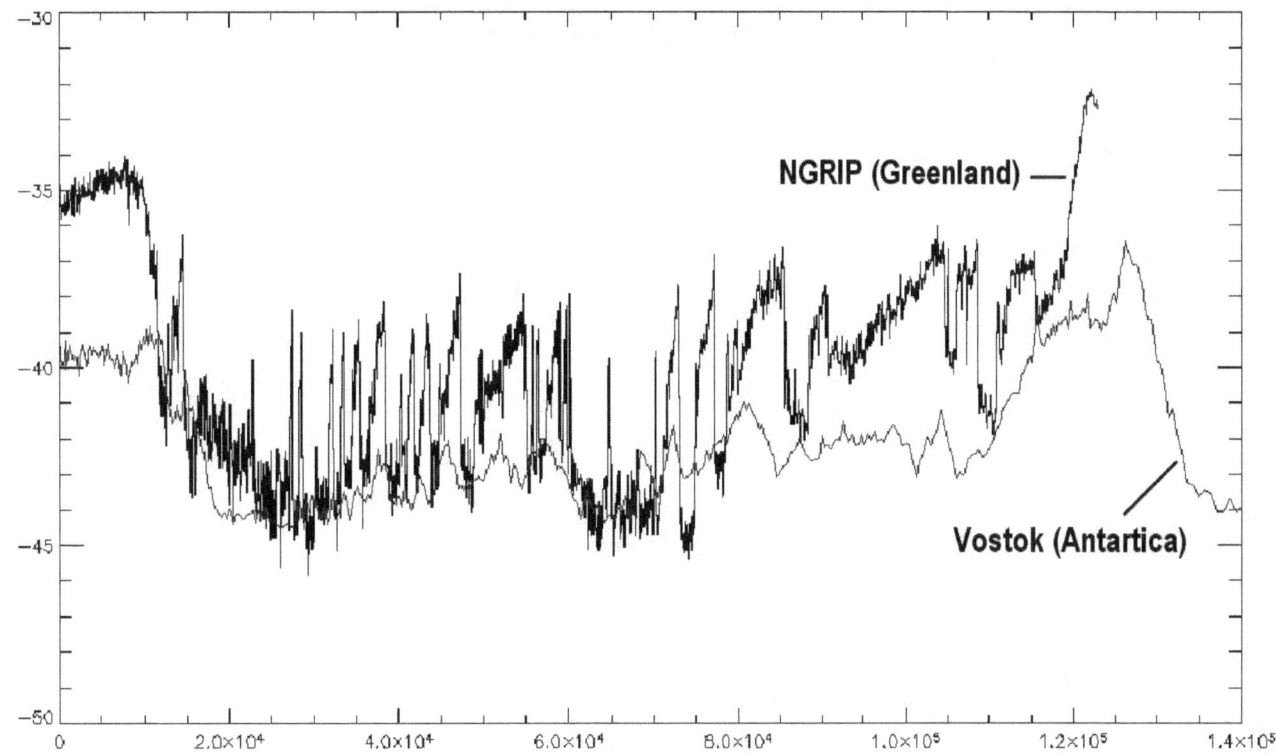

The type of rapid fluctuation that would be indicative of on-off transitions of the Sun would evidently be more strongly apparent in the ice core samples drilled from the ice sheets in Greenland, which is far from being a desert.

The temperature record that is shown here is from the North Greenland project where the temperature range gleamed from the ice samples is several times larger than the equivalent in Antarctica. In both cases the temperature gradient is gleamed from the ratio of the heavy oxygen isotope O-18 in the air, or the heavy hydrogen H-2 in Antarctica. This ratio is temperature sensitive. Colder temperatures produce a greater concentration of O-18.

Antarctica the washed out major trends

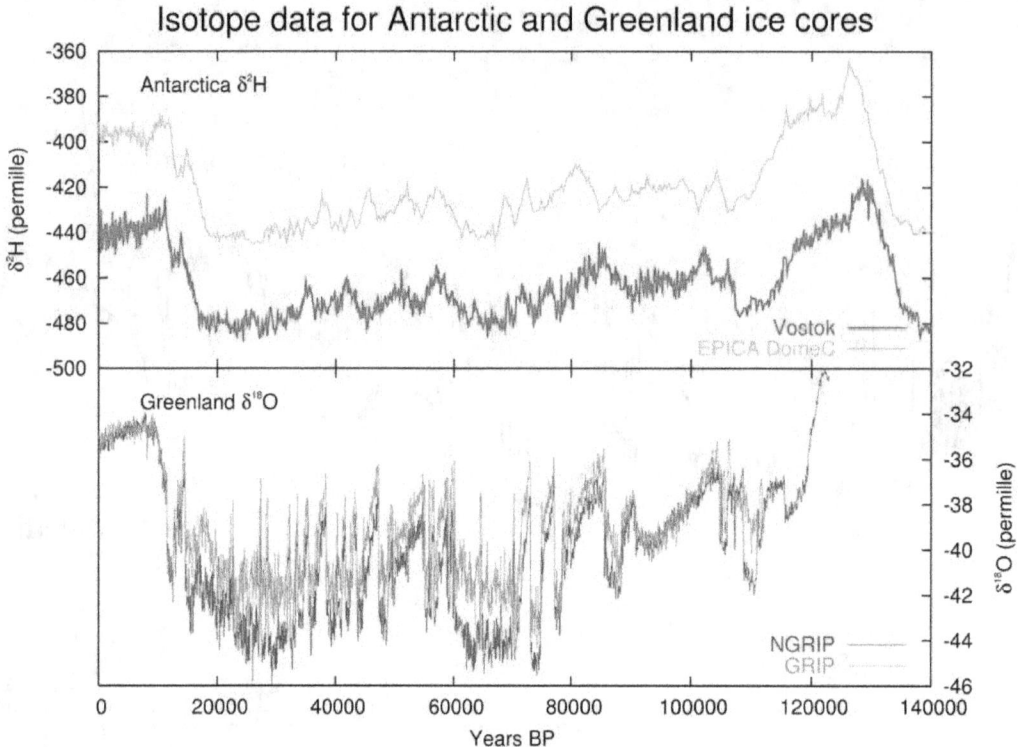

The baseline concentrations also vary somewhat between the drilling sites, both in Antarctica and in Greenland, but show the same pattern. What we see in the Greenland patterns presents strong evidence of the type that one would expect for solar on-off conditions, of which the Antarctic ice shows only the washed out major trends.

Rapid fluctuations in the Greenland ice

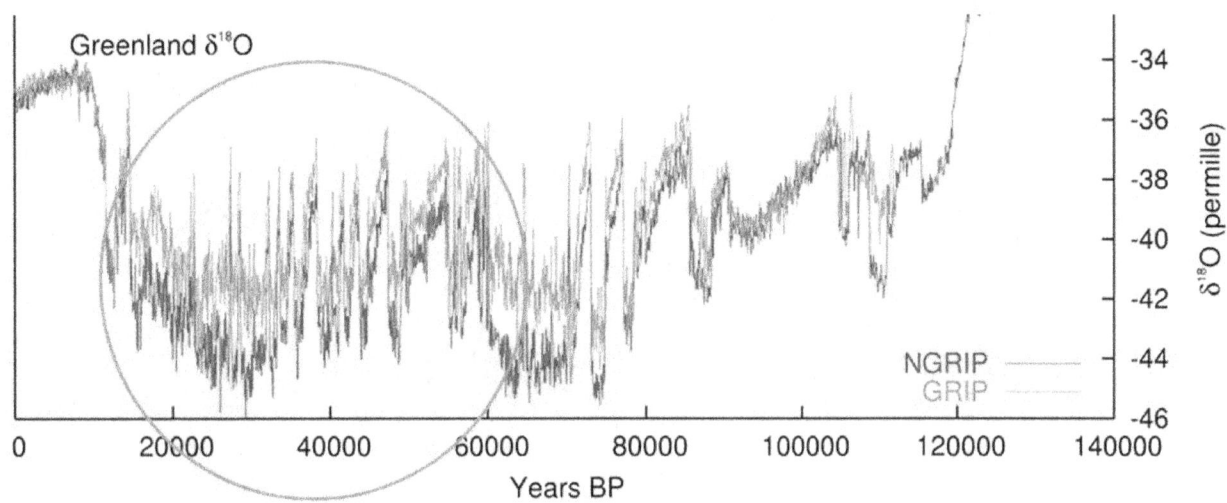

The rapid fluctuations that are detected in the Greenland ice, are called the Dansgaard Oeschger oscillations. These are transitions from deep glaciation conditions, almost all the way back to interglacial conditions.

Dansgaard Oeschger oscillations

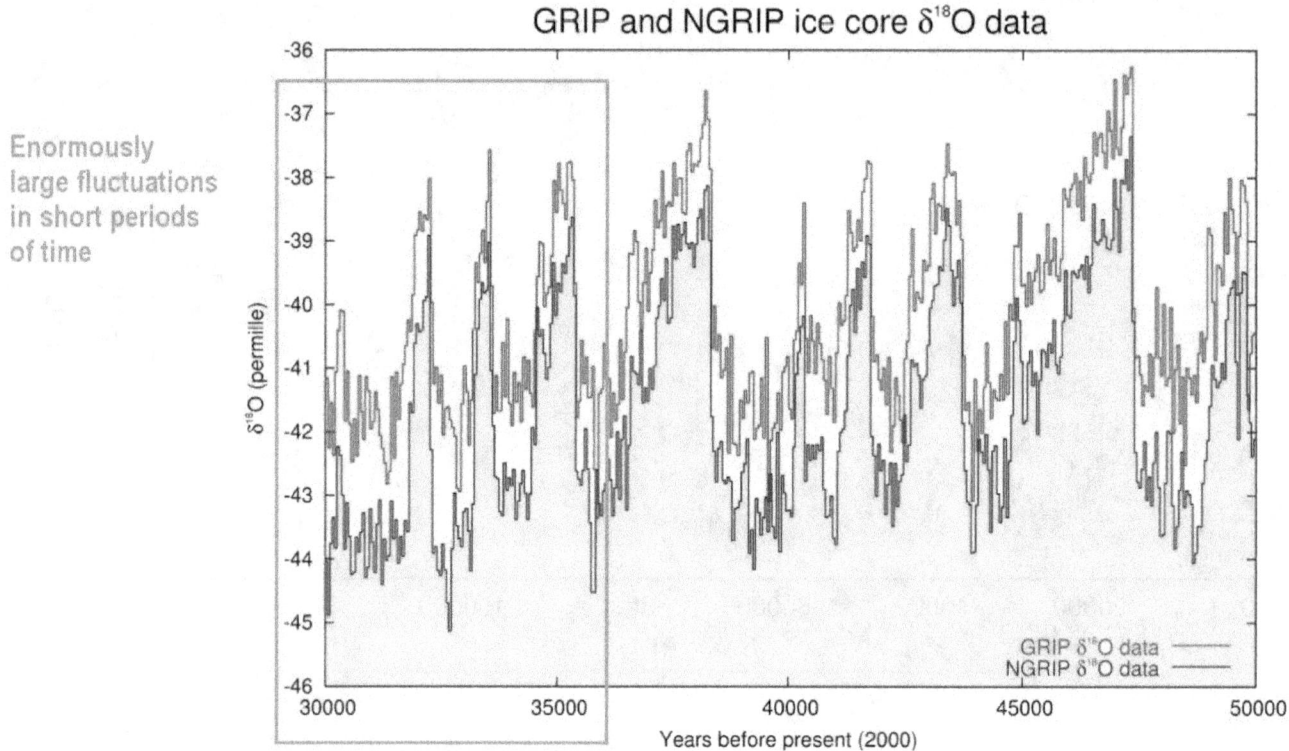

Enormously large fluctuations in short periods of time

These are enormous fluctuations. They tell us that the Earth heated up from full ice age conditions to near interglacial temperatures in twenty to thirty years, and then cooled down again, gradually, to deep glaciation conditions, in a few hundred years. This rapid warming, and gradual cooling, is evident in all of the big Dansgaard Oeschger oscillations. All told, 25 of these big fluctuations have been recognized.

Power-off of the Sun

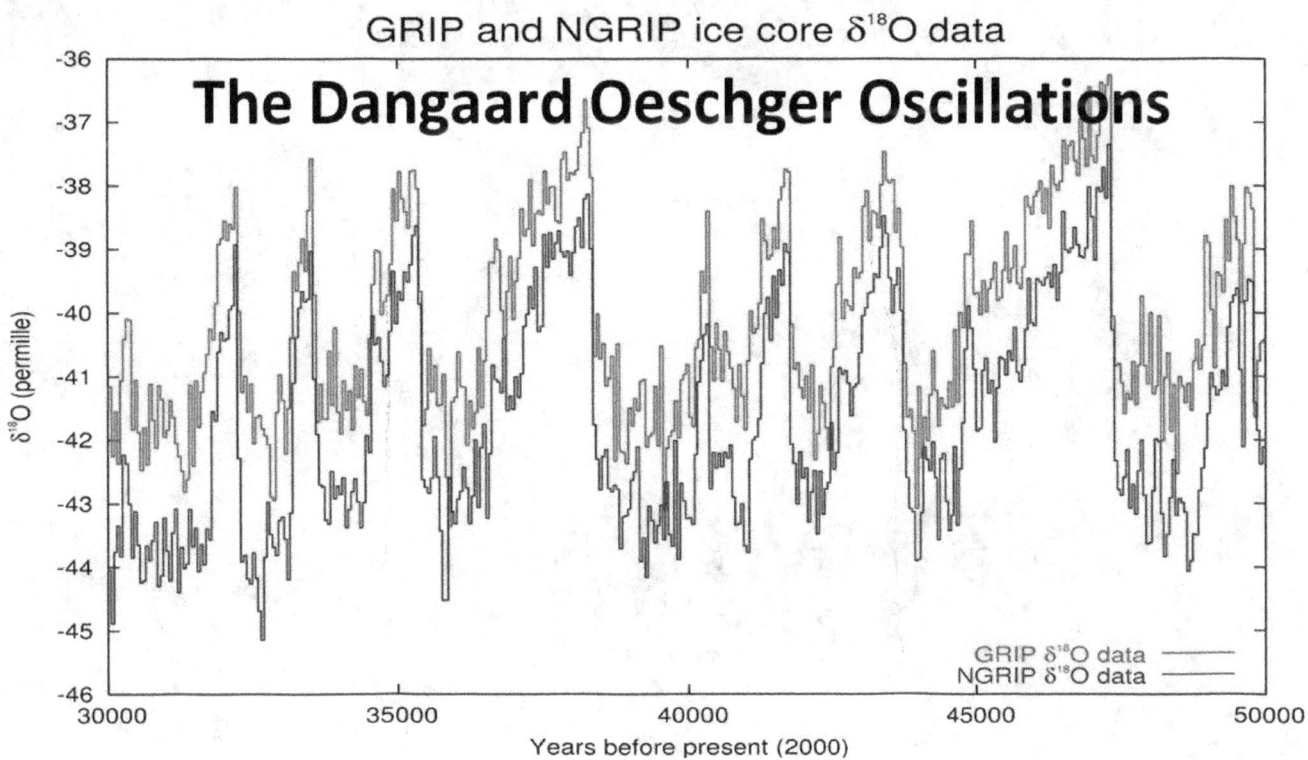

In some cases where the interval is long - where the gradual cooling spans a longer period - we see evidence of small, sharp, upwards spikes along the way, suggesting that numerous short bursts of the powered state of the Sun have occurred that have caused a periodic re-warming of the Earth, and also of the Sun itself, internally.

The evidence suggests that the last entire ice age, and those before, were created by a long series of the power-off state of the Sun interspersed with short periods of the Sun being fully powered. It appears that whenever the Primer Fields are established, the Sun is actively powered to roughly its full potential, and that both the Sun and the Earth cool down during the longer powered-off periods.

Giant red sprites

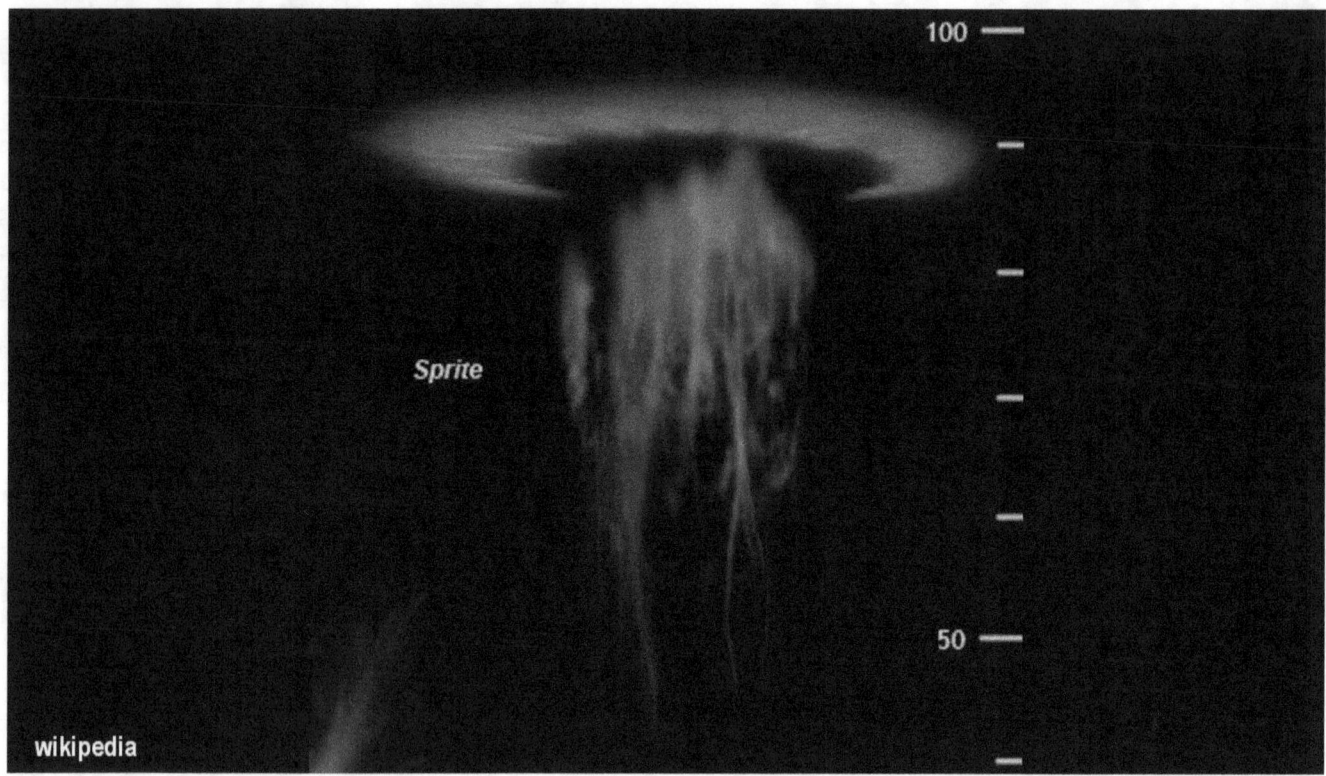

It may seem irrational to assume that such a giant system, as that which feeds our sun, can turn on or off in the space of the day. However, in electric systems, such rapid transitions are totally natural and are readily observed in the natural environment. When storm clouds reach high into the atmosphere, an electric field becomes established at times that extends into the stratosphere and causes a plasma-flow connection.
When this happens the basic pattern of the Primer Fields appear in the sky. They flicker on strongly in the shape of giant red sprites, and then vanish just as fast. They rarely last for more than a second. The Primer Fields that activate our Sun, evidently can 'flicker' on and off in a similar rapid manner.

After the Sun turns off

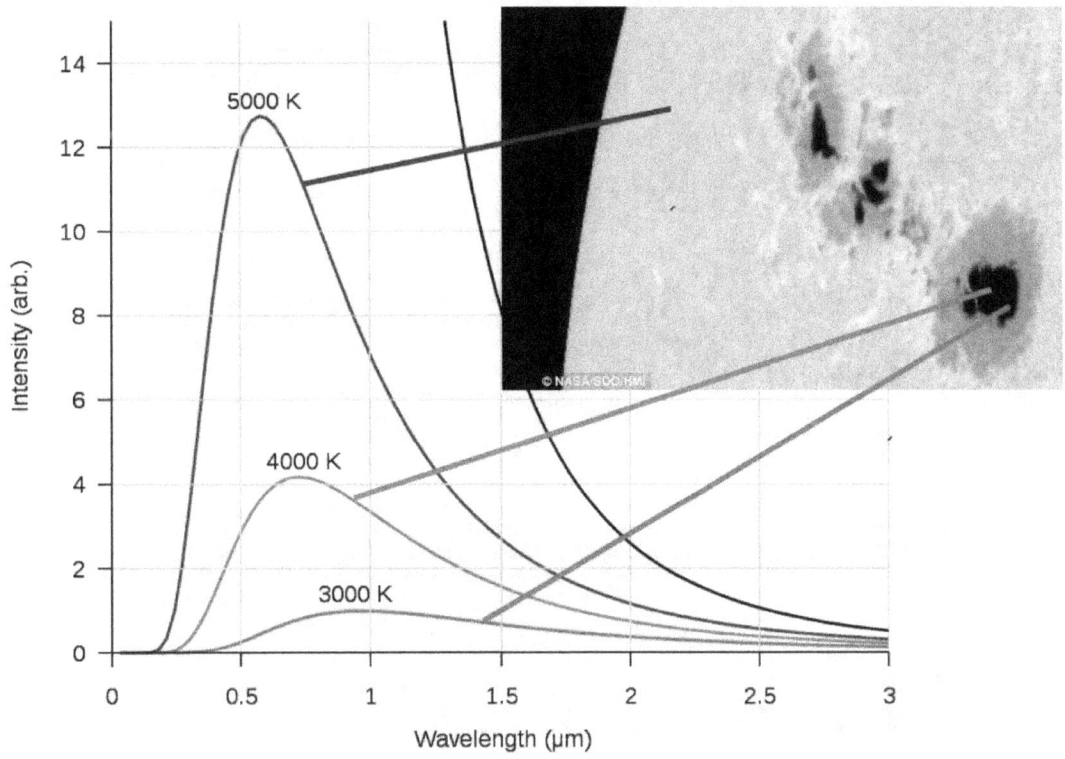

What we see in the resulting relationship between the on-and-off periods for our Sun has an enormous impact on what will happen to the agriculture that we depend on, the moment that the turn-off happens that marks the start of the next Ice Age. While we may see a gradual cooling of the climate of the Earth after the Sun turns off, the effect of the cooler Sun will have an immediate impact on agriculture. Not only will there be radically less energy available for the chlorophyll of the green plants to function, but the radiation spectrum will also shift away from the wavelengths of the visible light that are critical for the absorption by chlorophyll, without which plants cannot grow.

When the Sun turns off, we will see an immediate 70% reduction of the total energy coming from the Sun, and an immediate shift of the energy profile towards the red. We can compensate for the energy loss by placing our agriculture into the tropics, enhanced with artificial lighting, and by placing some parts of it directly into indoor facilities with 100% artificial environments.

Absorption spectrum of chlorophyll

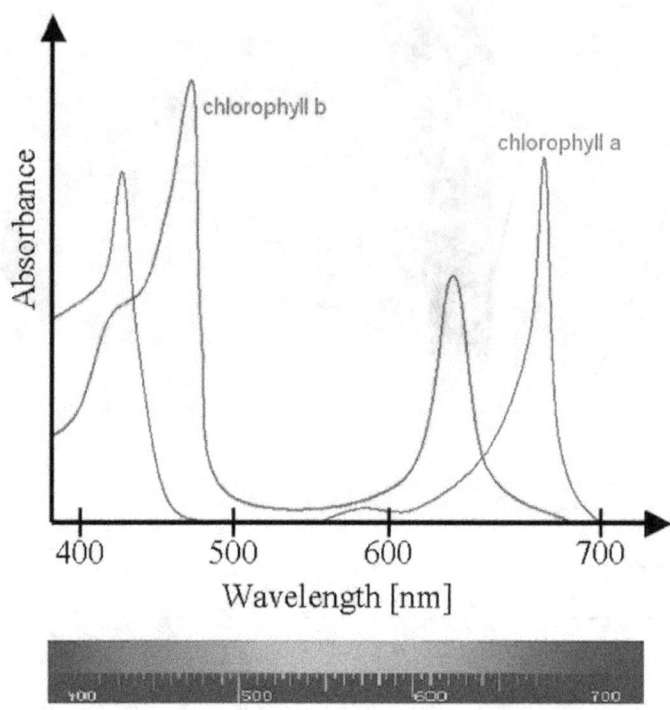

While the Sun becoming inactive poses some challenges, the challenges are not insurmountable. As you can see, both types of chlorophyll get most of their energy from the shorter wavelengths below 500 manometers, which the Sun provides even less of when it dims down. However, since the absorption spectrum of chlorophyll is narrow and specific, only small amounts of energy are required when the lighting is tuned to the absorption bands. At the present time, only 2% of the solar energy received is actually utilized by the plants. The entire amount then, can be provided with relative ease with the use of nuclear or more advanced types of electric power systems.

When the Sun enters its off-state

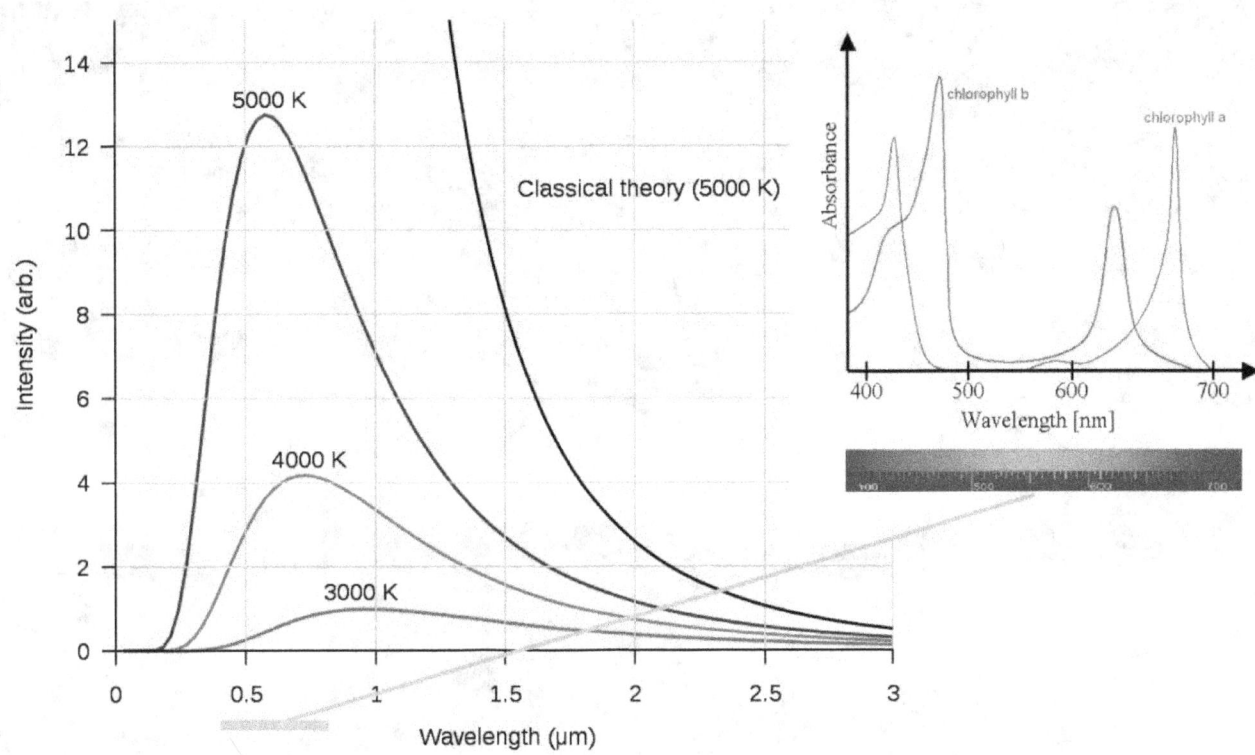

This means that when the Sun enters its off-state, that is its inactive residual heat-state, the remaining sunlight will be essentially useless for most types of agriculture existing today. This means that all the new agricultural platforms that need to be built to maintain our food supply, will have to be in place and be operating, before the day the Sun becomes inactive. This is the new reality. The transformation of the Earth will happen almost 'instantly,' possibly in the span of just a single day, or less.

When the solar off-transition happens, and it will happen, all of the temperate-zone agriculture, where presently nearly all of the world's food is produced, will be 'instantly' disabled. The transition won't happen gradually. It will happen 'instantly.' It will happen without warning. And when it happens, the consequences will begin immediately, on the very day. That's what we have to be ready for. That's what we must prepare for. The larger climate transition that unfolds along the way, in which the snows no longer melt, but increase, as big as this will all be, will actually be of lesser importance then.

Agriculture afloat on the equatorial seas

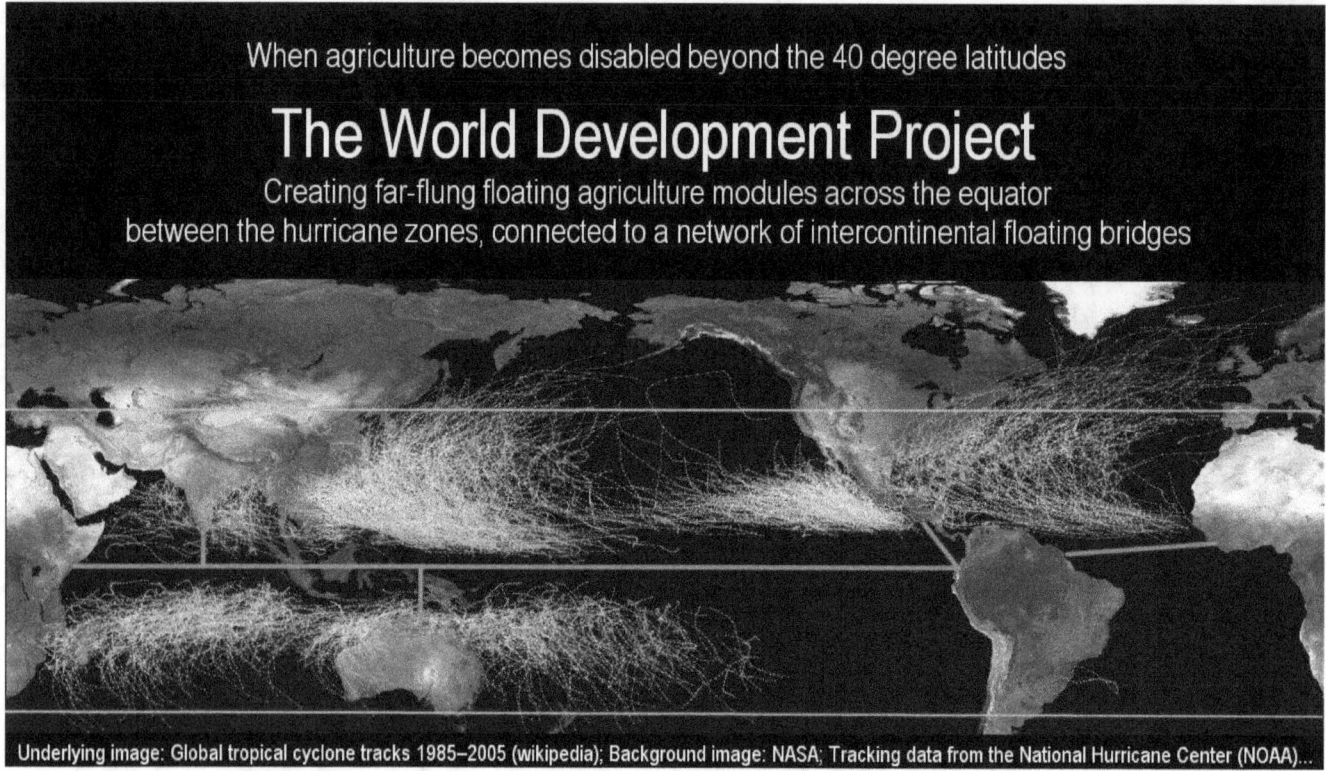

The impending ice age transition thus forces us to begin the greatest world development ever imagined, and to do this on a gigantic scale, like placing 90% of the world's agriculture afloat on the equatorial seas, all connected by floating bridges, and it being serviced by a new society living in floating cities along the way.

The requirements that we must meet for the future, demand us to start a new age of automated industrial production in the present - this means now and without fail - utilizing nuclear-powered high-temperature processes with the use of basalt as the feedstock. Basalt is none-corrosive and is lighter and stronger than steel, and is infinitely available. This also means that houses must be produced in automated industries, where they are produced at such a low cost in efforts that they can be given away for free as a part of the new infrastructures for living that must be produced, for the human journey to continue during the long nights of deep glaciation.

Worse than the effect of a nuclear war

If no preparations are made that compensate for the loss of the traditional food supply that results from the instant transformation of our planet when the Sun turns off, then most of humanity will simply starve to death. This isn't something to aim for, is it?

For this reason the floating agriculture will be built, with enhanced lighting, complete with floating cities to service them. These things will happen, because if we fail, the resulting effect will be worse than the effect of a nuclear war. The present global food reserves won't last for no longer than just a few months. It won't be a pleasant thing to watch seven billion people to starve to death. Only a few million made it through the last glaciation cycle alive.

We want to do far better this time around. And we will do better. The Primer Fields theory opens the door to understanding what we are up against, and what we must prepare for. This gives us an advantage that did not exist in the earlier times, but which exists now for the first time in the entire history of life on our planet. Likewise the technological power exists for this to happen, which makes it possible for the first time too, to build the infrastructures that we require to 'weather' an ice age with.

Where the sunlight is the strongest

To get there, our entire civilization will likely have to be rebuilt and set afloat in the tropics where land is scarce. This will happen. Shifting our agriculture into the tropics, where the sunlight is the strongest, will offset the early portion of the loss of the solar energy as the Sun is powered off. Some form of minimal artificial lighting will likely have to be added, and artificial climate control will likely have to be added even in the tropics, together with increased CO2 concentrations for increased plant growth.

These infrastructures will require some extensive scientific advances in plant biology and in physical engineering technology for the floating new environments, and the whole thing will have to be in place, and the infrastructures be operational, without fail, before the transition happens. This means we should get started now. We still have a chance to get this critical work done. We better not waste this chance before us, by doing nothing.. It may be our last chance.

The next deep glaciation to begin

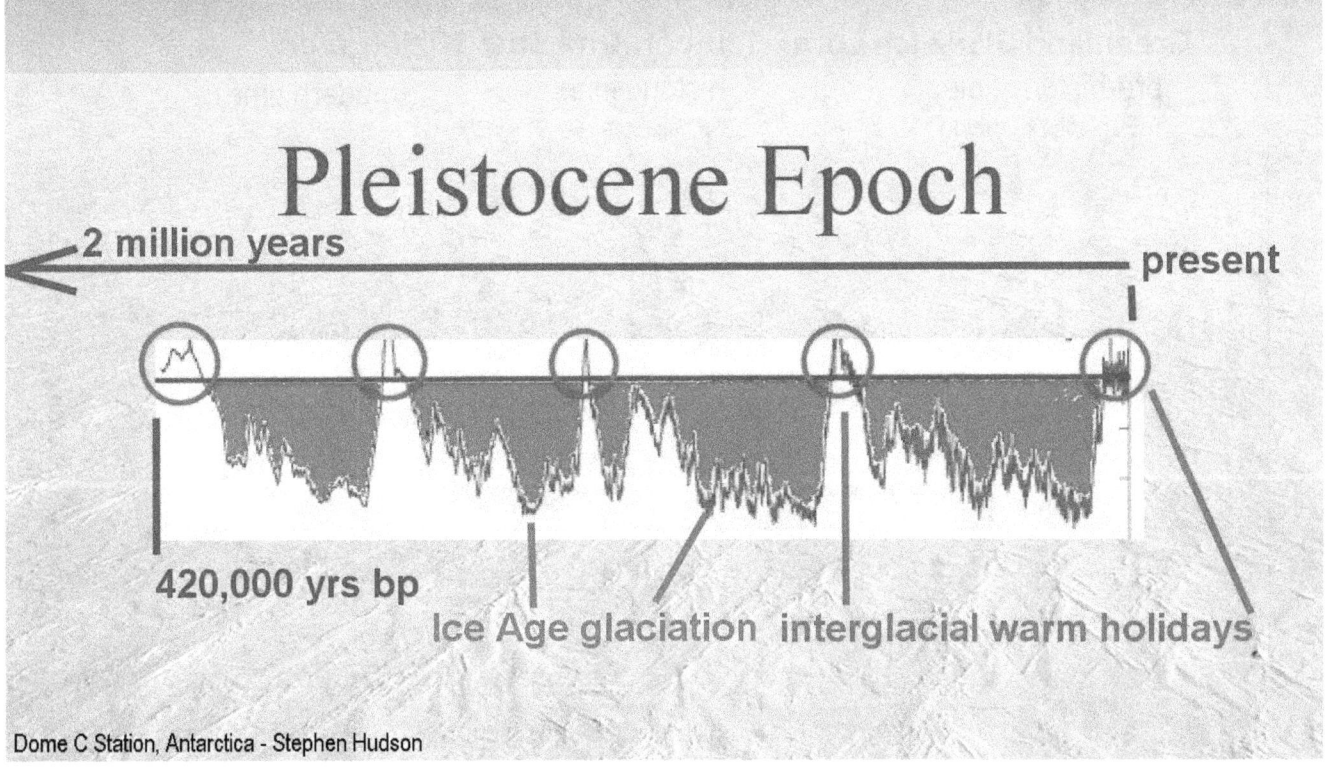

The ice age transition will happen without fail. We can count on that. We are close to the transition to the next deep glaciation to begin, though we don't know how close, close is.

The cut-off level

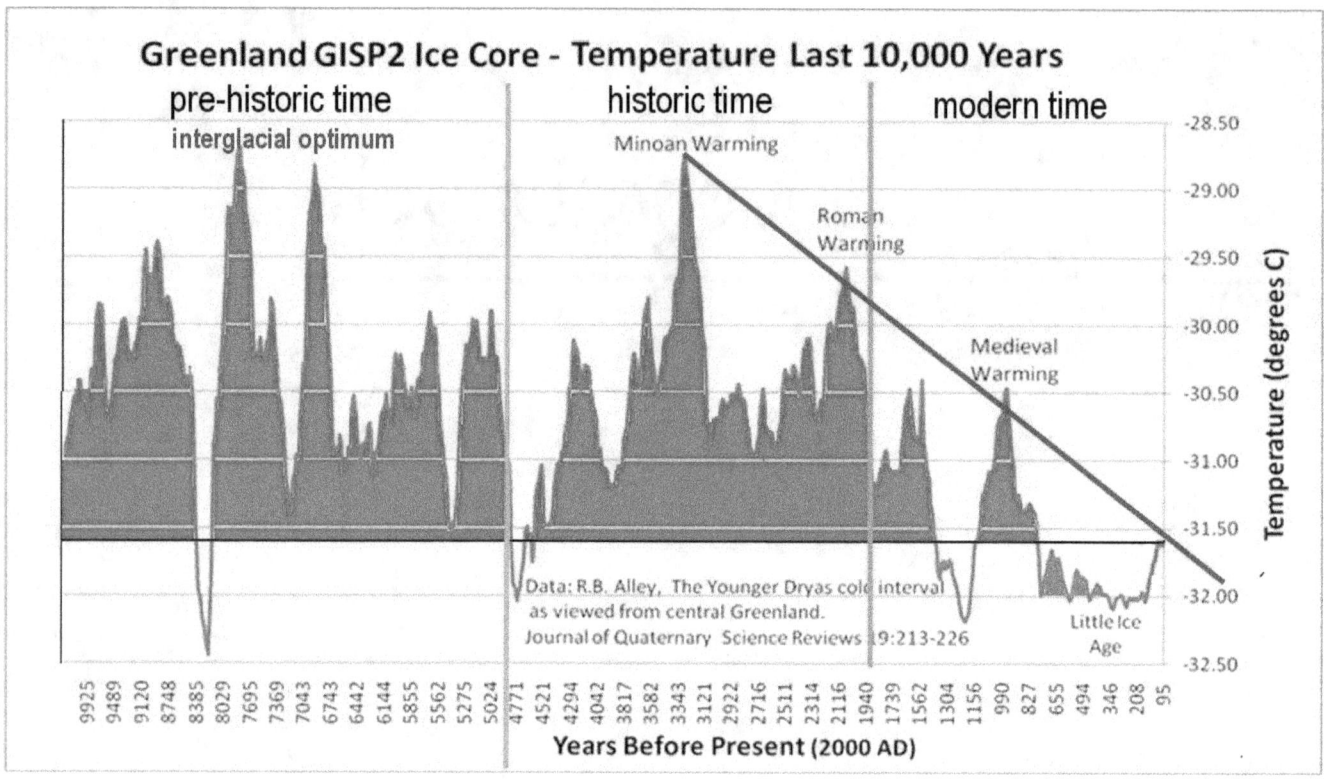

We also know that the trend towards the next glaciation has been progressing for 3000 years already, and is accelerating. We don't know exactly where the cut-off level is, beyond which the Primer Fields collapse and the Sun turns off. It may not be far below the level of the Little Ice Age. At the present trend, we may get to this point in twenty or thirty years, or fifty at the most.

NASA's Ulysses spacecraft

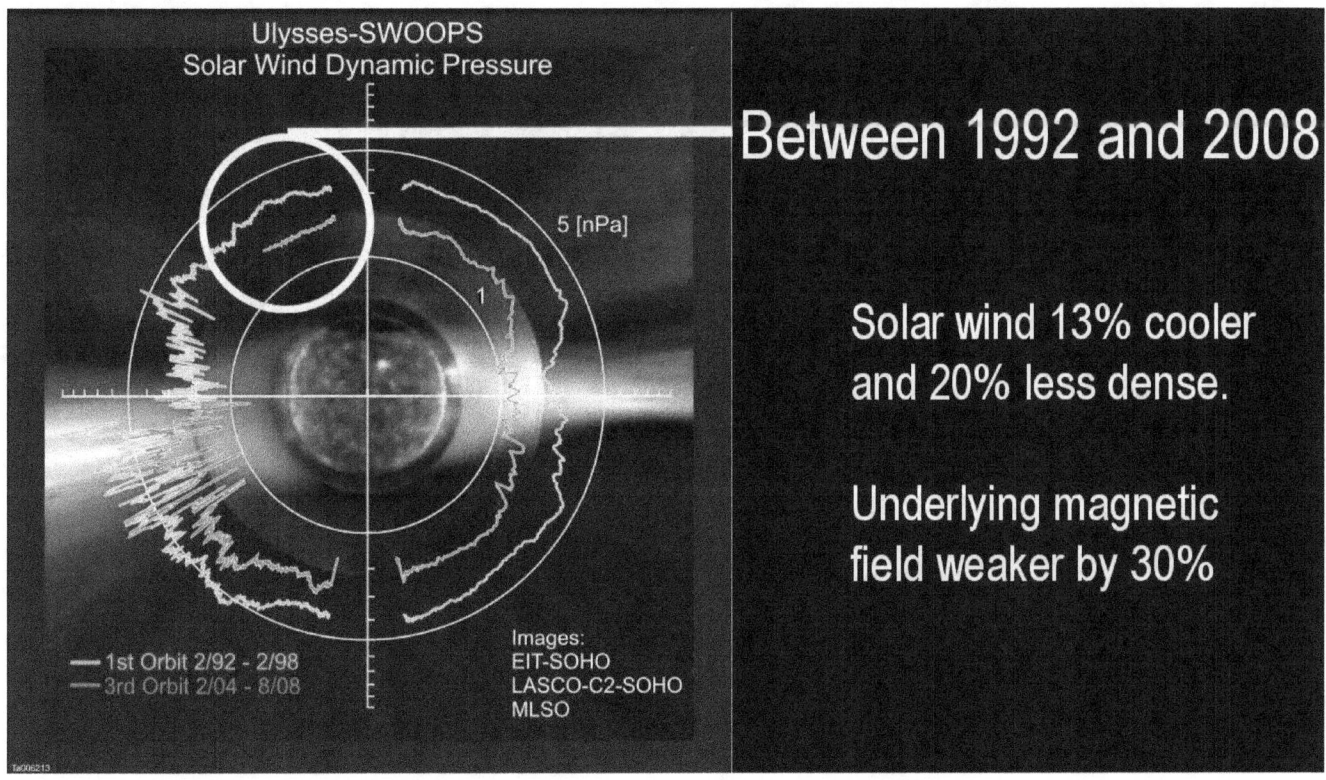

NASA's Ulysses spacecraft saw a 30% reduction in the solar wind pressure happening in just a single decade. That's not a small drop-off. And the trend is continuing. If our agriculture has not been transferred to the tropics before the high-power solar system stops, its game-over for humanity. Humanity will likely become extinct then, by the lack of food, except for a minuscule remnant of a few million that might escape the universal death. If death by starvation is what you wish for yourself and your children, then sit back and do nothing, because that way your desired fate will be assured with great certainty.

The brilliant life-giving 'fire' in the sky

The stepping away from this fate begins with the recognition that our Sun would not be the brilliant life-giving 'fire' in the sky without a dense plasma sphere surrounding it from which it draws its electric power that lights up its photosphere with electric plasma interaction.

If this realization is made, half the battle is won, because everything follows, and the needed steps become logical. And this realization shouldn't be hard to make.

It has been recognized a long time ago that our Sun is not the steady state nuclear fusion furnace, which it is widely said to be, but is powered by one of the vast networks of plasma streams that pervade the galaxies and the cosmos as a whole. Since plasma has mass, the plasma surrounding the Sun becomes attracted by the Sun's gravity and interacts with its outer atmosphere, the photosphere, which thereby becomes excited to 5,780 degrees Kelvin. It is that simple.

The Primer Fields will vanish in the near future

The Primer Fields prime the environment in which the simple process unfolds for as long as the fields exist. Once you realize that this presently operating system will vanish in the near future, and that solutions are possible for humanity to move forward in spite of the dimmer Sun, you may become inspired thereby to get out of the easy chair to gain a fuller understanding of how the critical process operates on which your continued existence depends.

Part 4 - The Primer Fields dynamics

The Primer Fields dynamics

Plasma sphere around the Sun

"The Primer Fields" lab experiments by David LaPoint on YouTube

The plasma physicist David LaPoint suggests that there is much more to the Primer Fields dynamics than what meets the eye. He suggests that our sun operates within a sphere of highly concentrated plasma that is generated by powerful magnetic fields, which he calls the " Primer Fields," and that the plasma sphere around the Sun is magnetically confined by these fields.

If this was not so, our sun would be but a dim speck in the sky, without the large magnetic fields acting on it that surround it with a sphere of concentrated plasma. The Earth would be a cold planet then, extensively covered with ice, as it once was 700 million years ago. But is David LaPoint correct? Do they Primer Fields really exist outside the laboratory environment? Can they be seen? And if they do exist and are visible, is it possible for these vital fields to collapse?

So, let's explore what stands behind it all.

David LaPoint uses two bowl-shaped magnets

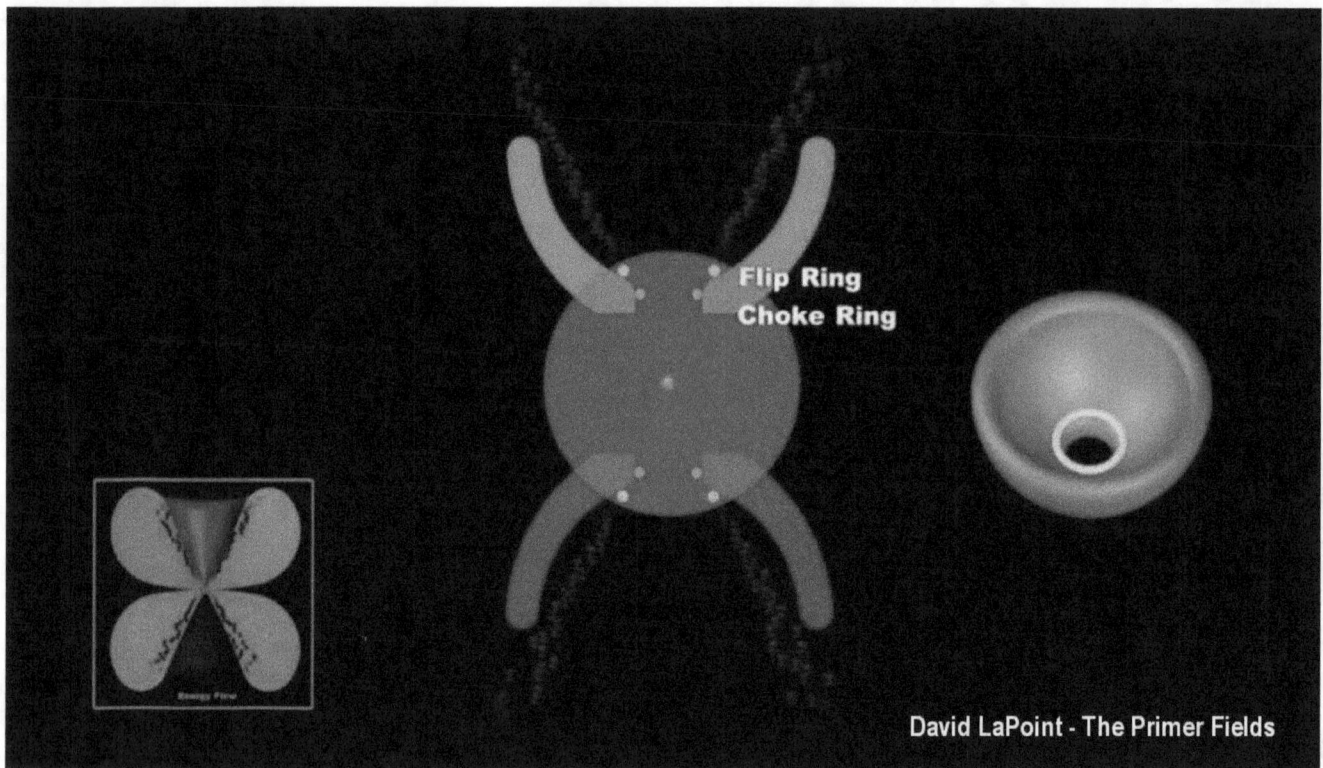

The key component of the Primer Fields theory is the existence of one or two bowl shaped electromagnetic fields. David LaPoint uses two bowl-shaped magnets of opposite magnetic polarity for his experiments conducted in a vacuum chamber.

The Zeta Pinch effect

The Sun at a plasma node

David LaPoint - The Primer Fields

In the environment of space, however, no bowl-shaped magnets hang in the sky. For the required magnetic fields to exist, they must be generated by the natural phenomenon of electricity flowing as plasma in space. And this is exactly what happens.

The term, plasma, refers to electrically charged particles that exist in free flowing form in space, primarily as protons and electrons, the stuff that atoms become made of when they are bound together. In space they are free flowing. However, flowing electricity creates magnetic fields, and by these fields the flowing electricity becomes pinched together.

When electricity is carried by two parallel wires, with the current flowing in the same direction, the wires are attracted to each other by the Lorentz force. The same happens in plasma in space where electricity is flowing freely without wires. Here the effect is called the Zeta Pinch effect.

Plasma currents in space become compressed

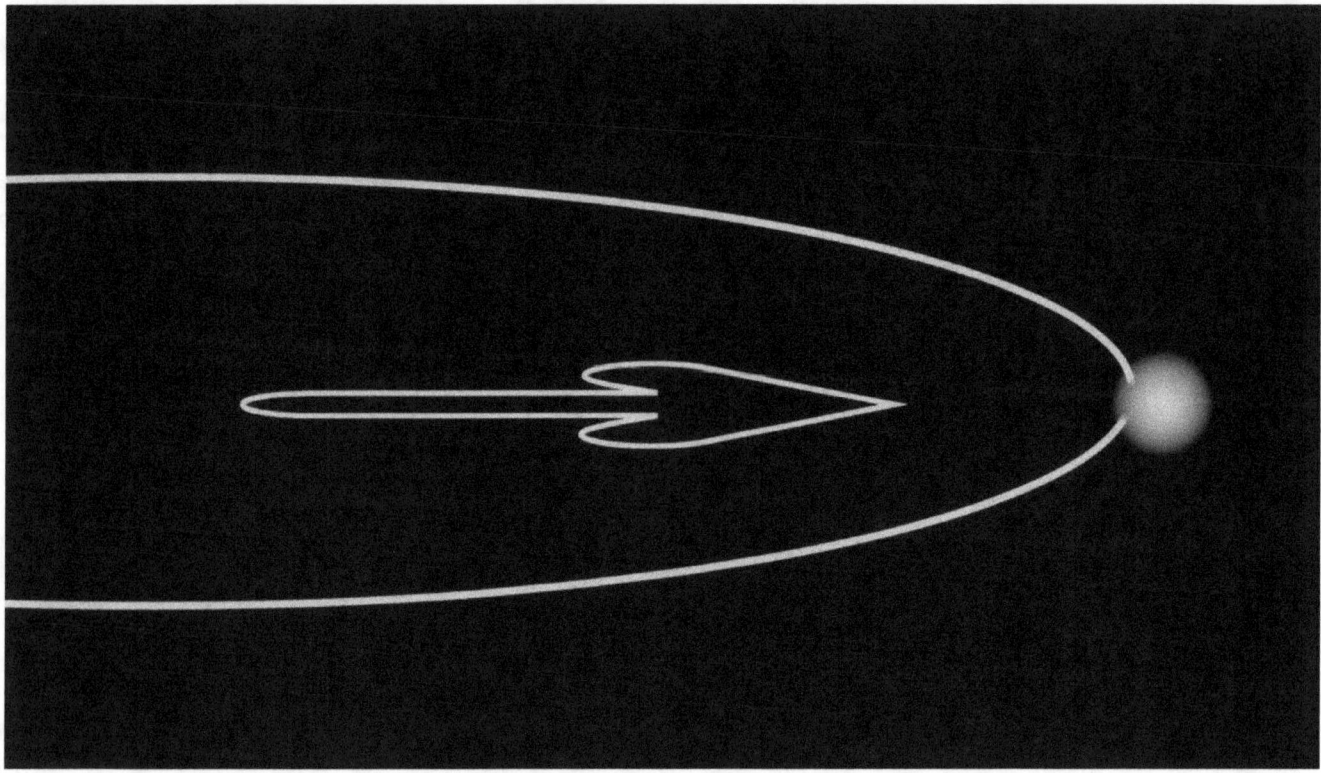

In space the flowing electric plasma particles are drawn to each other by the same magnetic forces that attract wires to each other. However while the wires remain physically fixed, plasma currents in space become compressed into ever-smaller magnetic confinement. By the confinement the current density is increased, which in turn pinches the plasma currents still tighter and tighter, forming a bowl-shaped magnetic field in the process at the very end of the pinched plasma stream.

Electromagneticly confined 'high-density' plasma On the platform of the Primer Fields

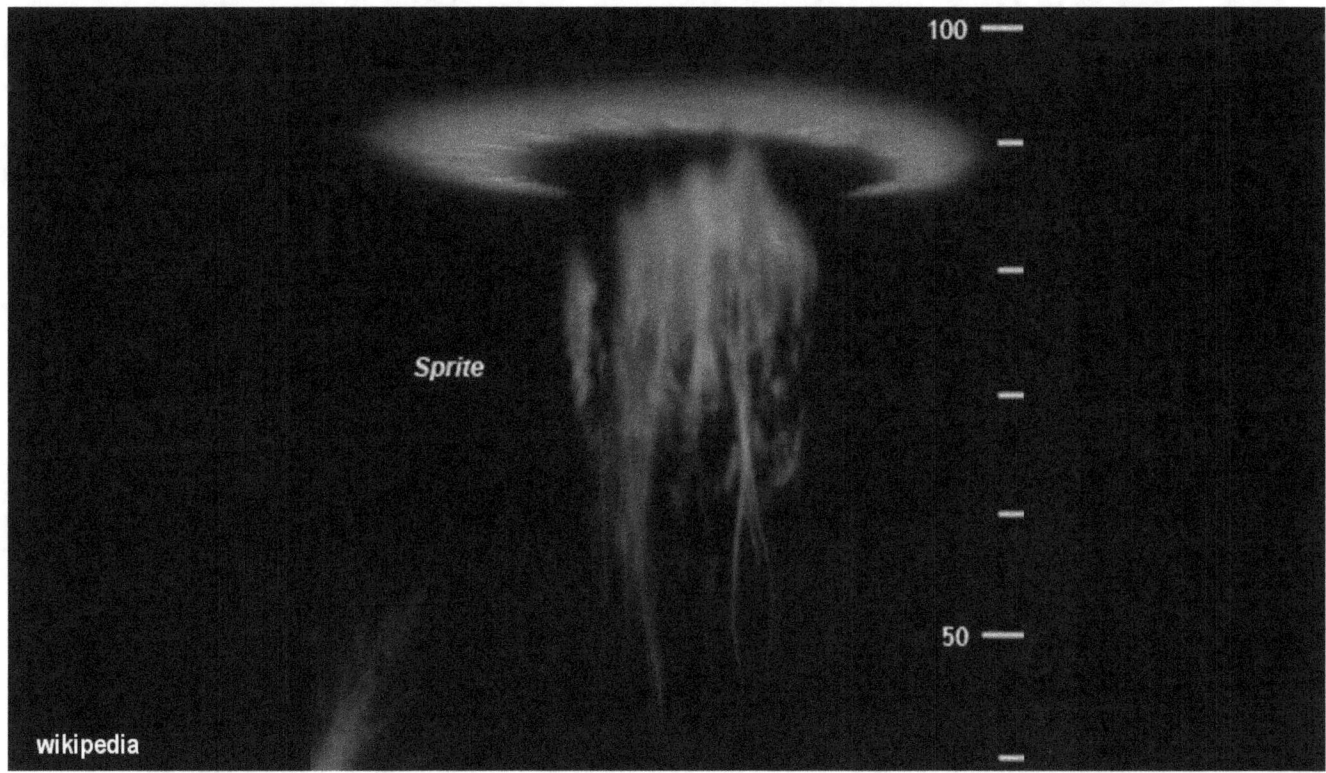

But something happens in the bowl when the plasma currents exceed a critical limit, as they converge to ever-tighter confinement. The currents and the resulting magnetic fields become unstable by the pinch effect, past a critical point. The currents become 'twisted into complex knots' whereby the bowl-shaped magnetic field that forms, opens up at the center. There, below the opening, an electromagneticly confined 'high-density' plasma stream is formed by a number of interacting effects.

On the platform of the Primer Fields the entire structure that we see that extends across 50 kilometers from top to bottom, forms in a fraction of a second. Typically the sprite remains active for about a single second, until the plasma flow becomes too weak to maintain the Primer Fields. At the point when the Primer Fields collapse, the entire structure simply vanishes.

Interglacial period of the Sun's active time

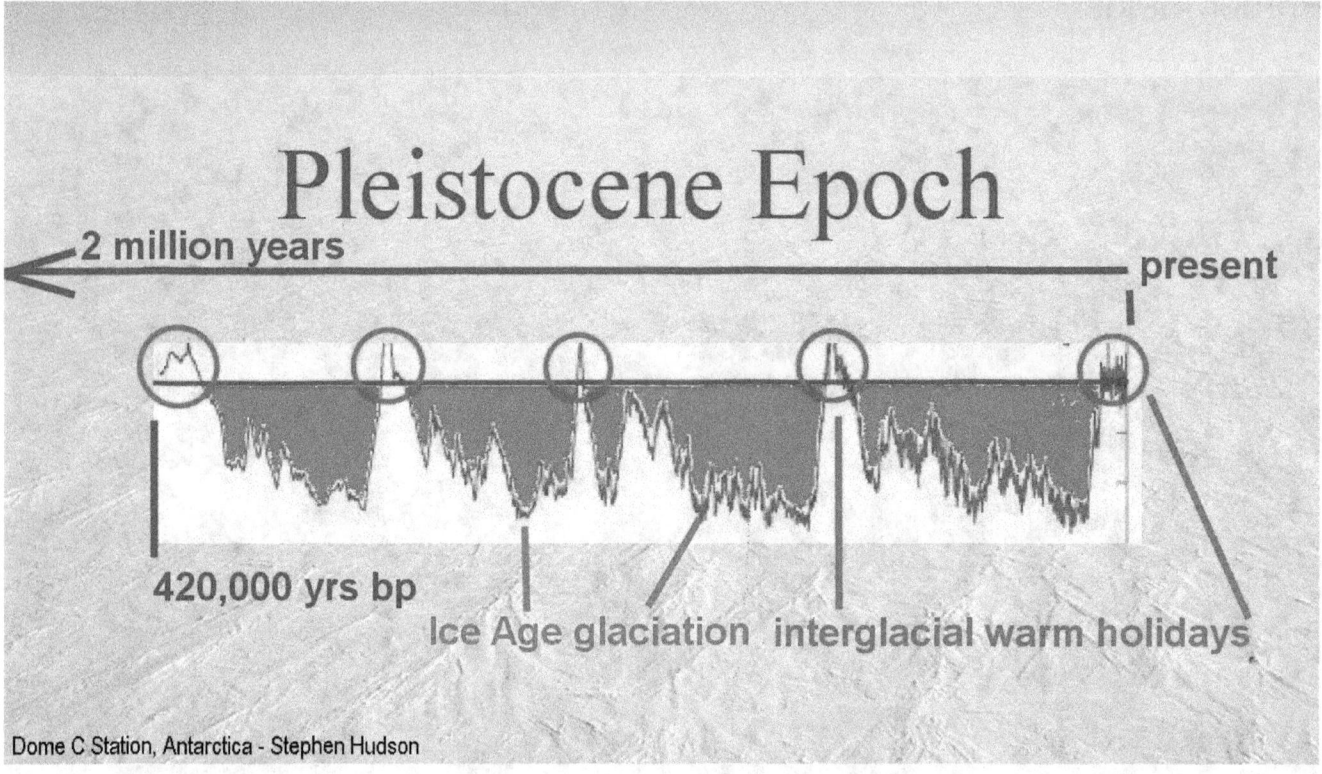

In comparing the sprite with the larger scale of the solar system, the sprite's one second active time is comparable to the solar system's interglacial period of the Sun's active time of roughly 12,000 years. The on-off process is the same in both cases, though different in scale.

Let me illustrate now how the process functions

Let me illustrate now how the process functions that forms the Primer Fields.

Dense Plasma Focus Device

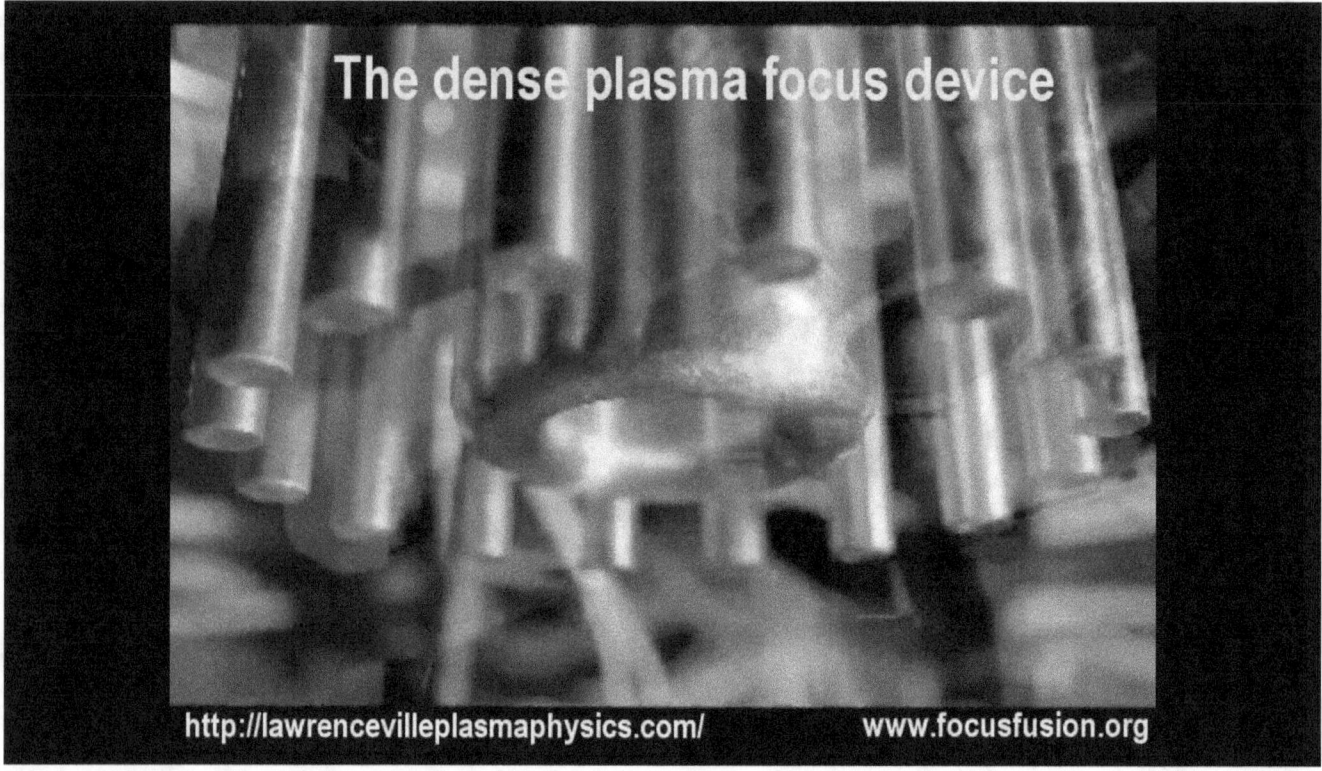

'This is best illustrated by looking at another lab experiment, that is carried out with an instrument called the, Dense Plasma Focus Device. '

A ring of electrodes

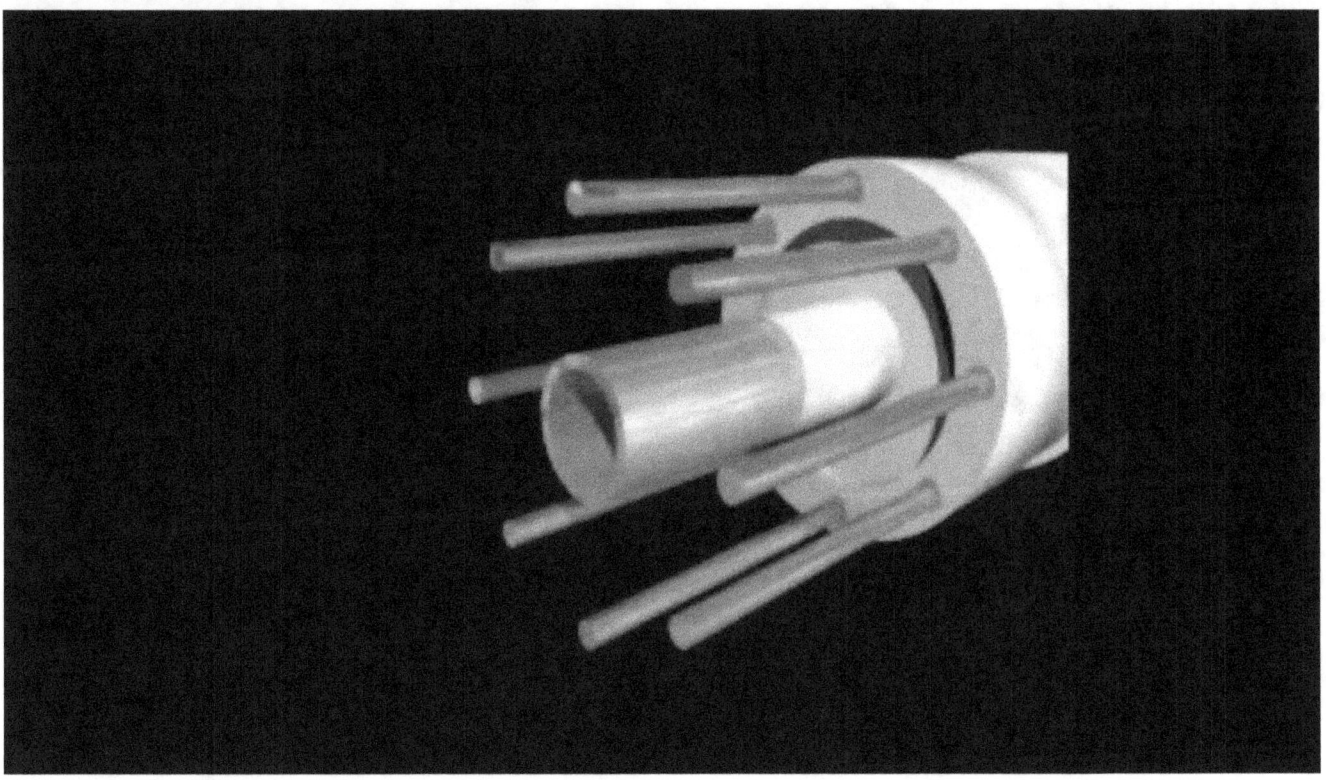

The device is made up of a ring of electrodes that surround a hollow electrode at the center.

A plasma sheet forms

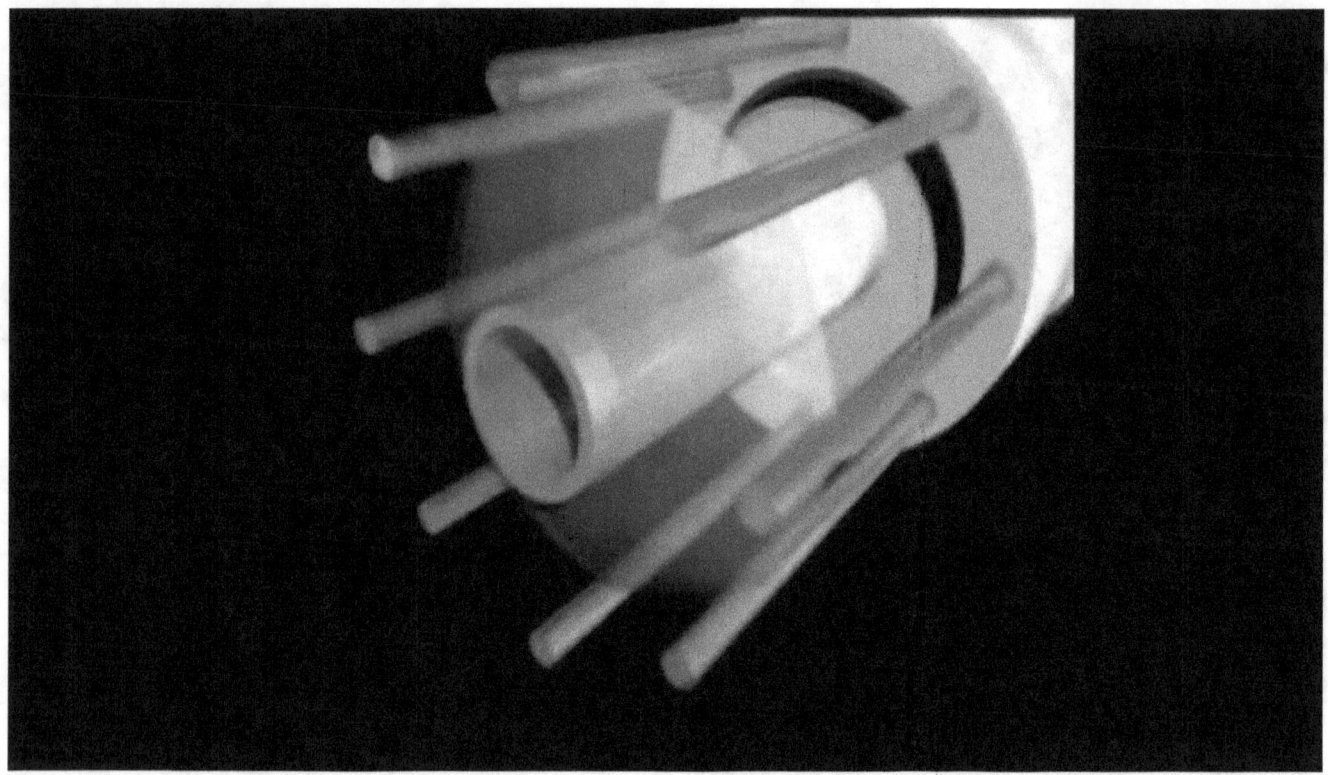

When an electric field is applied, a plasma sheet forms.

The plasma sheet

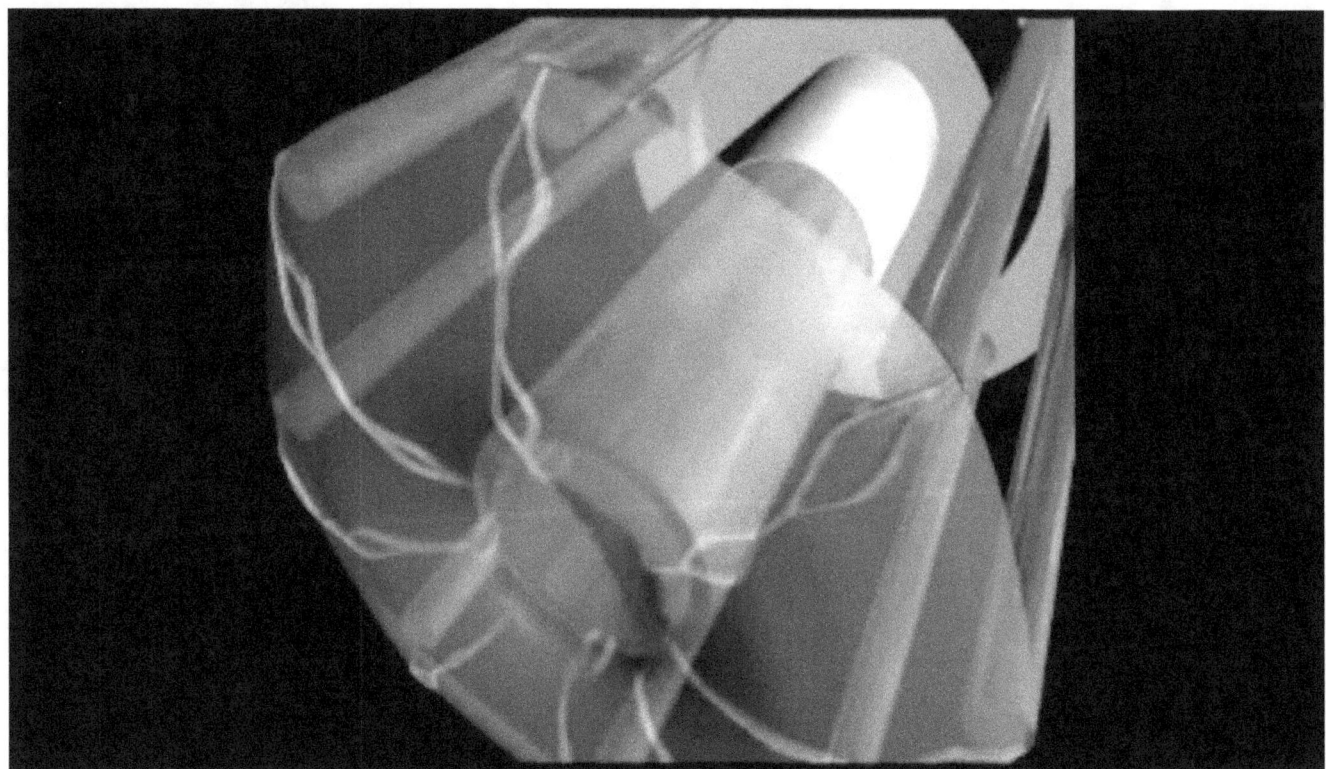

The plasma sheet is instantly drawn towards the opening of the central electrode, and is then drawn into it.

The plasma becomes extremely pinched

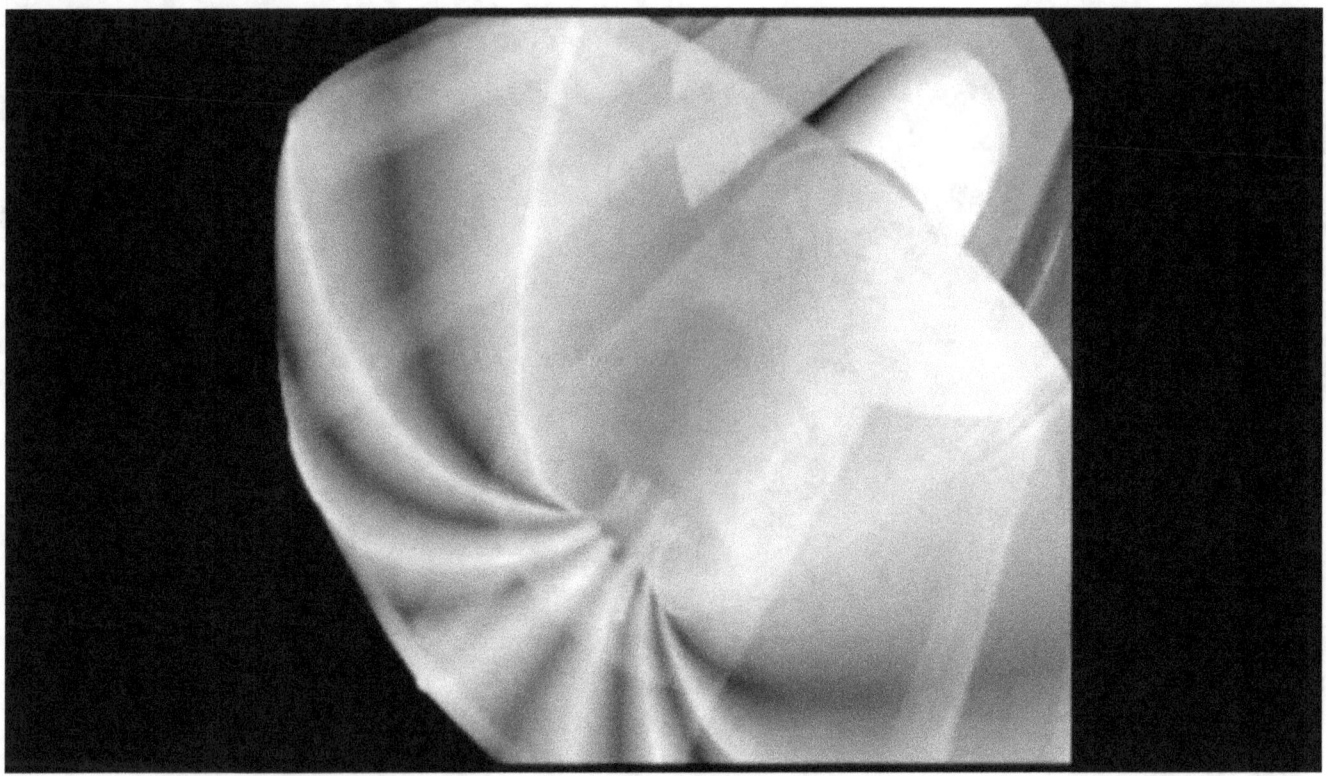

By it being drawn into the opening, the plasma becomes extremely pinched together.

It becomes unstable

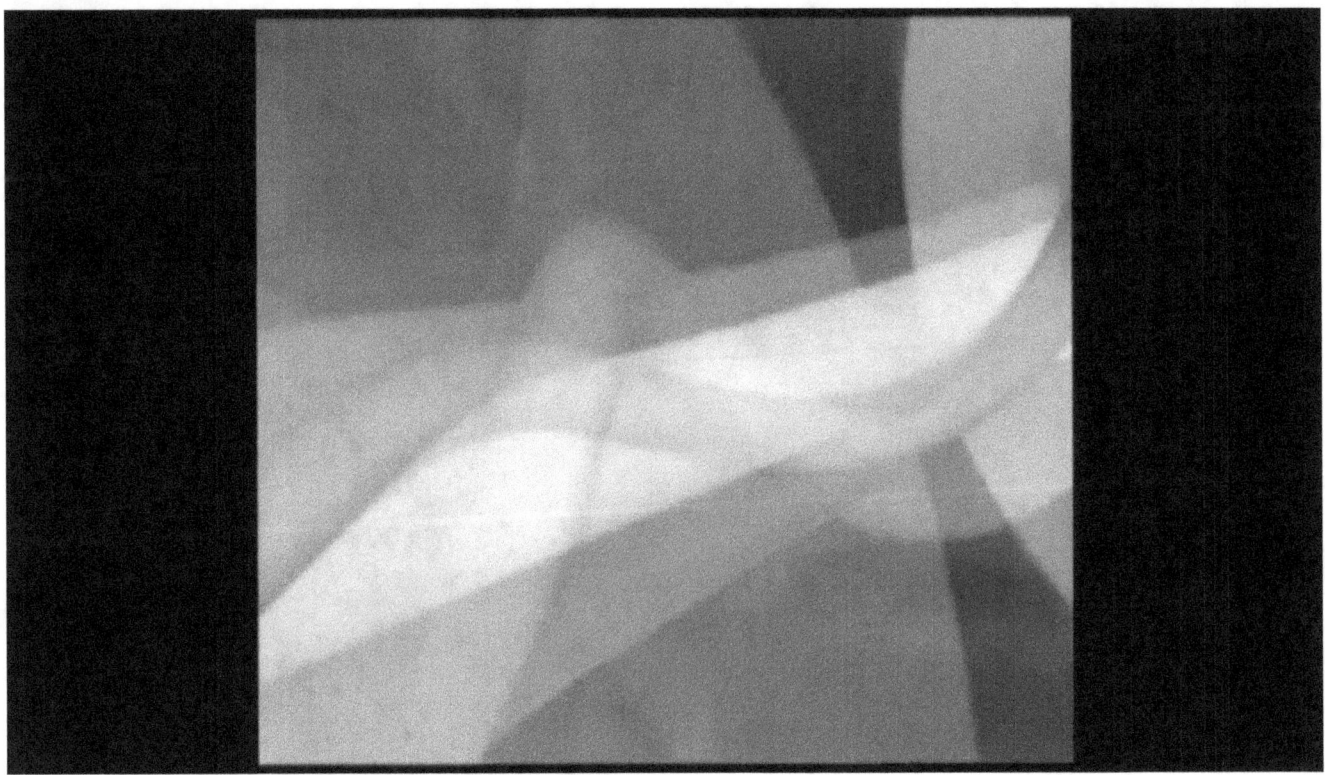

There, it becomes unstable and begins to twist.

The plasma twists itself into a spiral

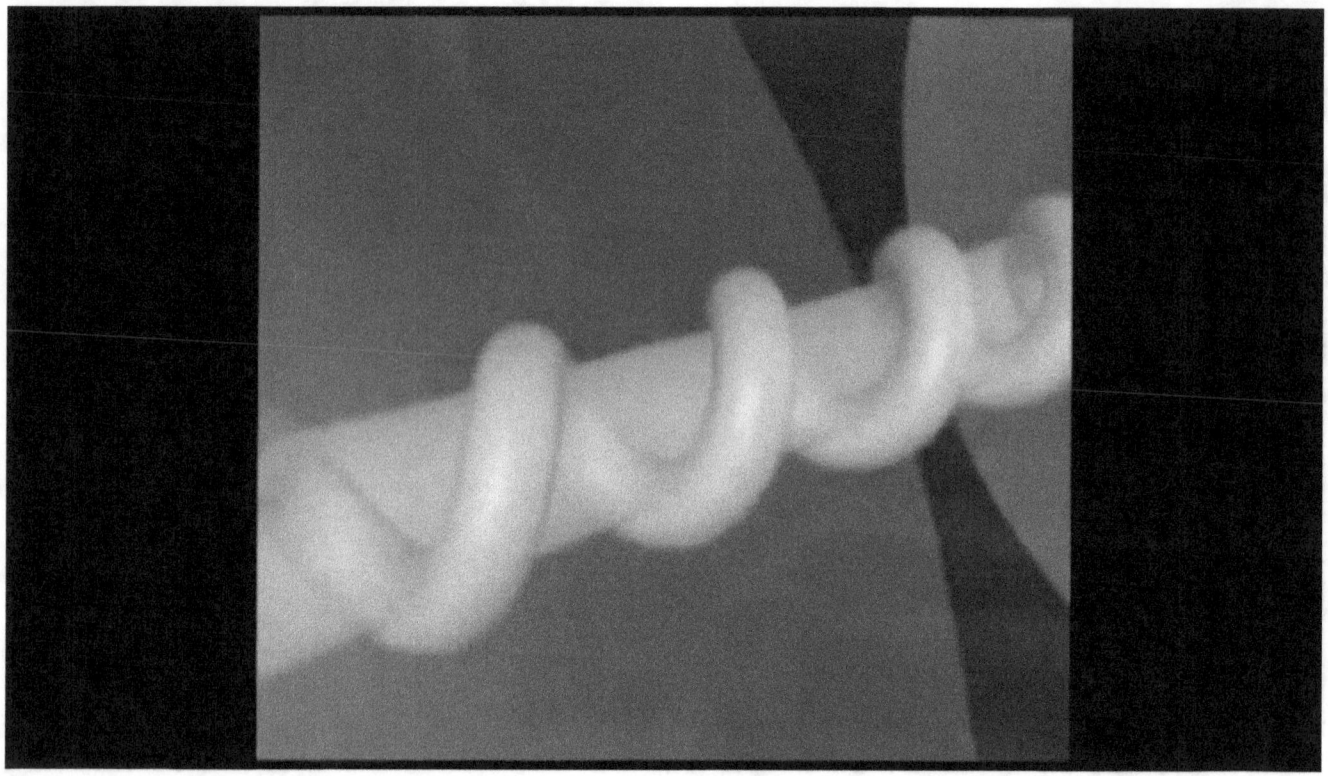

At first, the plasma twists itself into a spiral.

Then the spiral becomes compacted

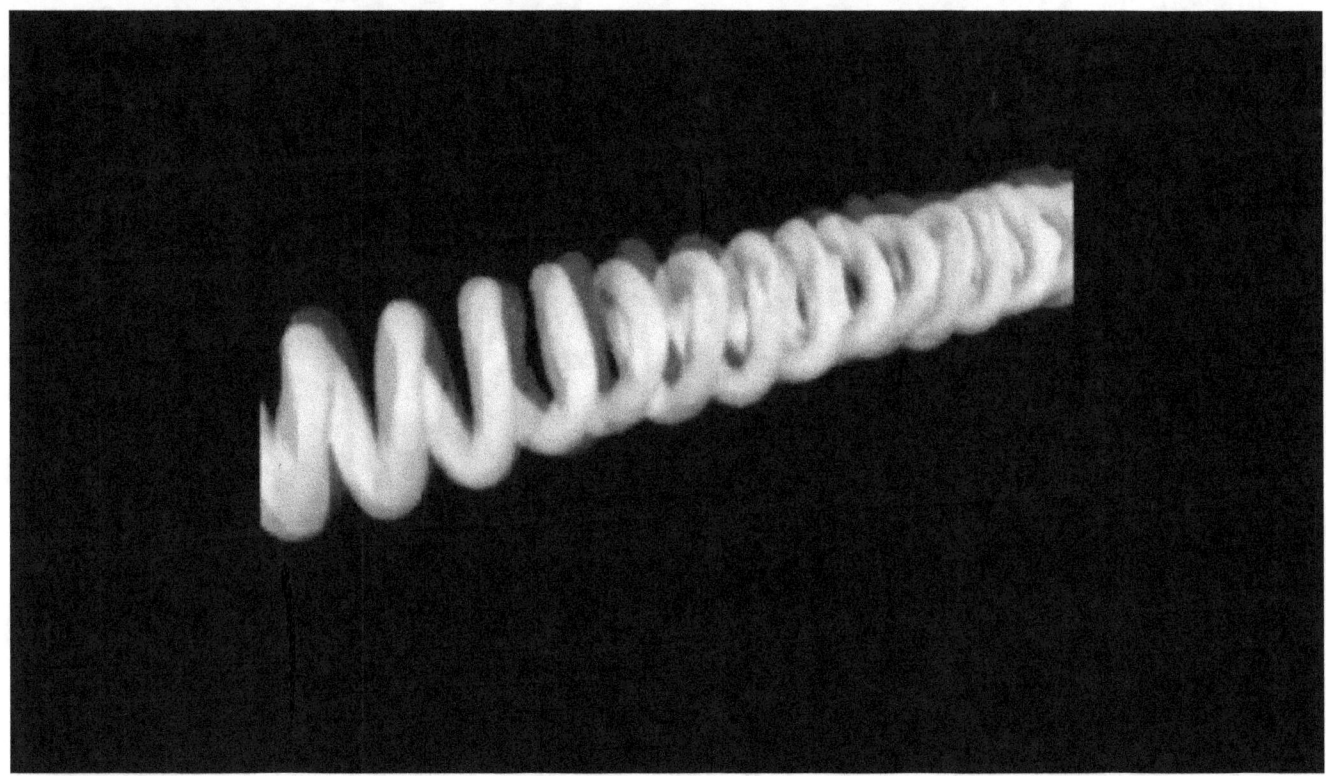

Then the spiral becomes compacted.

The more unstable the spiral becomes

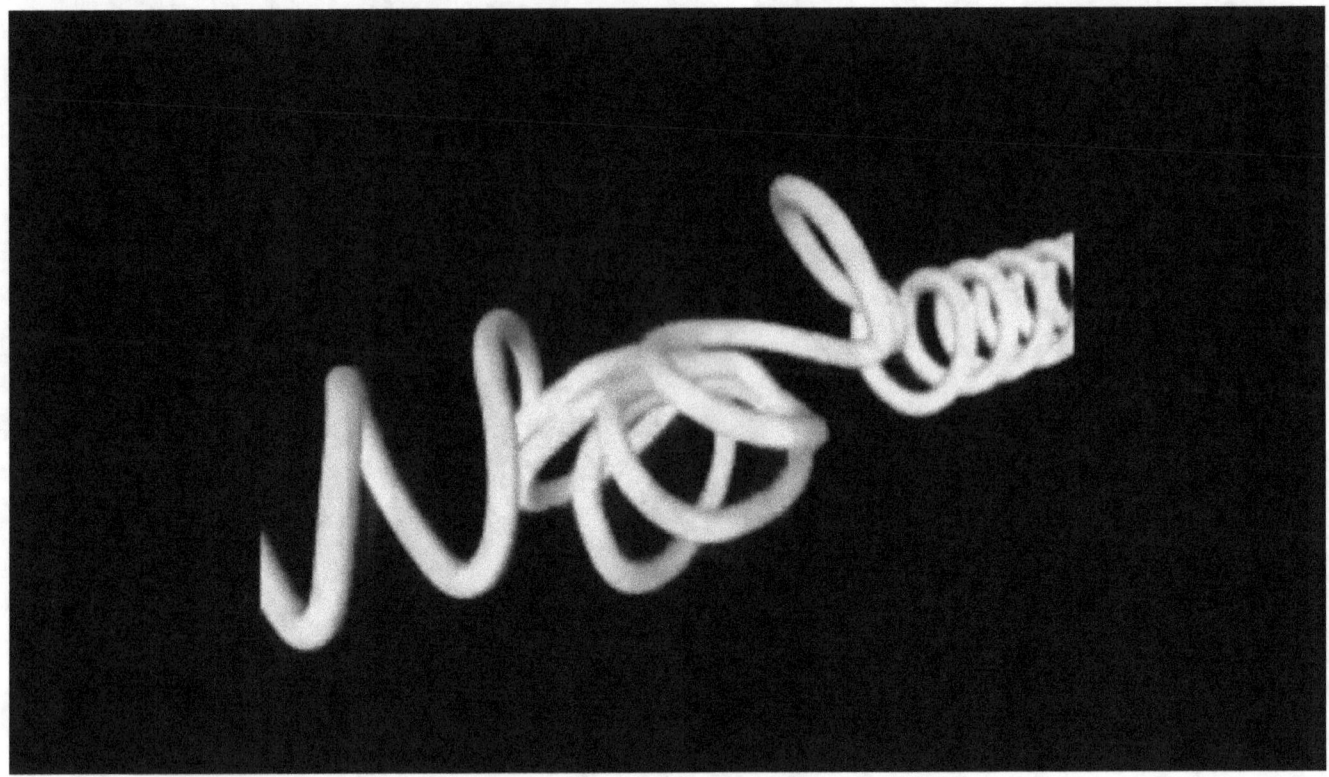

The more it becomes compacted the more unstable the spiral becomes, and becomes twisted.

The twisting forms a complex knot

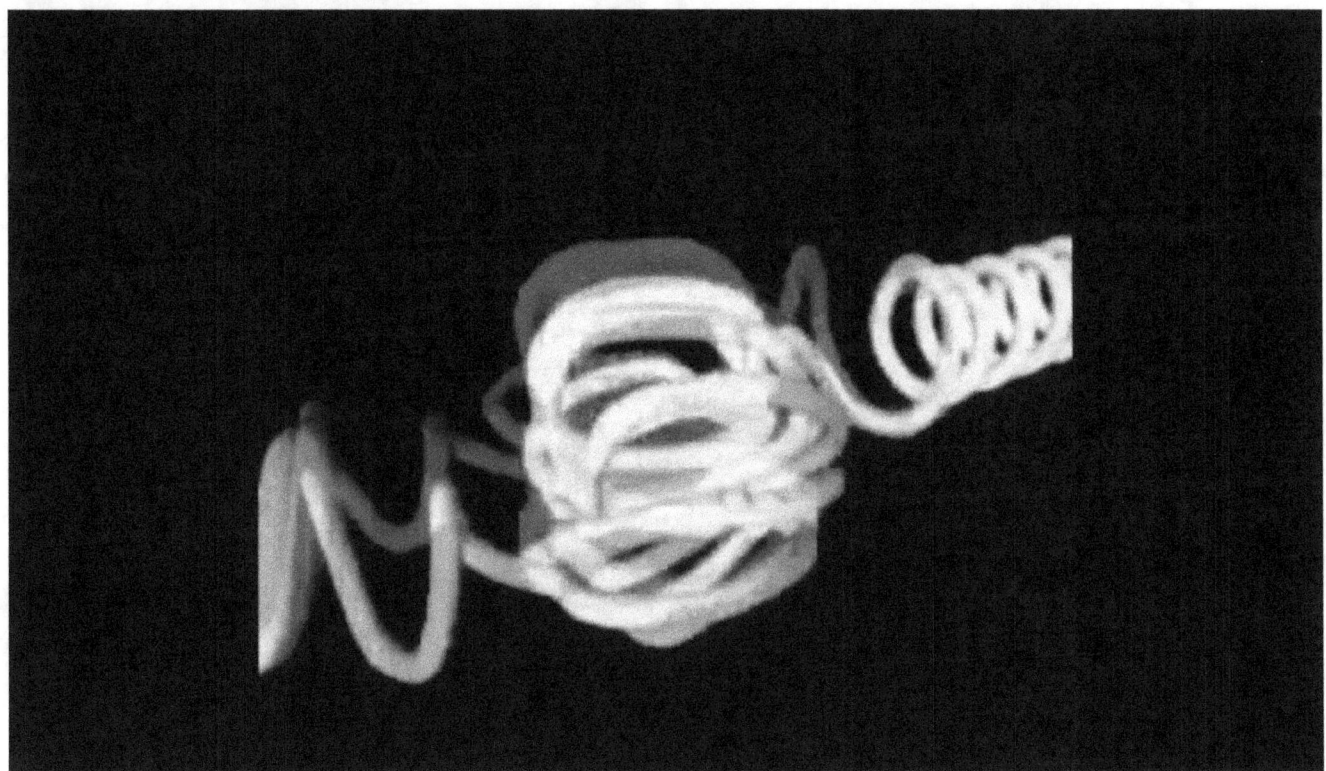

Eventually, the twisting forms a complex knot.

From a video about the Dense Plasma Focus Device

The illustrations are snapshots taken from a video about the Dense Plasma Focus Device.

A high-density plasma concentration forms

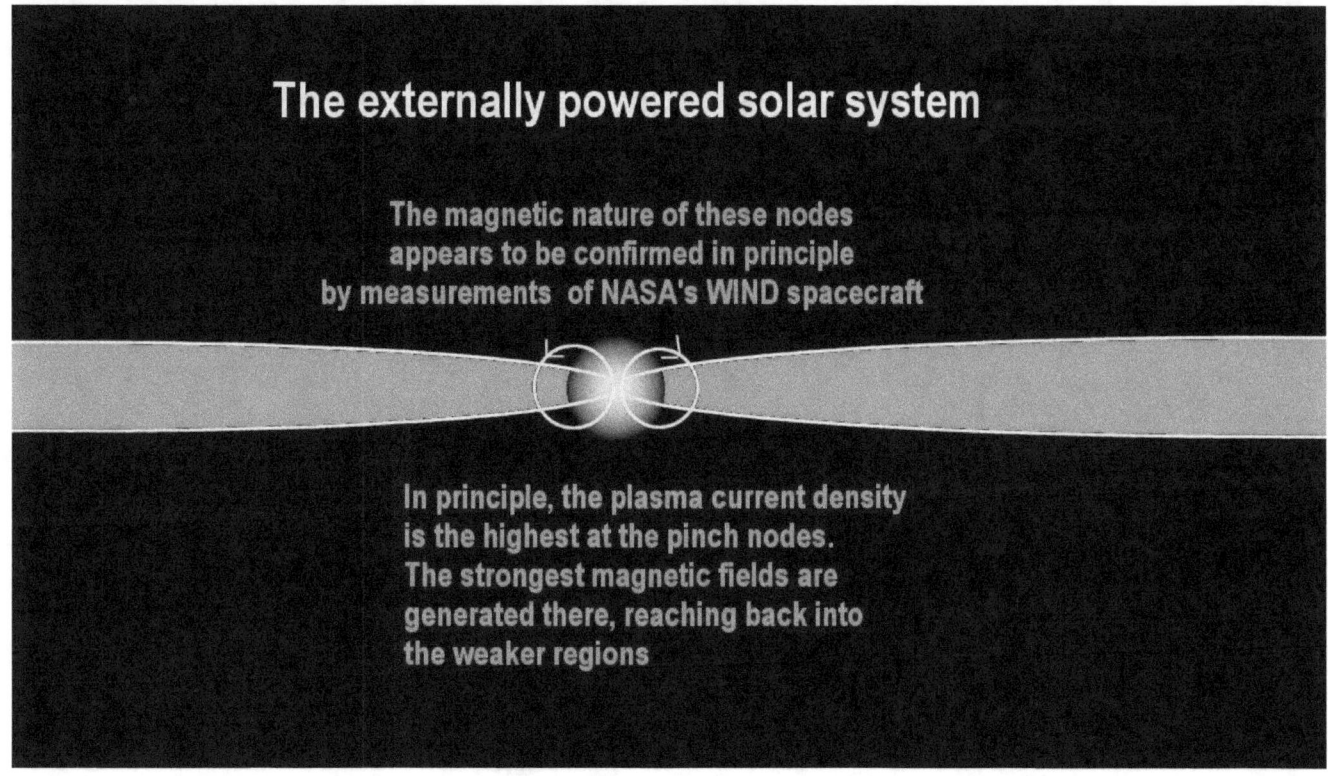

In space the plasma instabilities that form the complex knots open the magnetic bowls that form at the end of the concentrated plasma currents. As the process unfolds further, a high-density plasma concentration forms outside of the magnetic holes. That's where the sun is located in a solar system and is powered thereby.

When the plasma streams are too weak

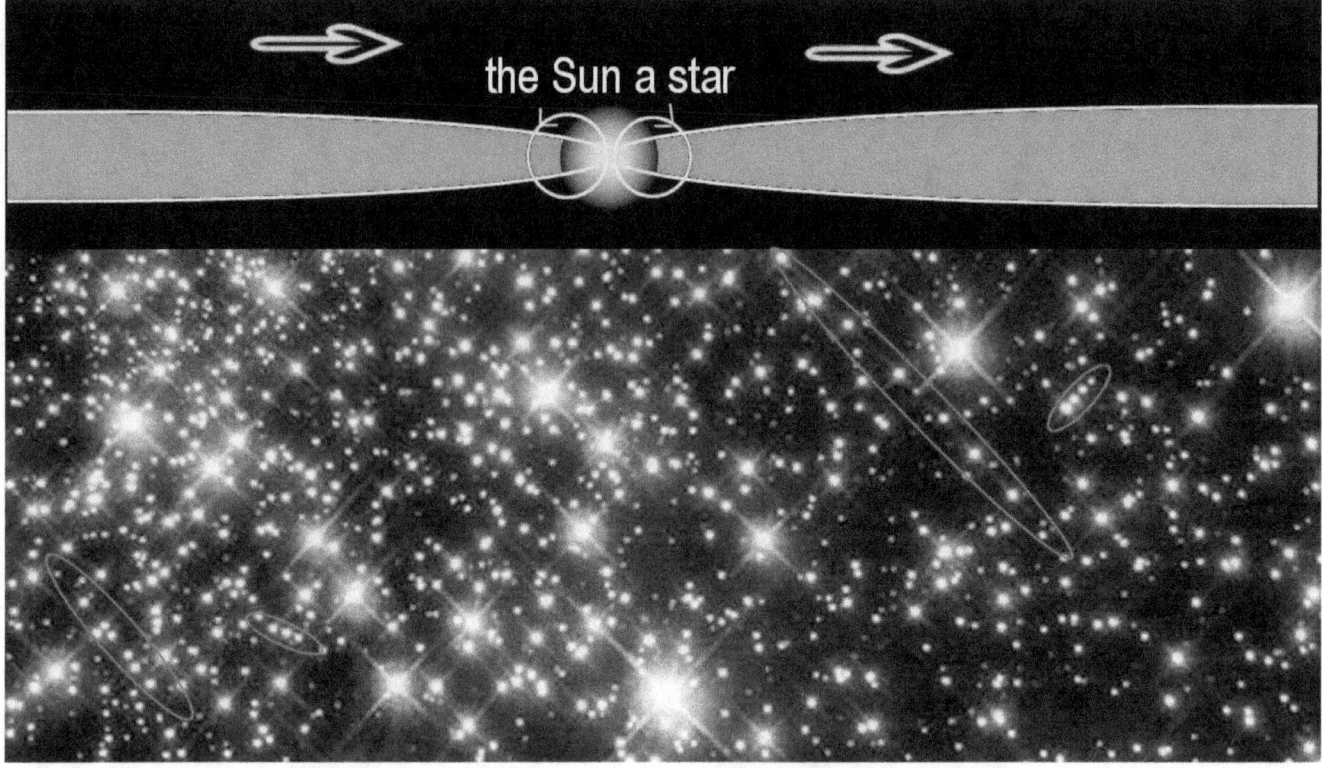

When, however, the plasma streams are too weak to cause an extreme pinch effect to happen, the plasma streams simply flow through the solar system without activating anything. The Sun thereby remains dim and not actively powered.

When the Sun is powered

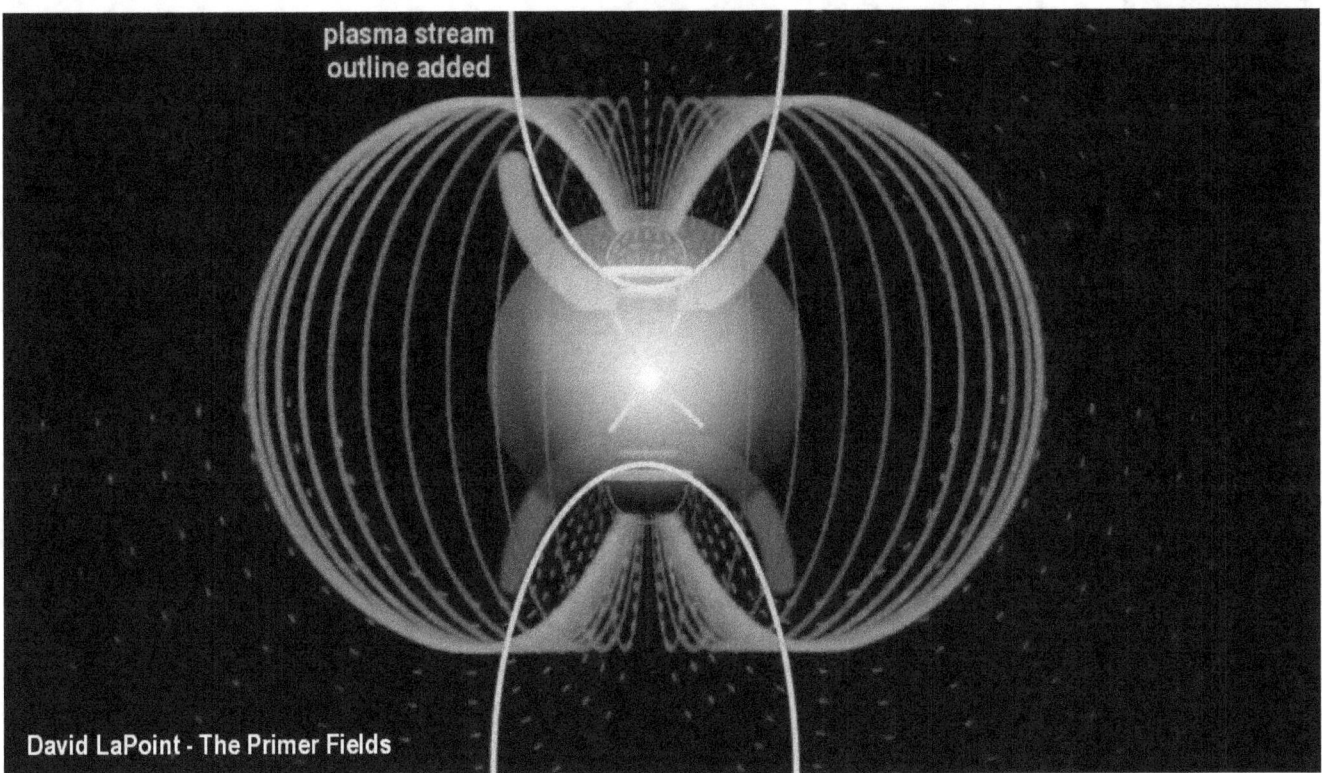

When the Sun is powered by a dense plasma sphere surrounding it, the flow-through process still happens. Plasma flows out of the plasma sphere, since the Sun utilizes only a small portion of it. In the outflow another magnetic field is generated, but with the opposite polarity. The evidence suggests that all large electromagnetic structures that exist in space, when they are drawn to a sun, or on the larger scale to a galaxy, whereby the Primer Fields form, exist typically in complementary pairs.

A number of interesting effects

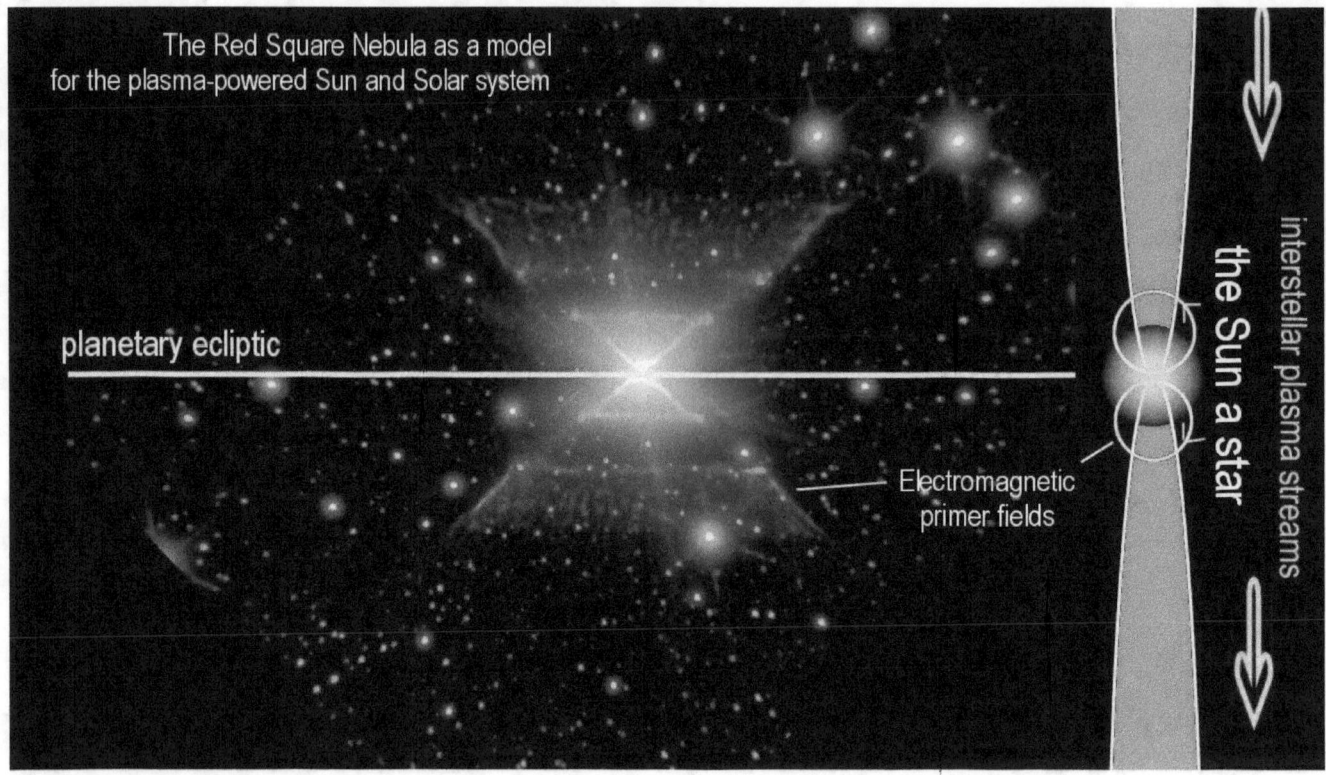

Between the two giant complementary electromagnetic bowls, a number of interesting effects come to light with interesting principles that are critical for the overall dynamic interactions.

In the narrow space between

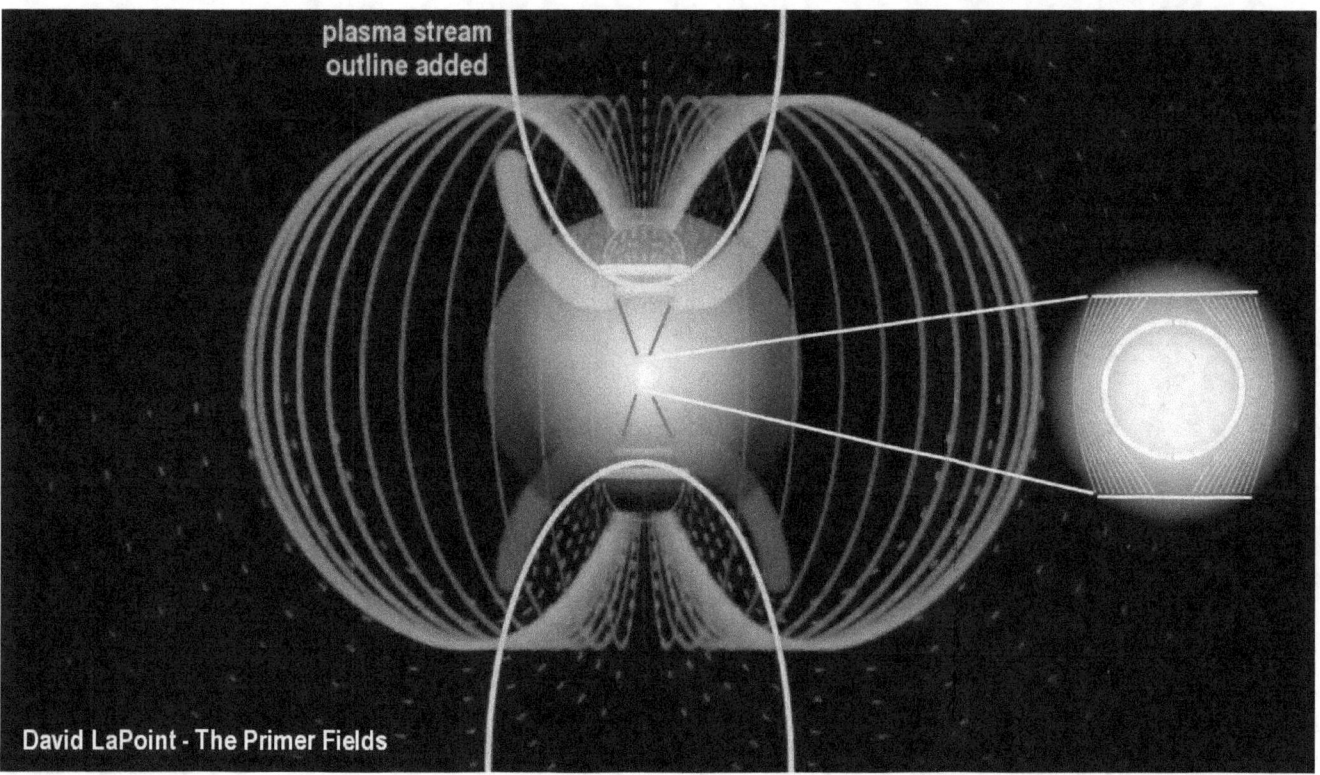

The magnetic fields are the strongest at the focal point of the bowl-shaped magnetic structures. In the narrow space between the two complementary structures lies typically a solar system with a sun, or several suns, on the central axis, and with the planets orbiting on a thin ecliptic plain in the space between the two electromagnetic bowl-type fields.

Two bowl-type electromagnetic fields work together

While the two bowl-type electromagnetic fields are each separate entities, they work together functionally as a whole.

Their function is to concentrate the plasma flows that pervade all space and focus them into a tightly confined sphere in which our sun is located and is powered by it.

A polarity flip point

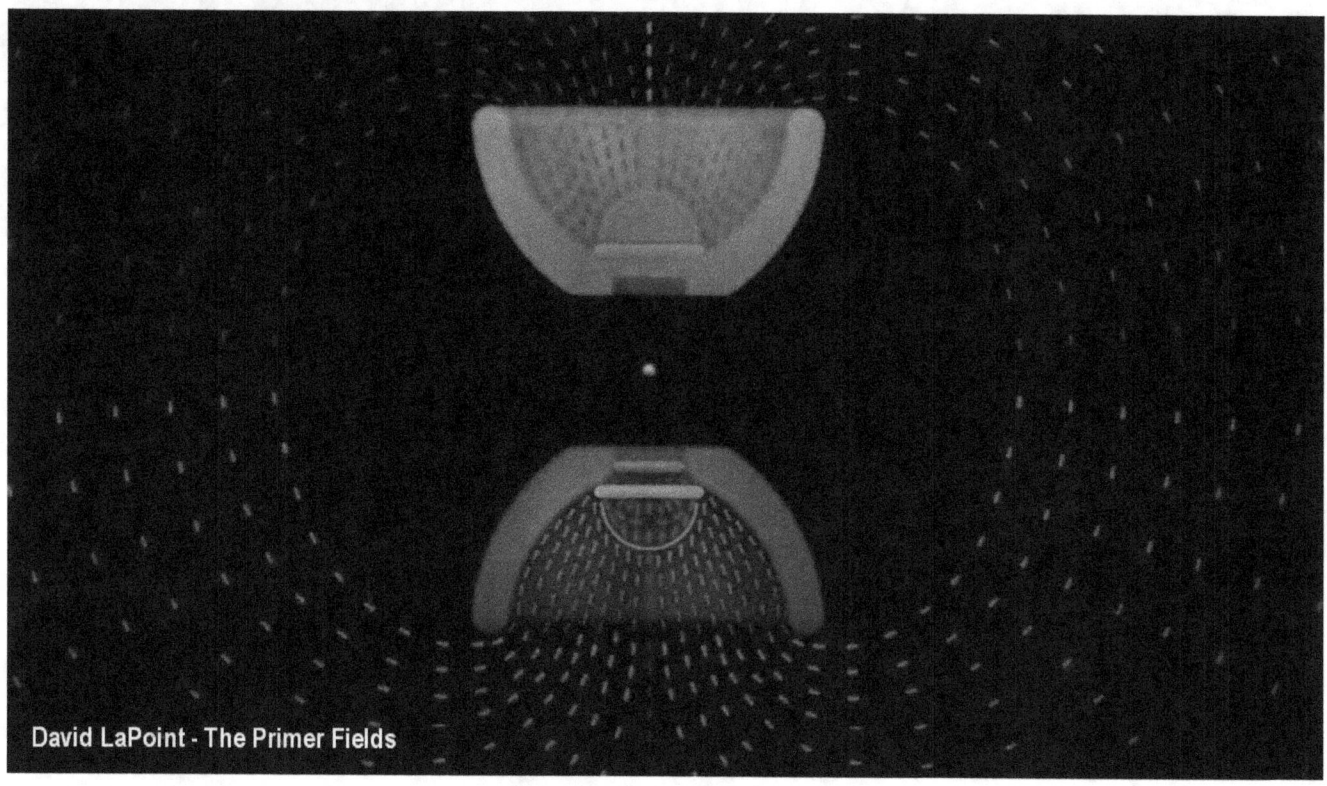

The remarkable concentration is accomplished in the laboratory by a set of two bowl-type structures, shown in red and blue, facing one-another with opposite (complementary) polarities.

The small point in the middle between the two magnetic bowl structures is where a polarity flip point is located. The flip point appears to be responsible for flipping the polarity of the Sun's magnetic field with every solar cycle.

The location of the flip point moves slightly when one of the two bowl structures becomes weaker than the other. This effect causes the polarity of the magnetic field of the Sun to assume the dominant polarity, and thus flip with the 11-year solar cycles that are simply resonance cycles between the complementary magnetic structures.

Three functional magnetic elements

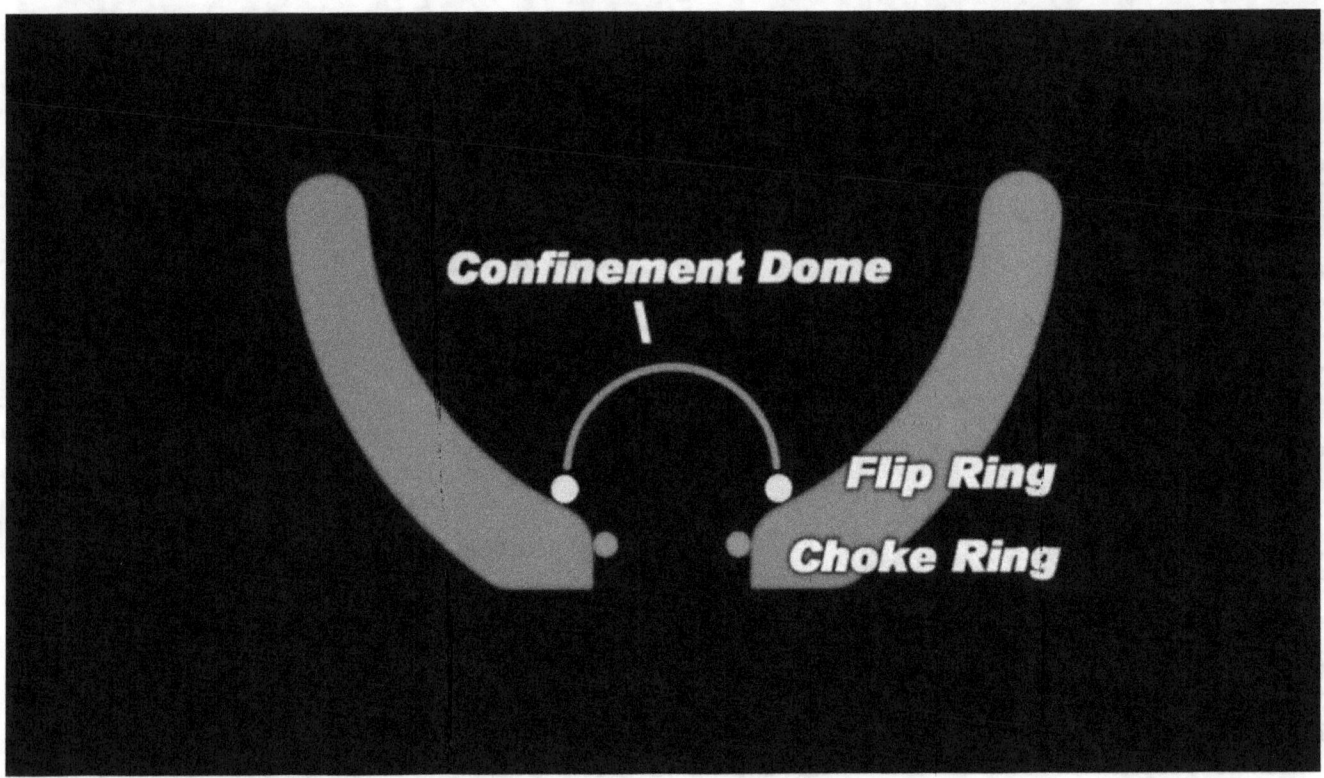

The plasma concentration that is required for the Sun to function is dynamically produced in the illustrated structure by the interaction of its three functional magnetic elements that are structured around the respective hole in the magnetic bowls.

Each has a specific function to fulfill

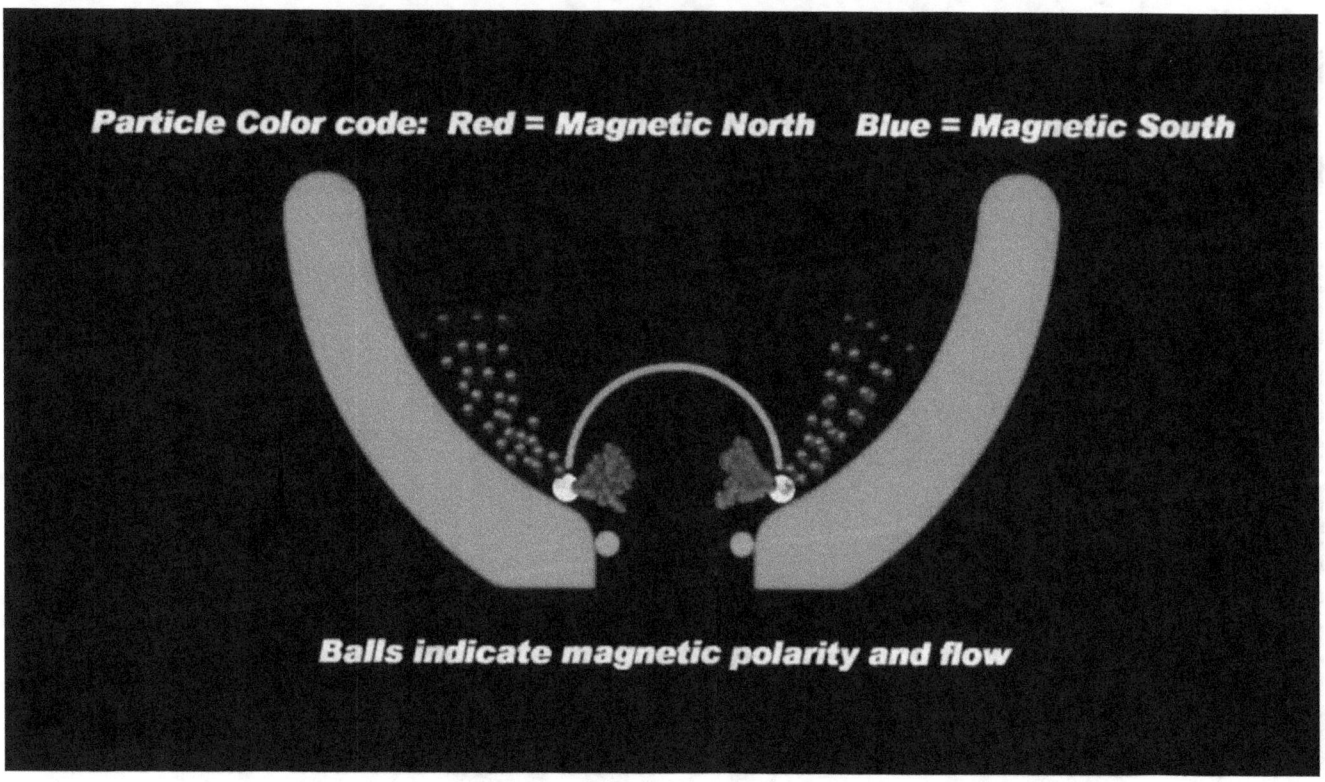

Each of the three structures has a specific function to fulfill. The flip ring flips the orientation of plasma. It flips it under the magnetic confinement dome, while the choke ring below helps to keep it there.

The flip ring

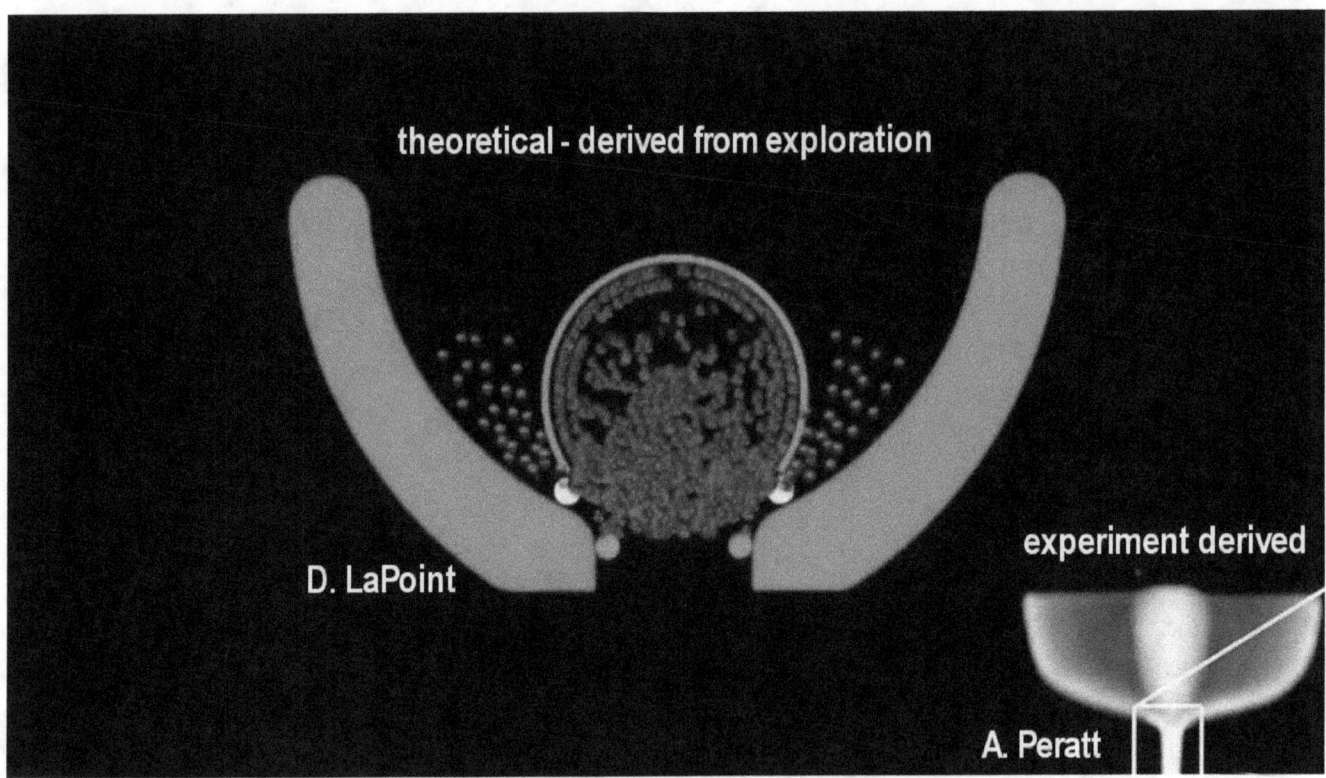

The plasma particles that flow into the big red bowl are flipped upwards as they pass the flip ring (yellow). They are collected together there into a massive accumulation. The plasma particles become concentrated by this process. The concentration creates a 'high-pressure' environment.

The magnetic choke ring

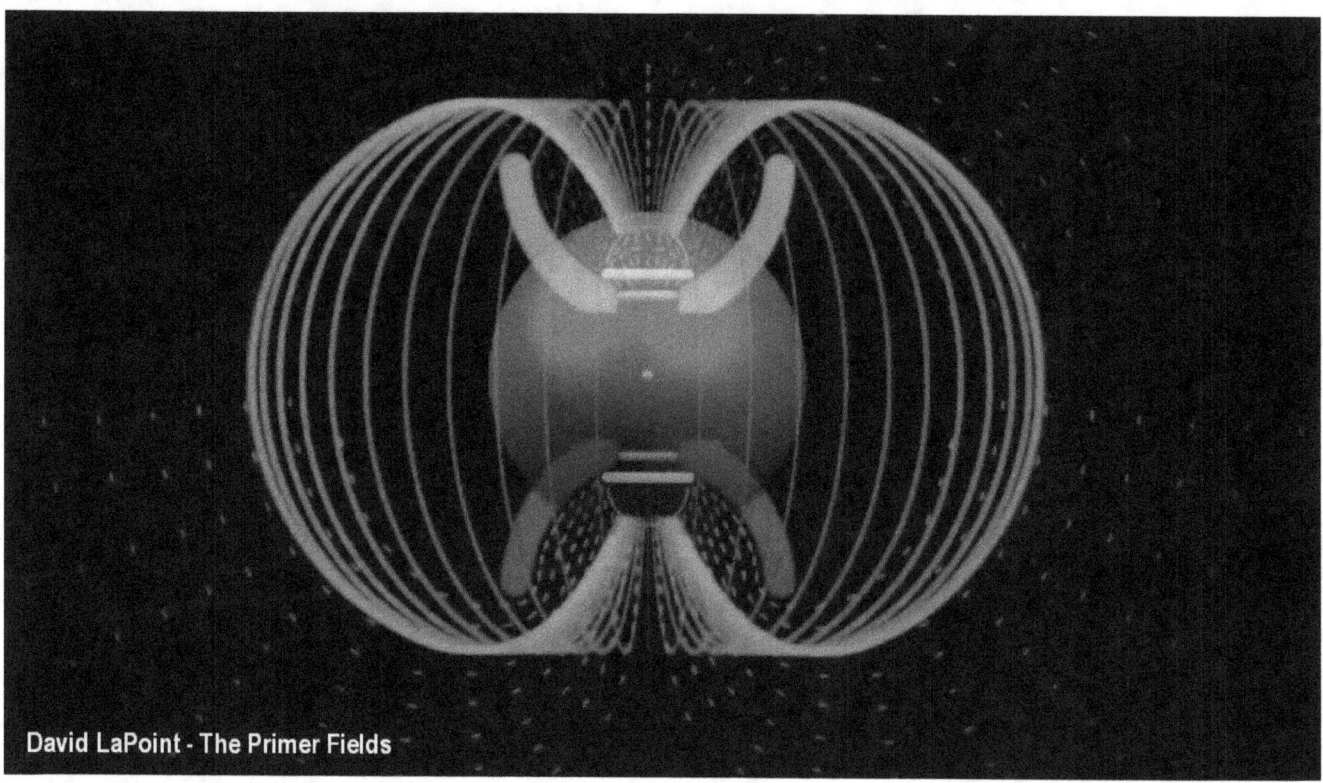

The magnetic choke ring, within the opening of the magnetic bowl, focuses the 'escaping' plasma flow into a tightly concentrated stream beneath the hole. In some lab experiments, the focused stream expands into a sphere.

The out-flowing stream

The Sun at a plasma node

David LaPoint - The Primer Fields

When plasma is drawn out of the sphere in the flow-through process, a complementary bowl structure is formed with opposite orientation and opposite magnetic polarity. In this structure the out-flowing stream of lesser density is drawn into the bowls where expands in the reverse process, reverting back to the 'normal' density of the prevailing plasma stream.

The process that is illustrated here can be verified with laboratory experiments.

A high-power plasma experiment

In a high-power plasma experiment of the type that is conducted at the Los Alamos National Laboratory, the complementary bowl structure that the plasma currents form by their own interaction, is clearly visible. Also the containment dome that forms inside the bowls is visible in the experiment that is illustrated here.

In this particular experiment the magnetically focused plasma stream that flows between the bowls, did not form a sphere, for which a catalyst would be required, but it did form a distinct plasma ring around the focused stream, centered between the bowls.

Archetypal drawings

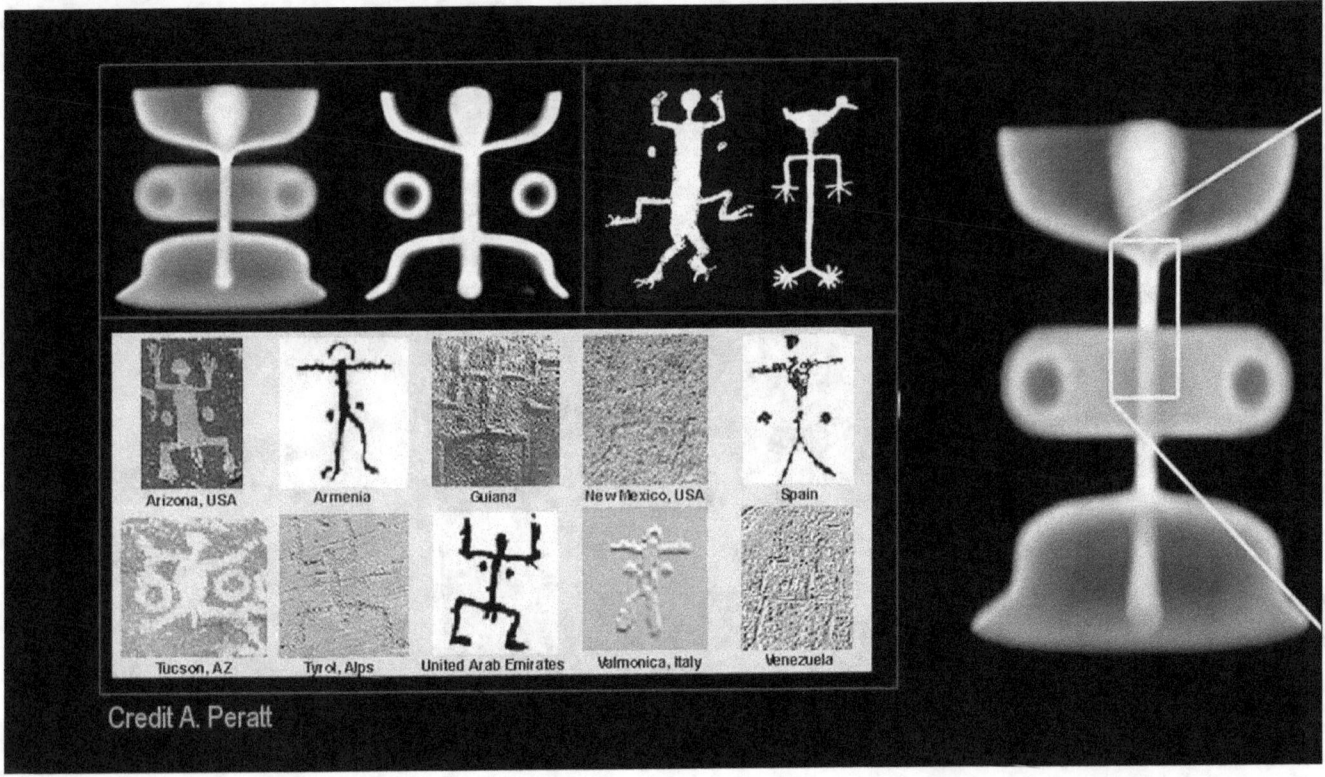

Evidence exists that the lab-created shape of plasma formed by the Primer Fields, was visible in ancient time in the sky. Archetypal drawings collected from widely separated regions on earth, show a remarkable similarity of their design with the lab-created plasma formations. The similarity suggests that the complementary plasma bowl shapes were common occurrences at one time, appearing and disappearing in the skies like so many UFO sightings today, or like the sprites still do under special conditions.

The plasma jets

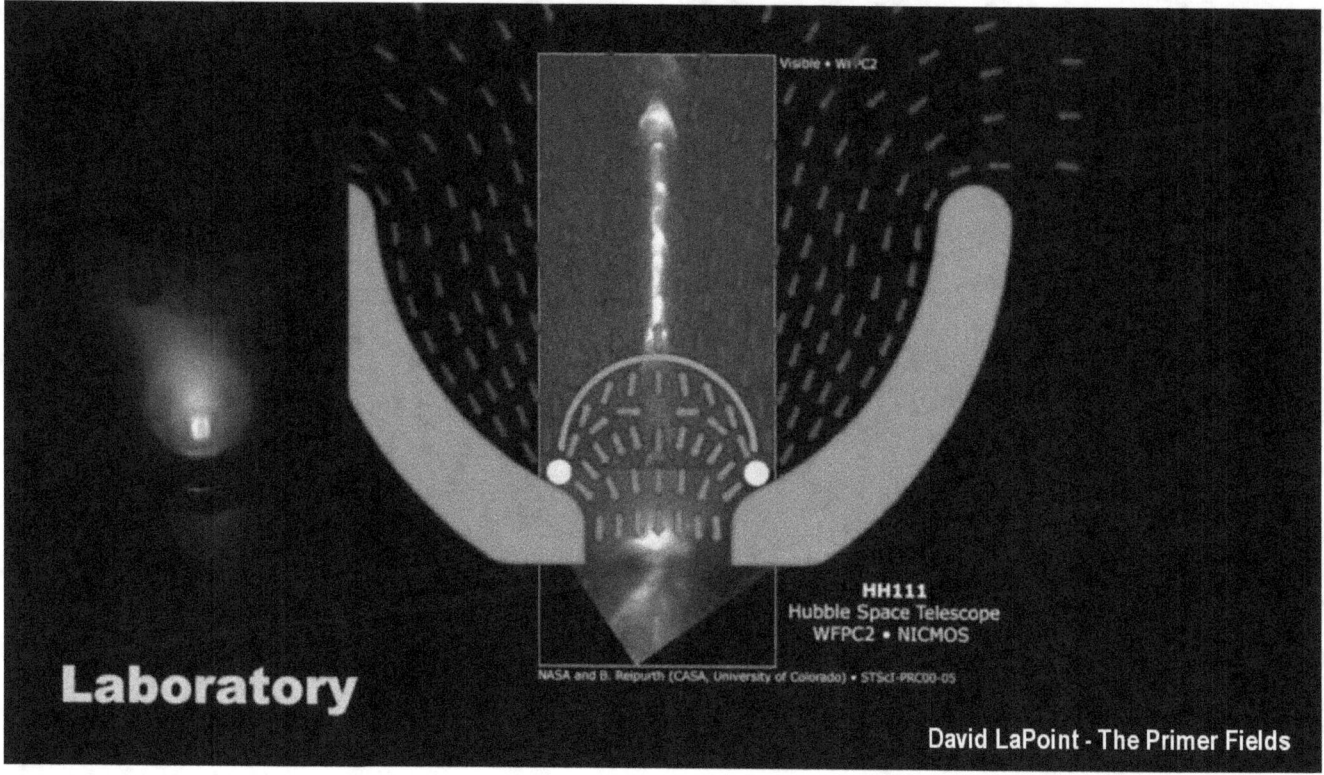

David LaPoint also discovered two more features of the Primer Fields, for which widely known evidence exists, which are the plasma jets and the magnetic flip point. He discovered that when the plasma pressure under the confinement dome becomes too great, the dome will rupture at its weakest point, by which excess plasma escapes in a burst until the rift closes up again under the resulting lower pressure. By this plasma venting process the magnetic strength of the respective bowl structure weakens somewhat. The resulting imbalance shifts the convergence of the magnetic fields, and with it it shifts a magnetic flip point that David LaPoint also discovered, forms below the opening of the magnetic bowl.

The 11-year solar cycles

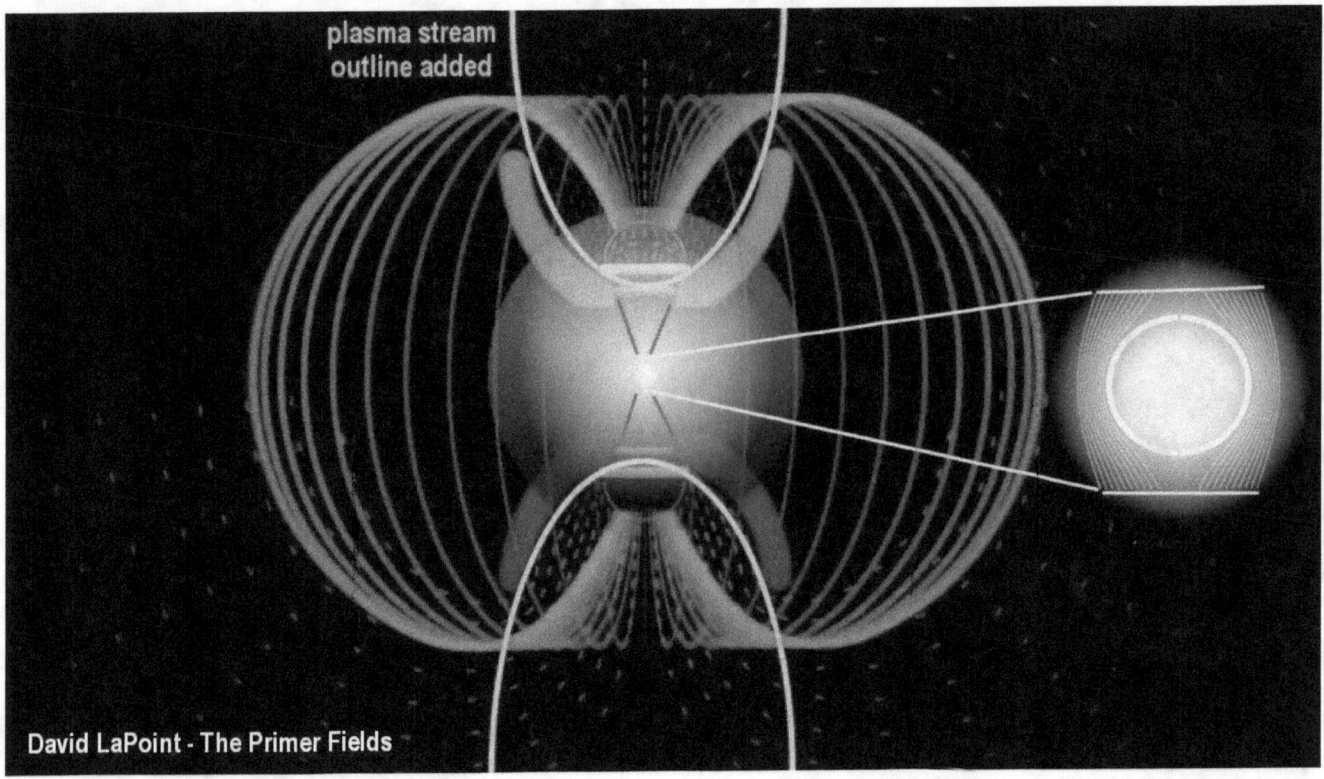

In the case of the solar system the resulting imbalance, which shifts the flip point, flips the Sun's magnetic field at the high point of the 11-year solar cycles are thereby recognized to be simply resonance cycles that oscillate between the two complementary electromagnetic structures of the solar system.

Operation of the Red Square Nebua

The Red Square Nebula

What the complete Primer Fields system that powers our solar system and the sun may look like in practice, can be seen illustrated in the operation of the Red Square Nebua that can serve as a model for this purpose. In this model we see all the essential features of the Primer Fields system clearly visible.

Two complementary bowl-type structures in operation

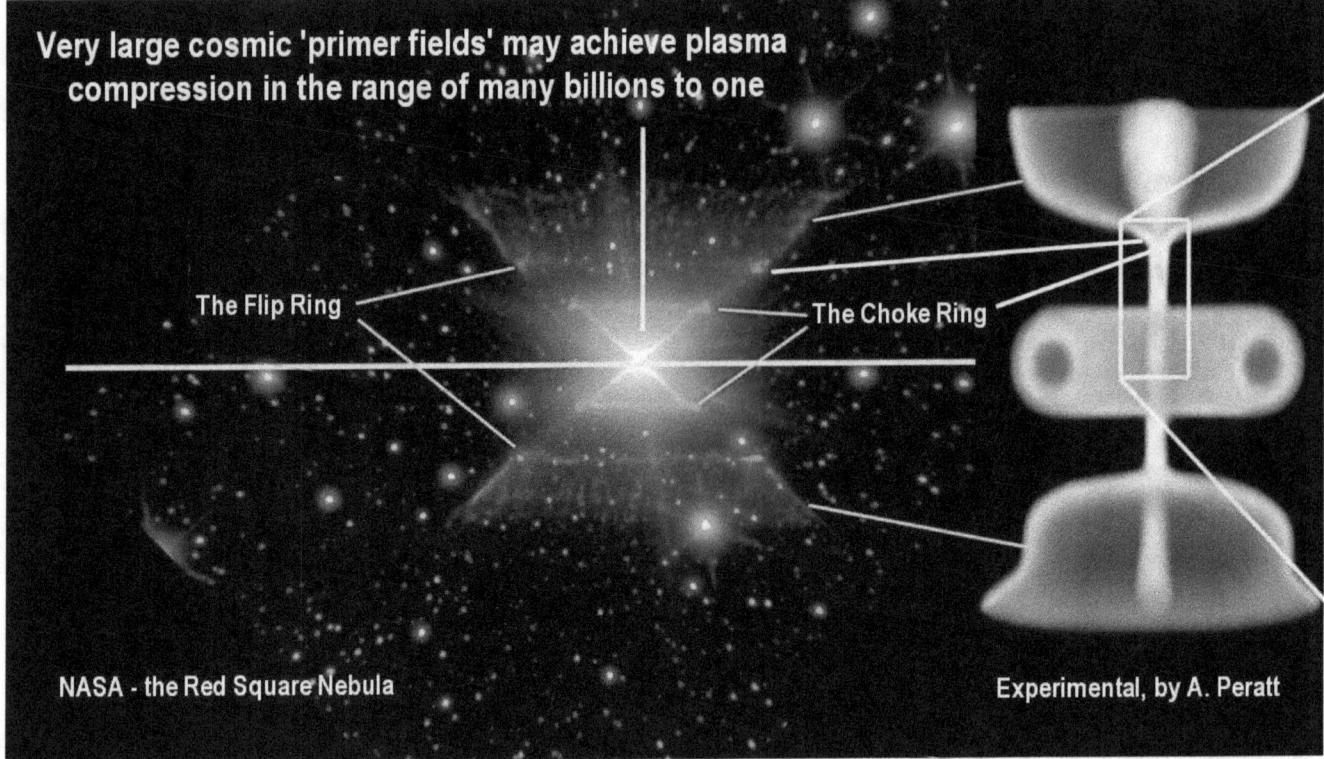

We see the two complementary bowl-type structures in operation. One concentrates the galactic plasma streams like a funnel. In the funnel we can see the flip ring, and below it the choke ring, below which a cone of concentrated plasma extends that focuses onto a sun, or a number of them. And we see the reverse happening for the outgoing plasma stream.

Observed in a laboratory plasma-flow experiment

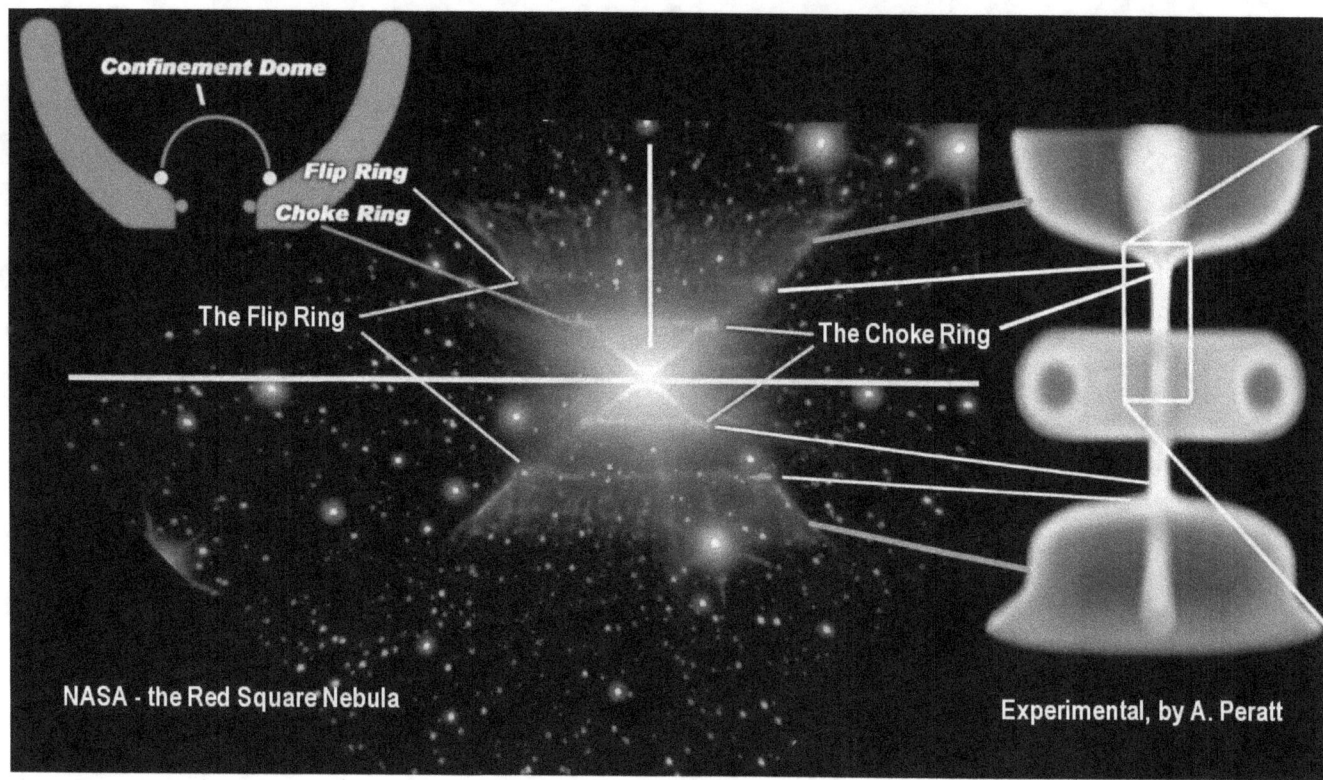

We can see all the essential features reflected here that have been observed in a laboratory plasma-flow experiment.

Our galaxy, when it is observed as a whole

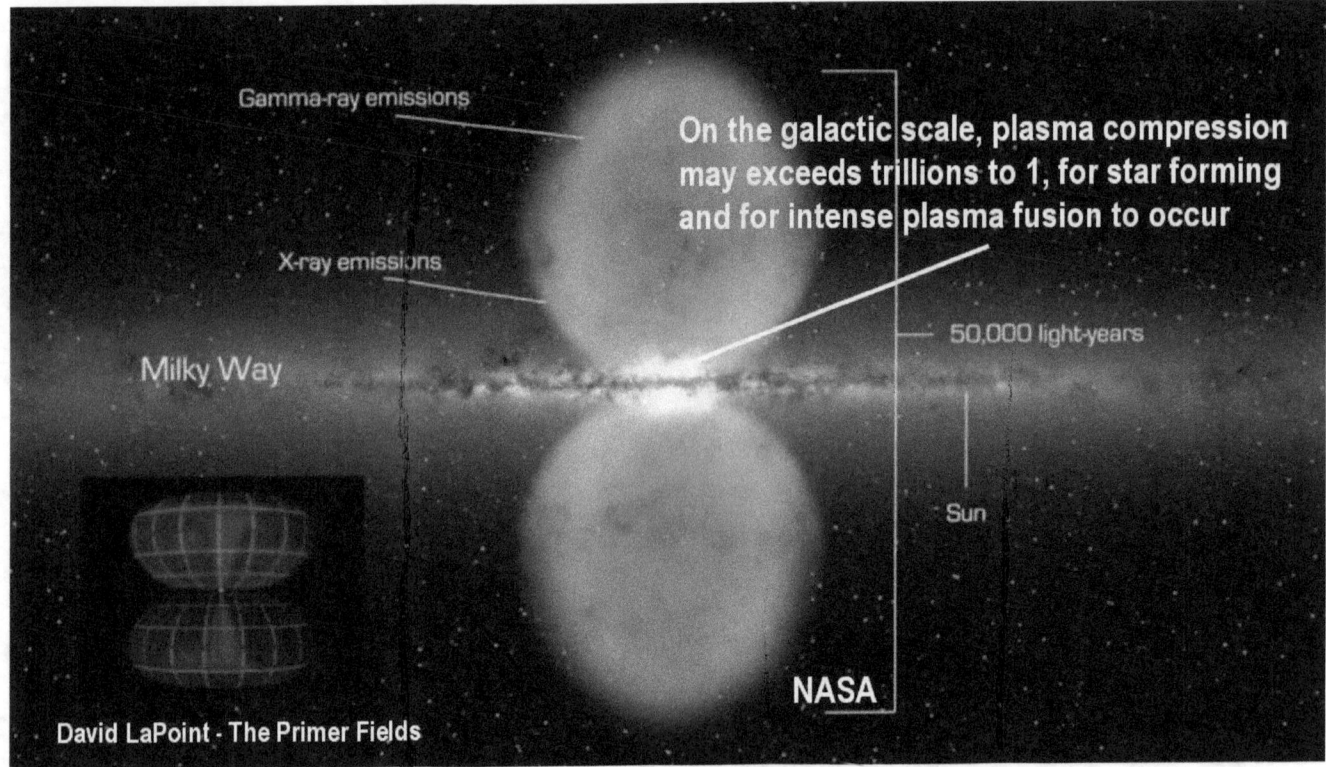

Our galaxy, when it is observed as a whole, also operates on essentially the same dynamic platform, though on a vastly larger scale than a solar system.

The center of the galaxy

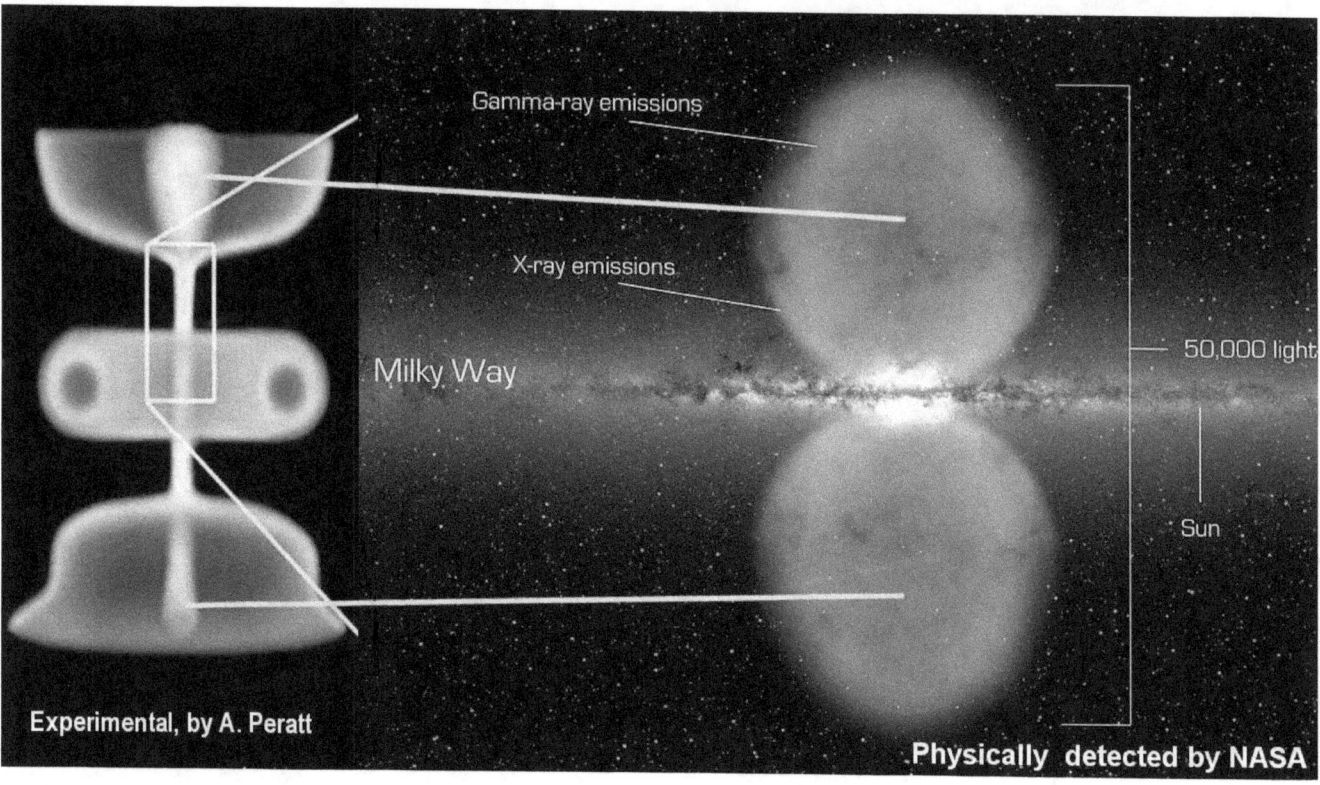

But here too, we see unmistakable evidence of two complementary electromagnetic bowls that form highly condensed plasma concentrations under their respective confinement domes.

The concentration is visible in x-ray and gamma-ray emissions. We can also see the the extremely concentrated plasma sphere below the confinement domes and between the two bowl structures that are indicated by the existence of the confinement domes. The concentrated plasma sphere is also the center of the galaxy.

Large intergalactic plasma streams

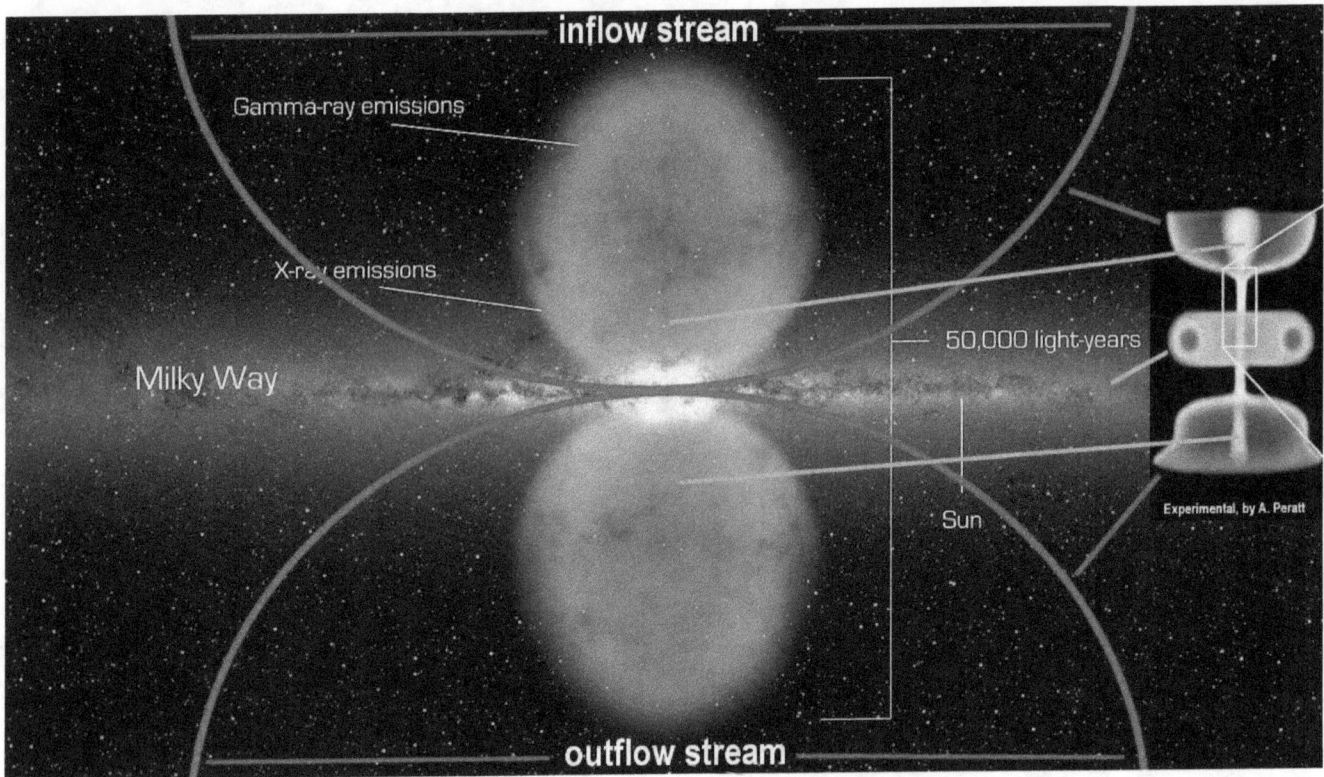

Large intergalactic plasma streams feed into and out of the Primer Fields that form the confinement domes.

Two long intergalactic connecting streams

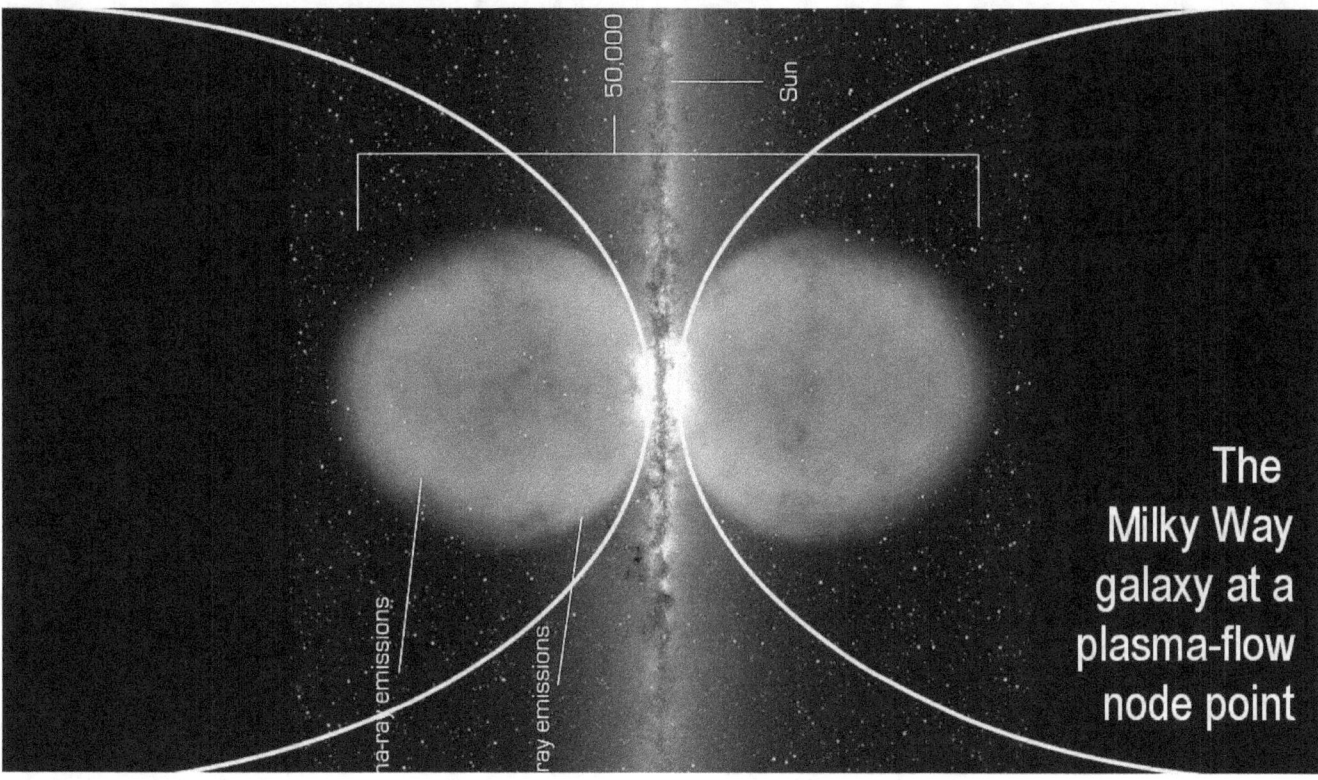

These two long intergalactic connecting streams, one incoming and one out-going, both have very long resonance cycles.

very long electric resonance cycles

Galaxies at node points

These very long electric resonance cycles that correspond to the long distances between the galaxies, evidently affect the strength of the Primer Fields that power the galaxy, and are thereby the cause for the two long climate cycles that have been observed on Earth.

Very long electric resonance cycles

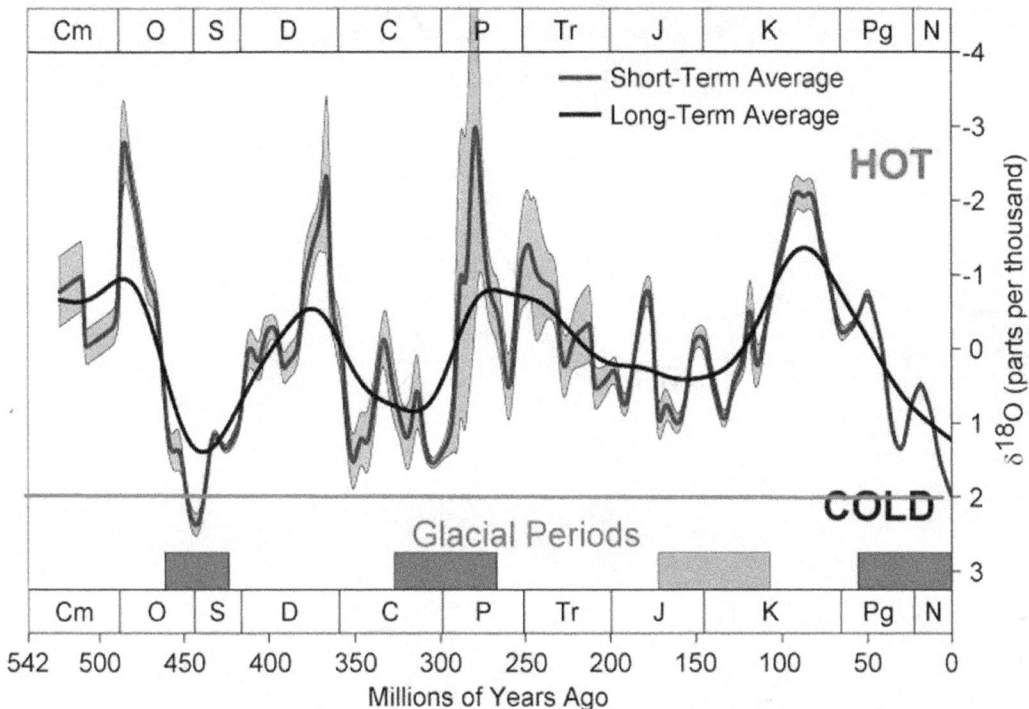

These very long electric resonance cycles - the sixty-two million year cycle, and the hundred-forty million year cycle - show up as long climate cycles that have been preserved in sediment records that enable us to look back in time more than 500 million years.

Both near their minimum point

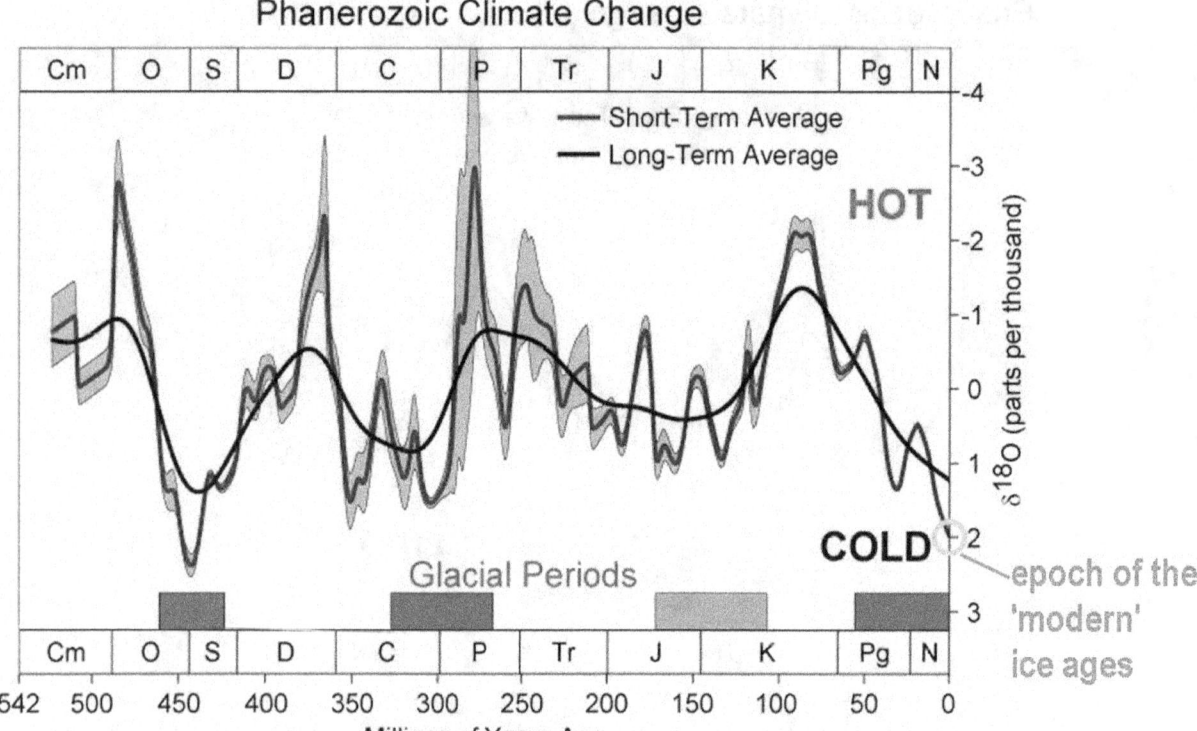

Presently, the two very long climate cycles are both near their minimum point, whereby the weak plasma conditions have been created that have gripped the Earth for the last two million years, in which the ice ages happened.

Breakdown of the Primer Fields

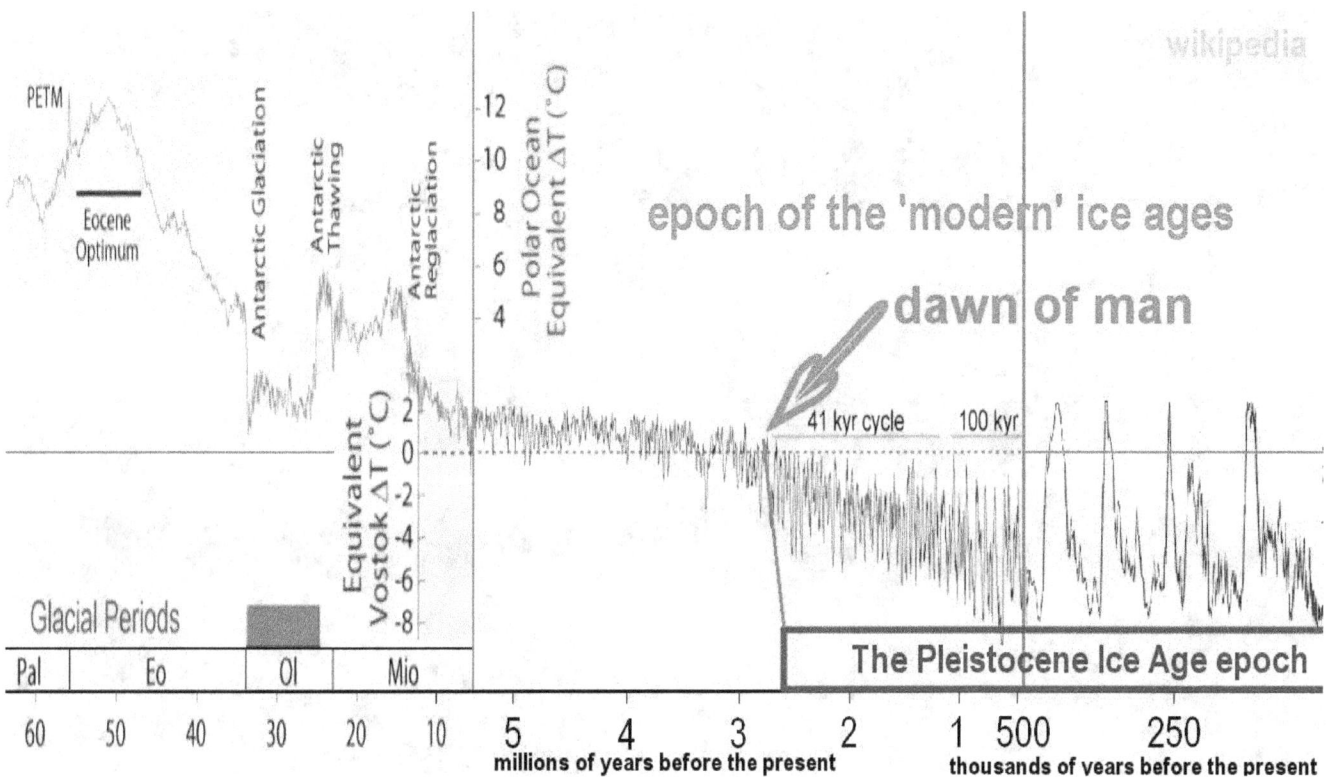

During these weak conditions that have been slowly developing for the last five million years, which are getting still weaker, the breakdown of the Primer Fields that power our solar system has become a regular occurrence during the weakest times.

The Primer Fields cannot form

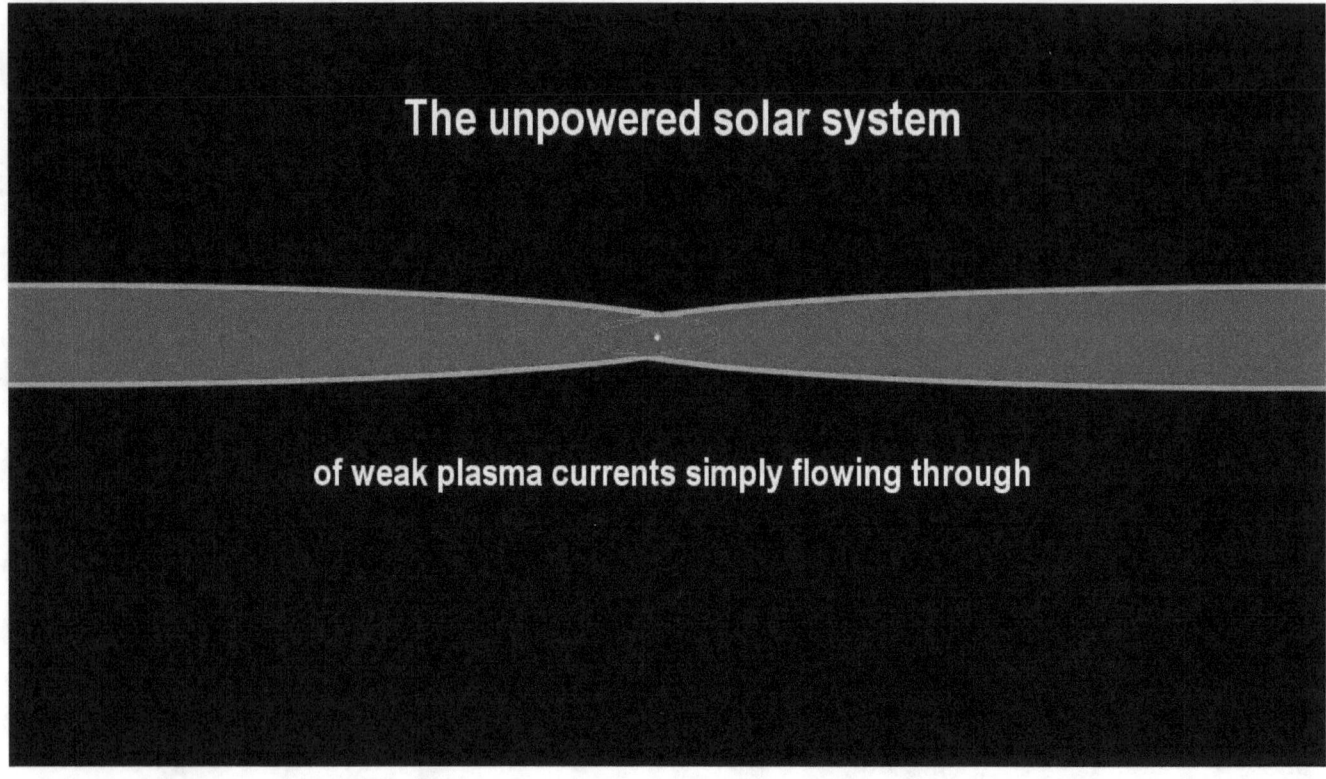

Then, when the plasma streams that flow through the solar system become so weak that they drop below a threshold, a point is reached when the pinch effect in the plasma streams is no longer strong enough to twist the plasma currents into knots to create the bowl-type magnetic structures with the void at the center that make up the Primer Fields. And so the Primer Fields cannot form. What once existed, suddenly exists no more.

The Sun becomes inactive, dim, and cold

The plasma currents become no longer concentrated then, but simply flow through the solar system without becoming focused around the Sun.

At this point the Sun simply turns off. It becomes inactive, dim, and cold. An Ice Age begins.

The Pleistocene Epoch

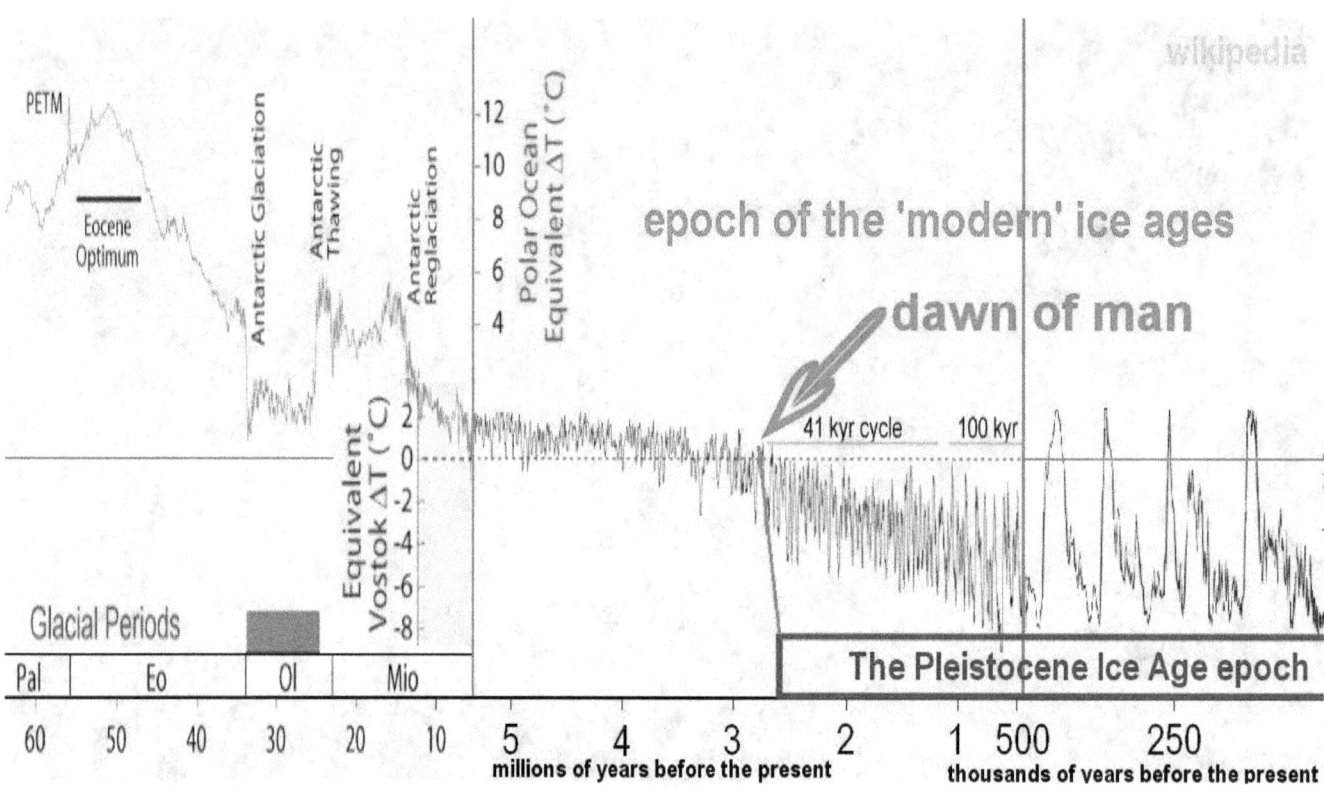

The ice ages didn't last as long in the earlier phase of the weakening conditions. They lasted only 41,000 years then. As the general weakening continued, the 100,000 years long ice ages began. This became named the Pleistocene Epoch.

Throughout the ice ages, in which the Primer Fields fail, the Sun becomes inactive for long periods. It becomes a dim yellow star that glows mostly by its stored up energy and whatever nuclear decay may be ongoing within it.

The Sun remains not totally shut down

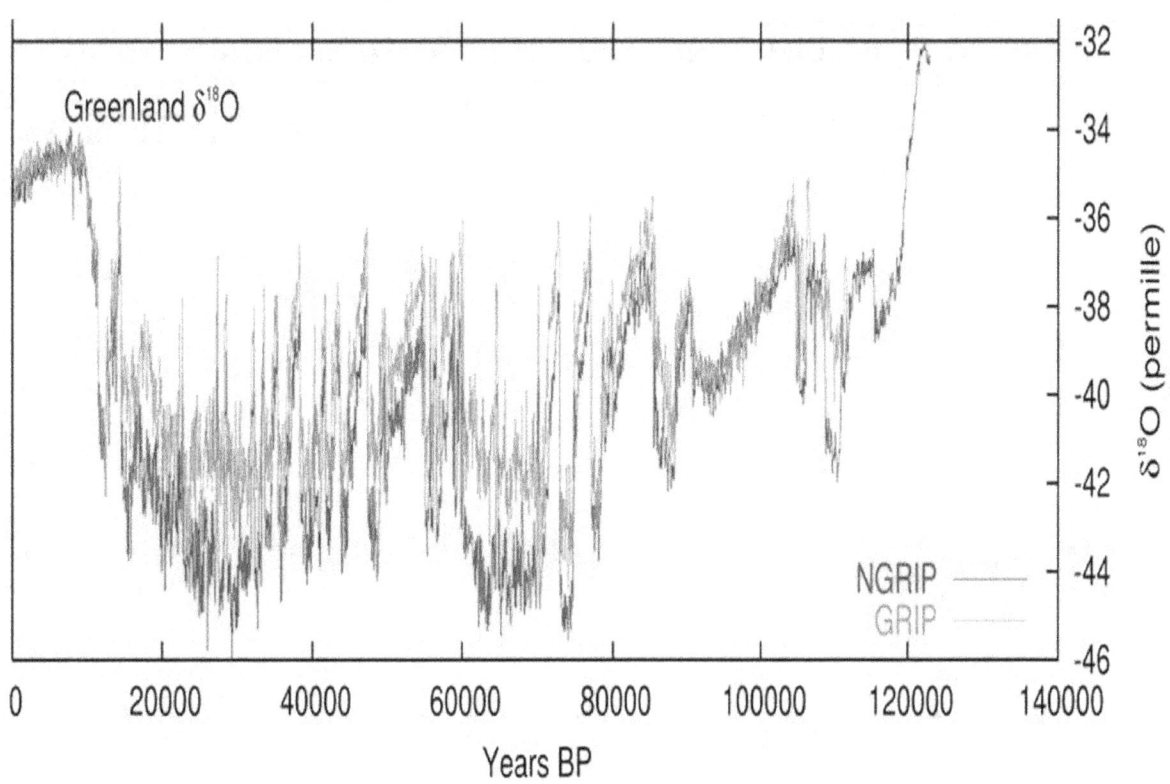

However, evidence exists that the Sun remains not totally shut down during the long ice age glaciation periods, which typically last 100,000 years. The Earth would turn into a snowball if the Sun would remain inactive for 100,000 years. Fortunately, this doesn't happen. Periodically short pulses of high-density conditions do occur during the ice ages, which re-invigorate the plasma streams that enable the Primer Fields to form anew, whereby the Sun to becomes powered again. Unfortunately these pulses are short in duration. The Sun remains powered by these pulses for only a few decades, and then turns off again.

The Dansgaard-Oeschger oscillations

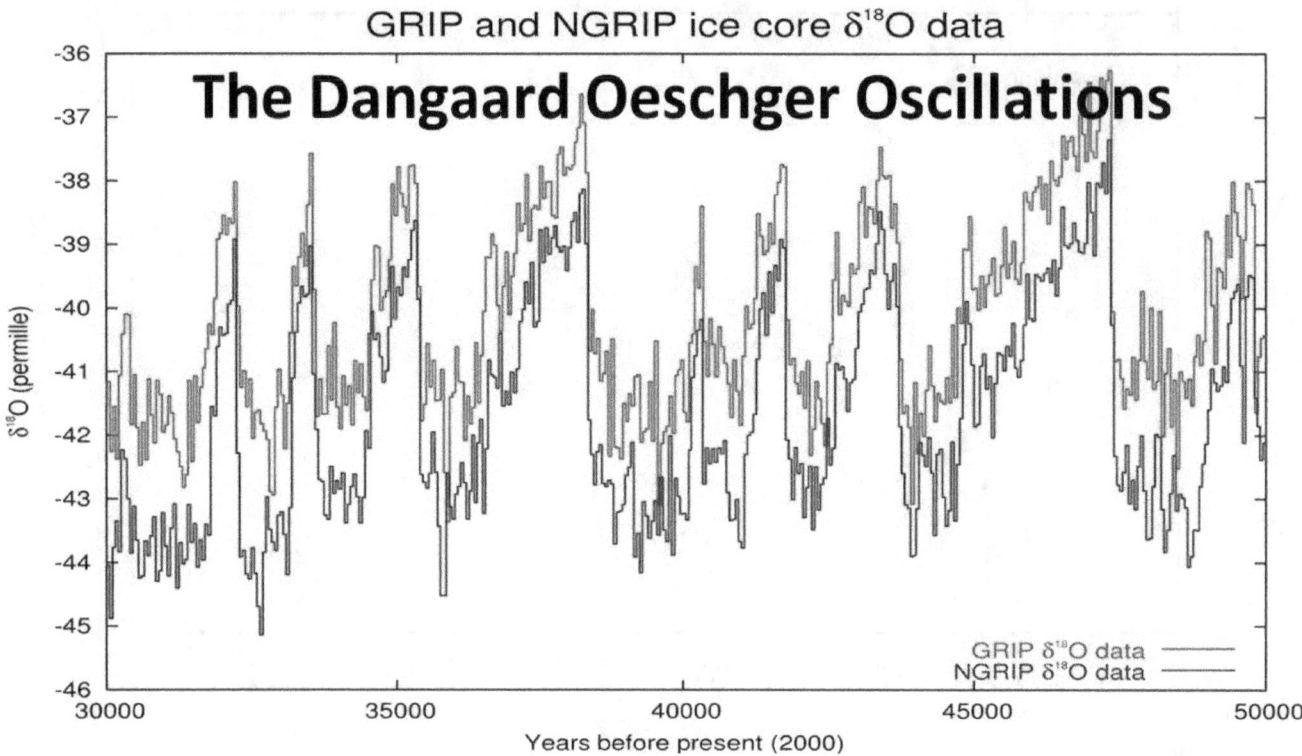

Evidence exists that these pulses occurred on a fairly regular basis. Their occurrence has created large climate oscillations. Evidence has been detected in ice core samples drilled from the Greenland ice sheets that these short periods when the Sun becomes active again, have occurred in intervals of 1470 years. The resulting oscillations have been named, the Dansgaard-Oeschger oscillations.

Part 5 - The Dansgaard Oeschger oscillations

The Dansgaard Oeschger oscillations

During the extremely weak conditions in the galaxy

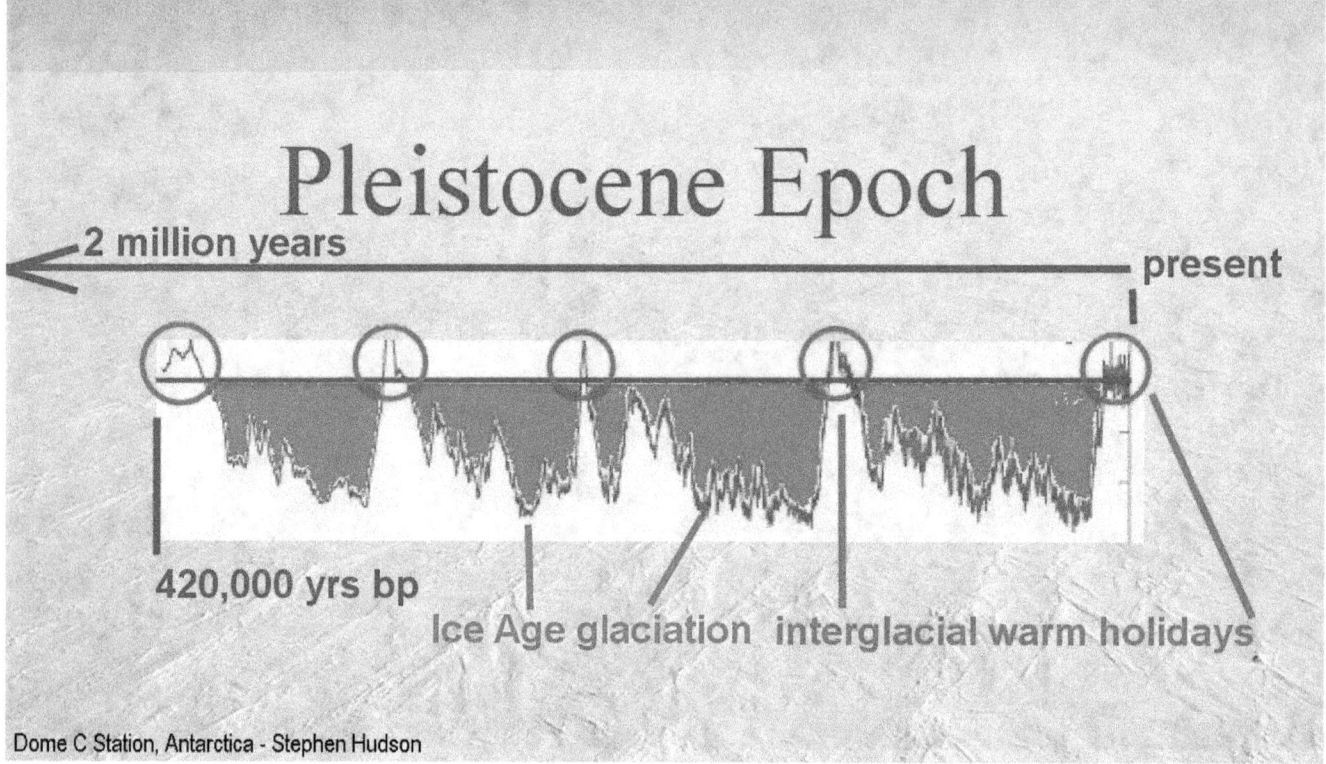

During the extremely weak conditions in the galaxy that enable the ice ages on earth, the Sun remains constantly active only during the timeframe of the interglacial pulses. In the interglacial timeframe the solar system is in its high-power mode.

During the long glacial periods

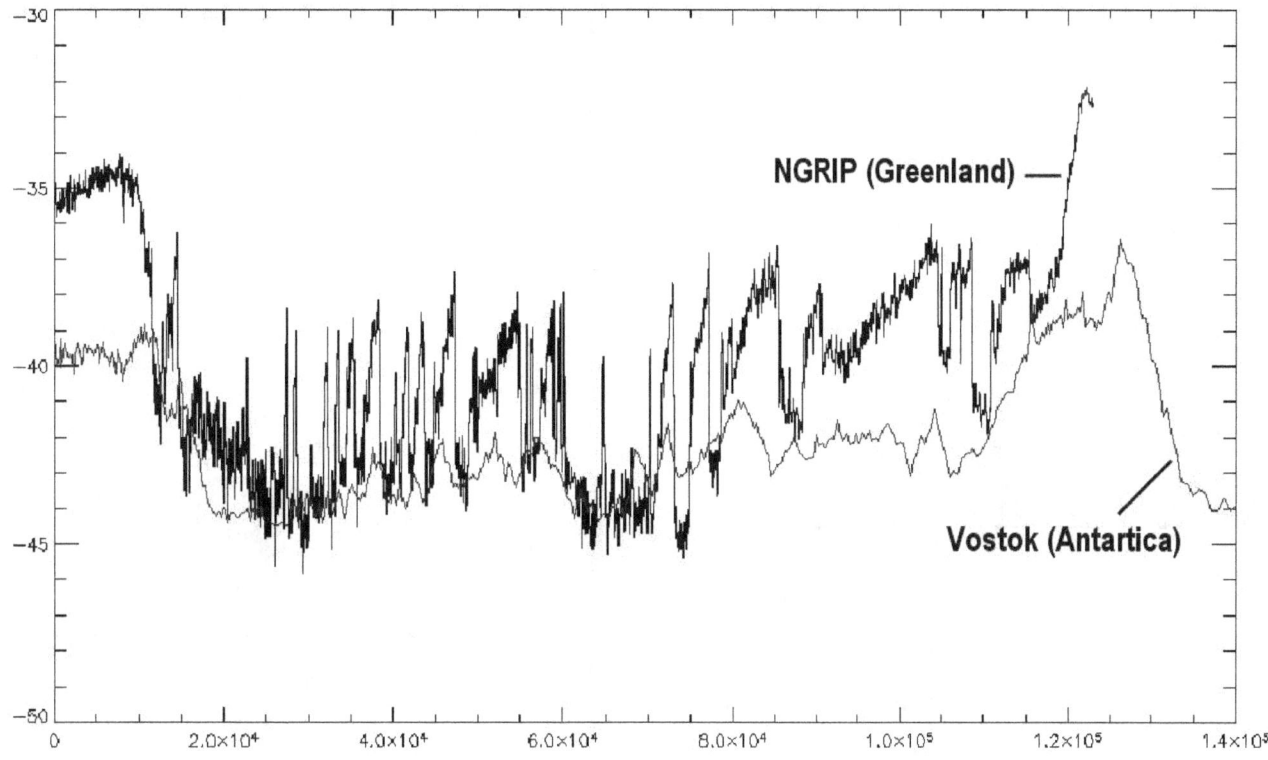

During the long glacial periods, however, between the warm and sunny interglacial pulses, large climate fluctuations have occurred that are named Dansgaard-Oeschger oscillations in honour of the discoverers of them.

These are large climate fluctuations, in the range between 20 to 40 times that of the Little Ice Age cooling. These large oscillations dominate the climate landscape all the way through the Ice Age glaciation periods. These enormously massive climate events were evidently caused by the Sun turning on and off periodically, rather than by ocean current fluctuations.

The Dansgaard-Oeschger oscillations

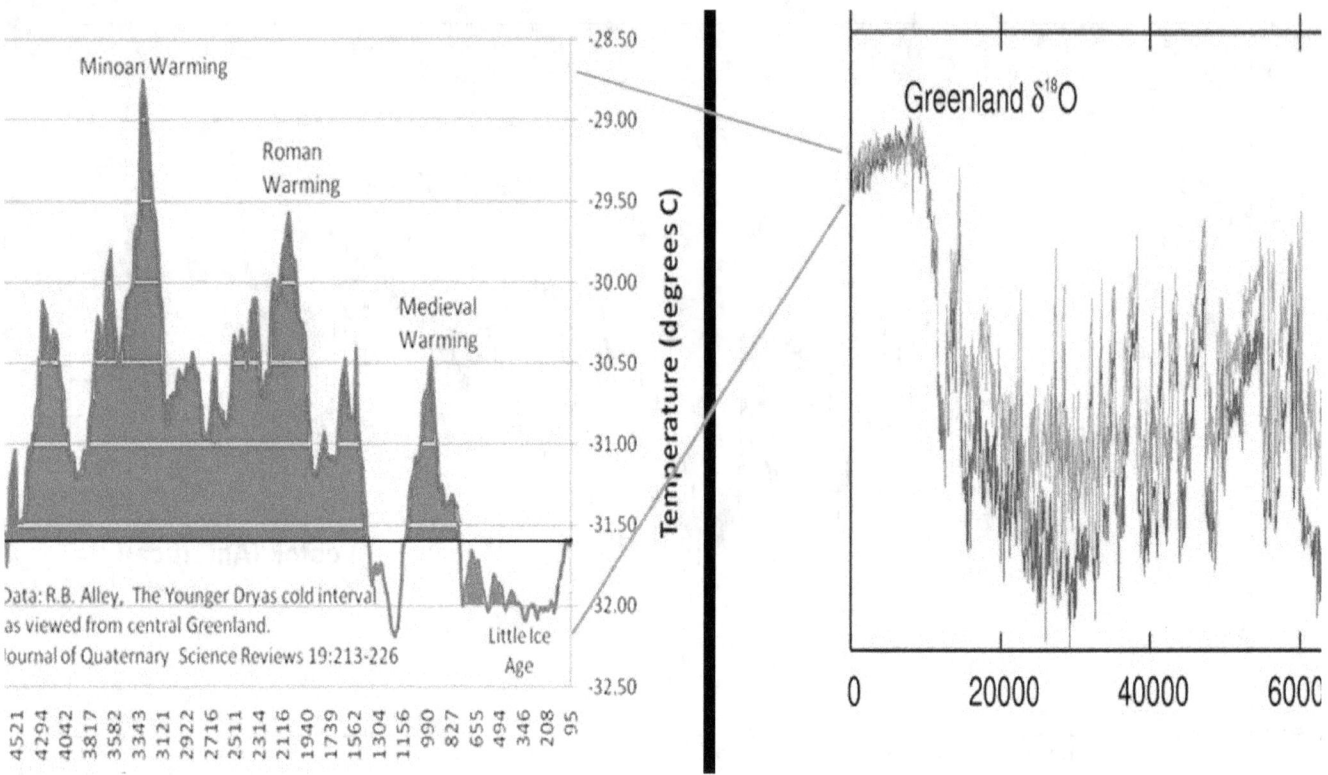

The Dansgaard-Oeschger oscillations, which have been discovered in ice core samples in Greenland, were apparently gigantic events that pale all the climate events in experienced history into insignificance.

The oscillations that have apparently spanned all the modern ice ages, may have prevented the Earth from freezing up completely, as it may have had in distant time roughly 700 million years ago, when the entire Earth became a snowball and remained frozen for tens of millions of years.

The snowball-Earth concept is just a theory

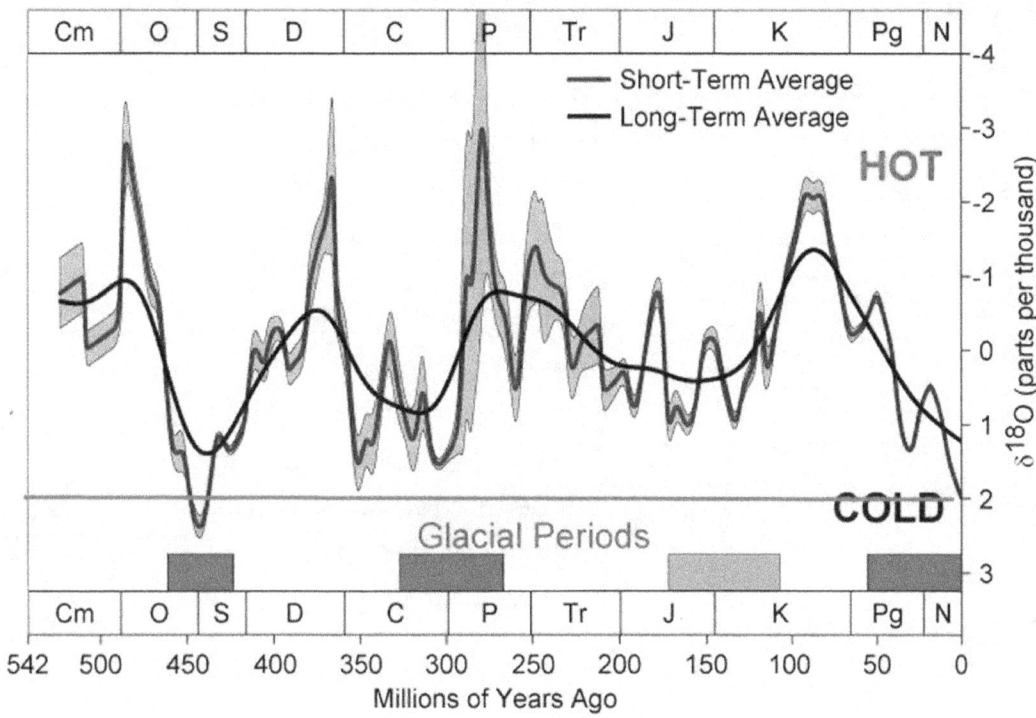

While the snowball-Earth concept is just a theory based on records laid up in fossils, the last four glaciation periods, that were discovered with oxygen-18 isotope measurements, coincident with data in fossils, were evidently real. It is known, for example, that the deep glaciation that occurred around 445 million years ago, was so severe at the time that half of all species of life that were known to exist at the time, became extinct. The extinction occurred not on land, but in the oceans. Life on land did not exist at the time. It evidently takes enormously massive climate events to affect the oceans so severely that half of all species living in the oceans were unable to survive. The result is referred to as the Ordovician extinction event.

The enormous rise and fall of sea-levels that occur during glaciation conditions, which had repeated exposed and flooded large areas, may have sufficiently altered the habitats to exterminate many a long-established species.

Ocean levels dropped 400 feet

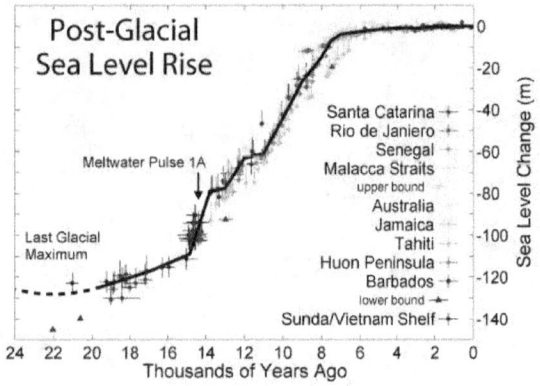

Expect a near-term sea-level reduction

The danger is that we will experience a massive reduction in sea level in the near term as the re-glaciation begins with the Ice Age Transition now in progress.
Picture a loss of only 20 meters
A powerfull new renaissance will be needed to meet the physical challenge for infrastructures

Wikipedia

In modern time, during the last glaciation period, which is generally called the Ice Age, the ocean levels dropped 400 feet as the water became piled up on land in the form of Ice, over 10,000 feet deep, and remained frozen there. No evidence is known that the Dansgaard-Oeschger oscillations had caused sea-level fluctuations.

The Milankovitch Cycles

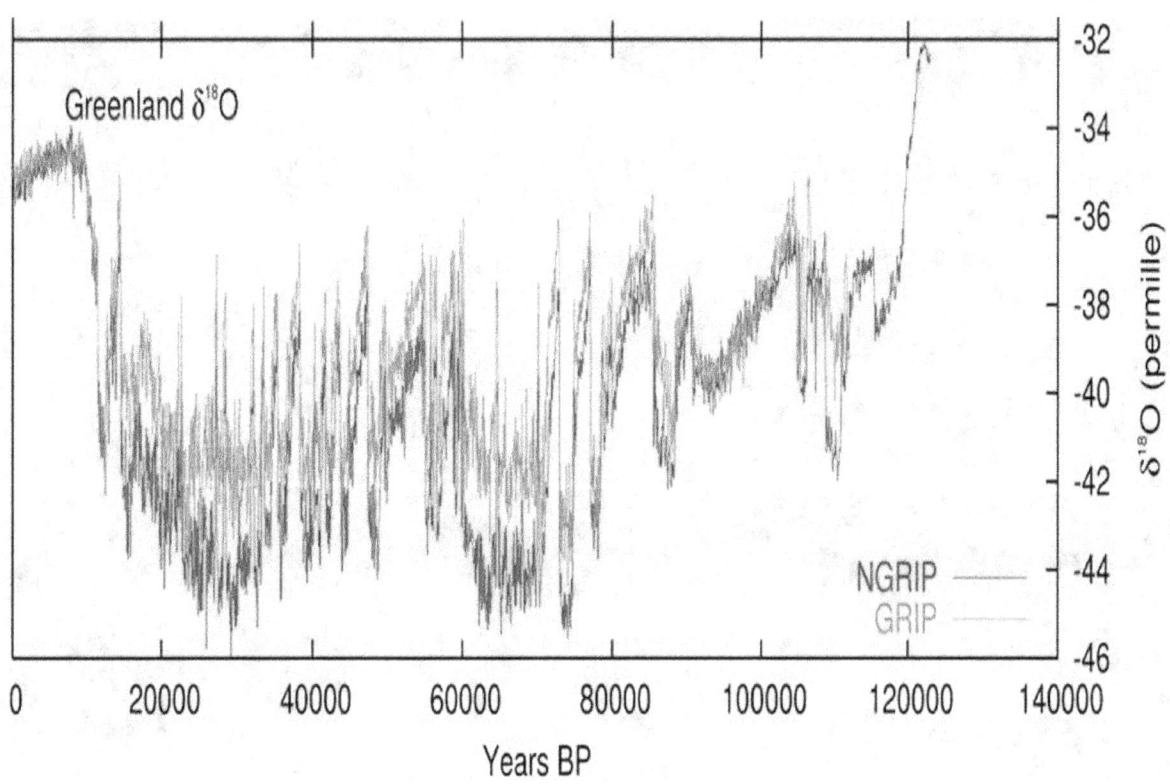

The oscillation pulses were evidently too short in duration to cause significant melting. They were primarily fluctuations between extremely cold climates and not quite as cold climates. It appears that some of the pulses came close to the interglacial level, but remained still below it. The cause for the pulses remains largely a puzzle, as their existence doesn't fit into the widely accepted Ice Age model where climate fluctuations are believed to be caused by three overlapping, long-term fluctuations of the orbit of the Earth around the Sun, which are termed the Milankovitch Cycles in honour of the man who pioneered the concept.

The Serbian geophysicist and astronomer Milutin Milankovich

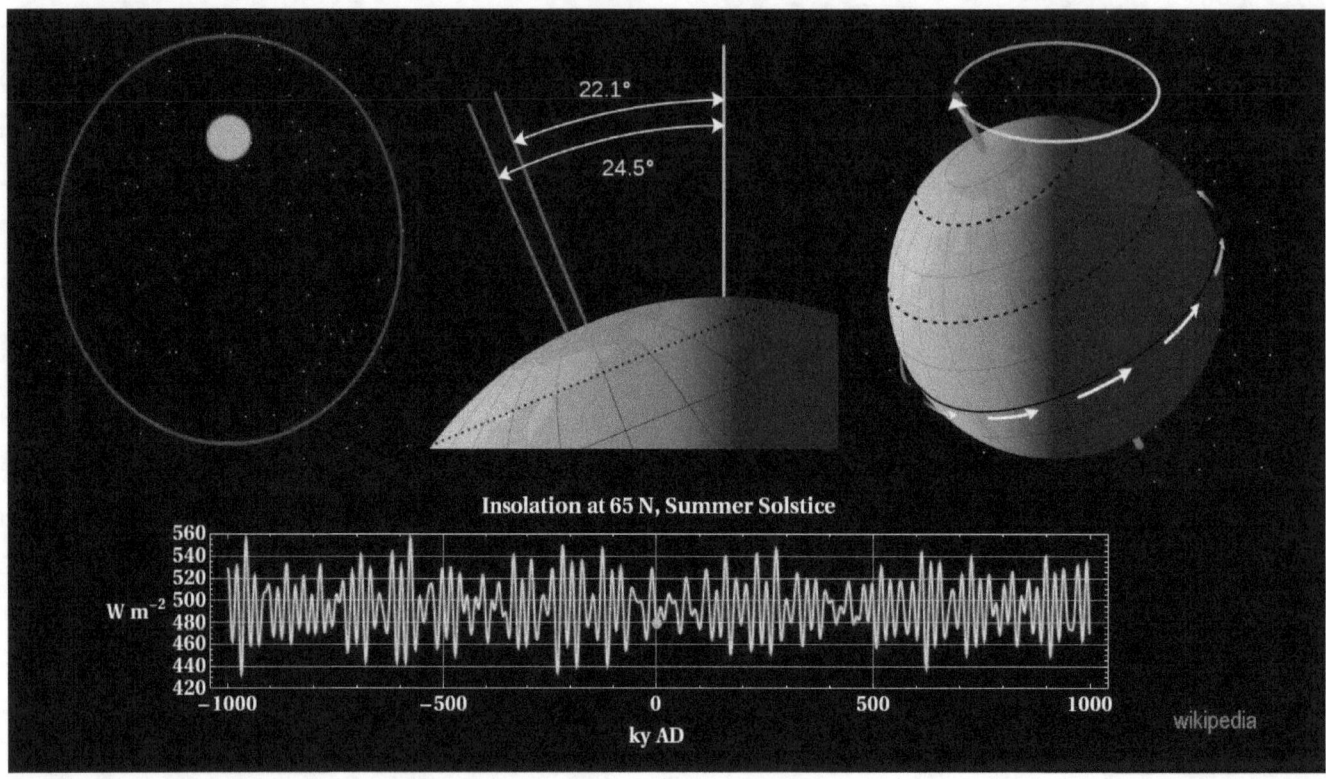

The Serbian geophysicist and astronomer Milutin Milankovich combined the 26,000-year precession cycle of the spin axis of the Earth, and the minute 41,000-year shifting of the spin-angle, with the 100,000-years cycle of the shifting eccentricity of the Earth orbit around the Sun, and theorized that the overlapping of these cycles, which cause minor variations in the seasonal and hemispheric distribution of the radiation of the Sun, causes Ice Ages to occur. This is the most elegant theory that has ever been put forth to explain the occurrence of ice ages under a constant Sun.

The problem with the theory is that the total amount of solar energy received remains always the same, no matter how the Earth's spin axis is shifting, and the eccentricity of its orbit varies.

Theatrical cause, doesn't match the measured ice core data

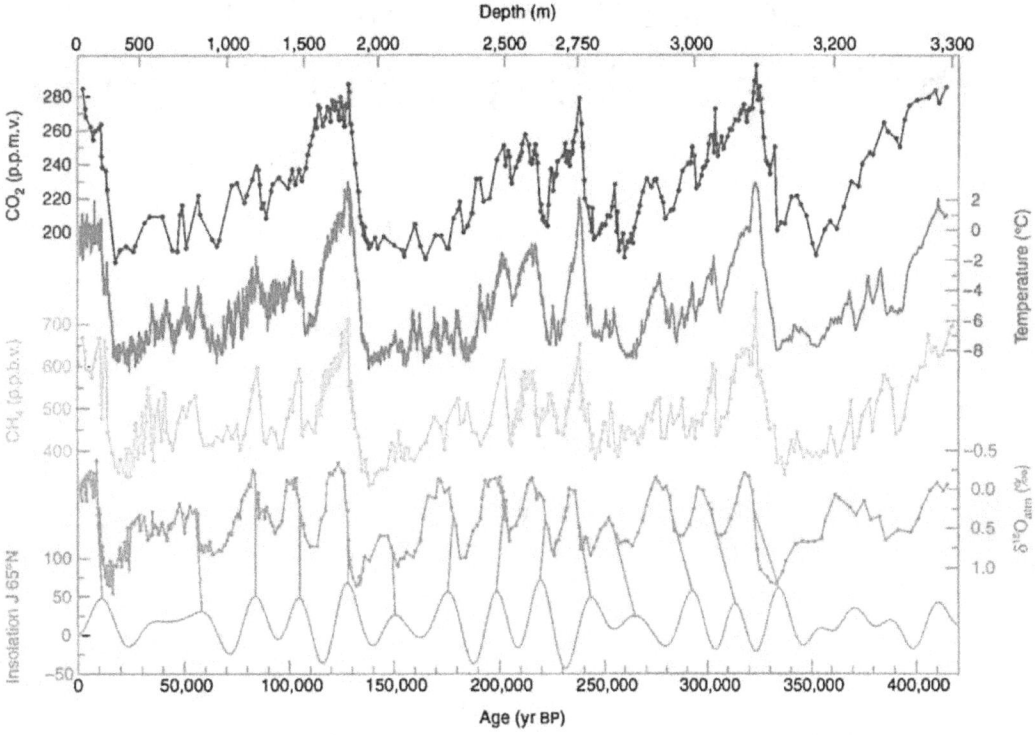

Another problem with the theory is, that the mathematically computed theatrical cause, doesn't match the measured ice core data, which renders the cycles to be subsequent phenomena of the astrophysical dynamics that cause the ice ages, rather than being causative for them. Sometimes the computed cause lags the events by 10,000 years, and sometimes precedes it by 10,000 years, as shown in the deviations presented in brown.

When it comes to the big Dansgaard-Oeschger oscillations

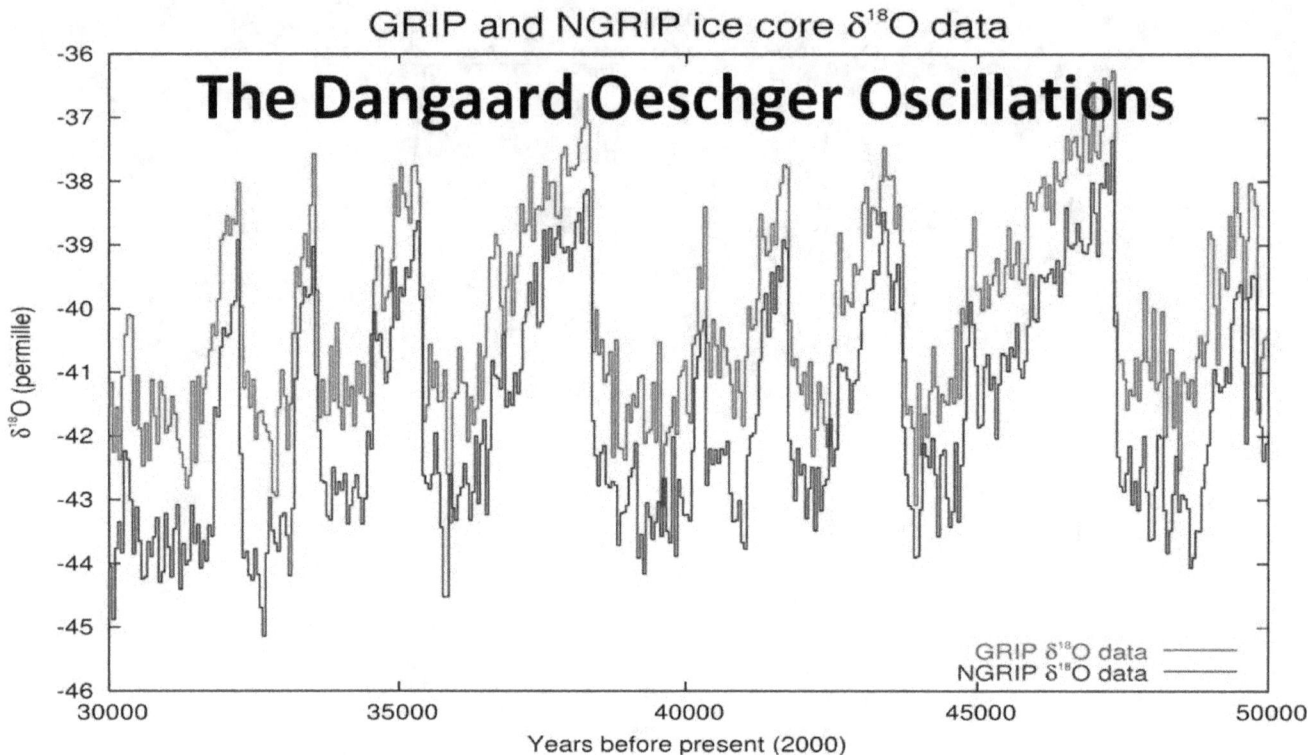

Of course, when it comes to the big Dansgaard-Oeschger oscillations that are able to spike in the range of decades and occur in intervals spaced 1470 years, the orbital cycles theory falls totally flat. A total of 25 such events have been identified in the ice core data. Unfortunately the Dansgaard-Oeschger oscillations hadn't been discovered at the time when the Milankovich cycles theory was invented. The Dansgaard-Oeschger oscillations were only recently discovered, when the very-deep drilling through the Greenland ice sheet was undertaken.

These drilling projects are not small undertakings

"Gripdome". Licensed under CC BY-SA 3.0 via Wikipedia

These drilling projects are not small undertakings. The Greenland Ice Core Project, named GRIP, was a multinational European research project, organized through the European Science Foundation. The funding came from 8 nations, Belgium, Denmark, France, Germany, Iceland, Italy, Switzerland, and the United Kingdom, and from the European Union.

The GRIP project drilled out a 3028-metre ice core

"Gripcor1". Licensed under CC BY-SA 3.0 via Wikipedia

The GRIP project drilled out a 3028-metre ice core, from a summit of the ice sheet in Central Greenland, to bed rock. It took 4 years to drill this ice core, from 1989 to 1992. The oscillations were discovered in this ice core, with obviously no small surprise.

The drilling was repeated with another 4-year effort

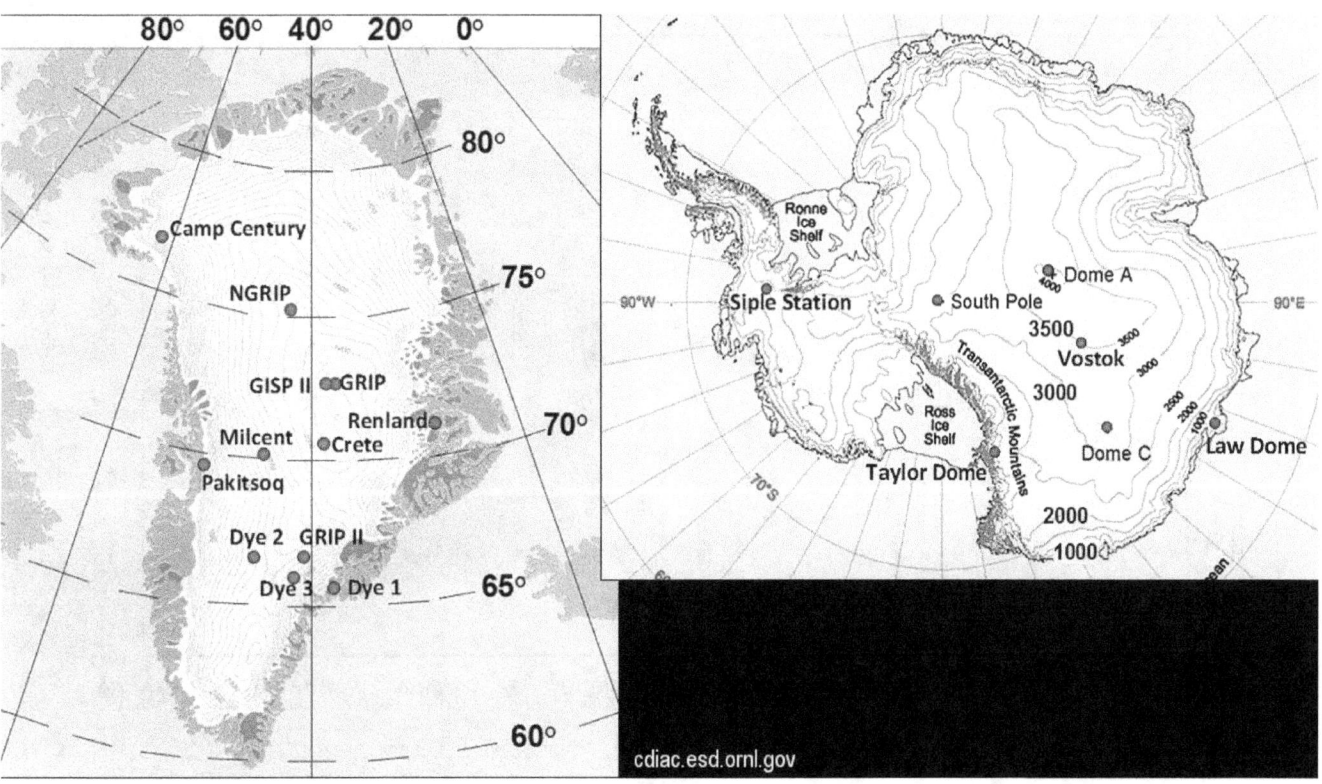

For numerous reasons the drilling was repeated with another 4-year effort. Another drilling was started a long distance further North, near the 75 degree latitute. It was started in 1999. Bedrock was reached in 2003 The drilling site and the new project was named the North Greenland Ice Core Project, or NGRIP for short. The drilling at this new site is significant in that it penetrated deeper and thereby into earlier ice, reaching back all the way into the previous interglacial period.

The new drilling confirmed

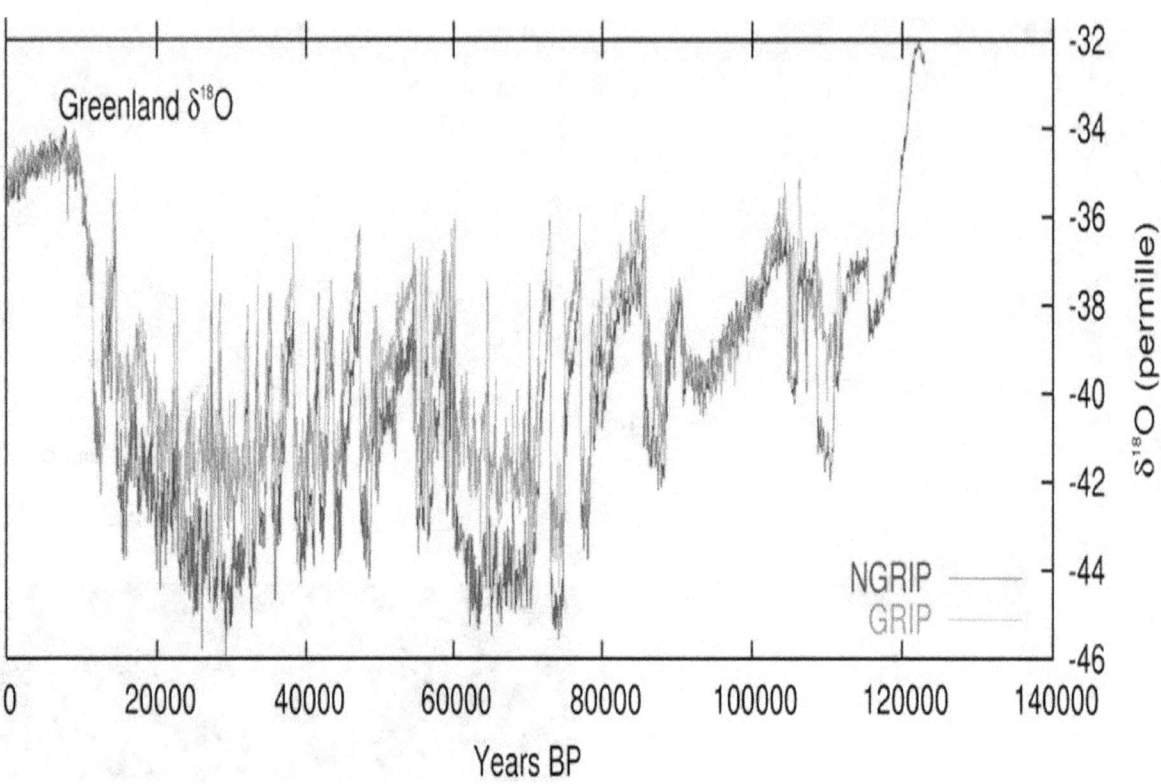

The new drilling confirmed that the Dansgaard-Oeschger oscillations were not a freak anomaly. Both the GRIP and NGRIP results are shown here, combined in different colors.

When one overlays the Greenland ice core data

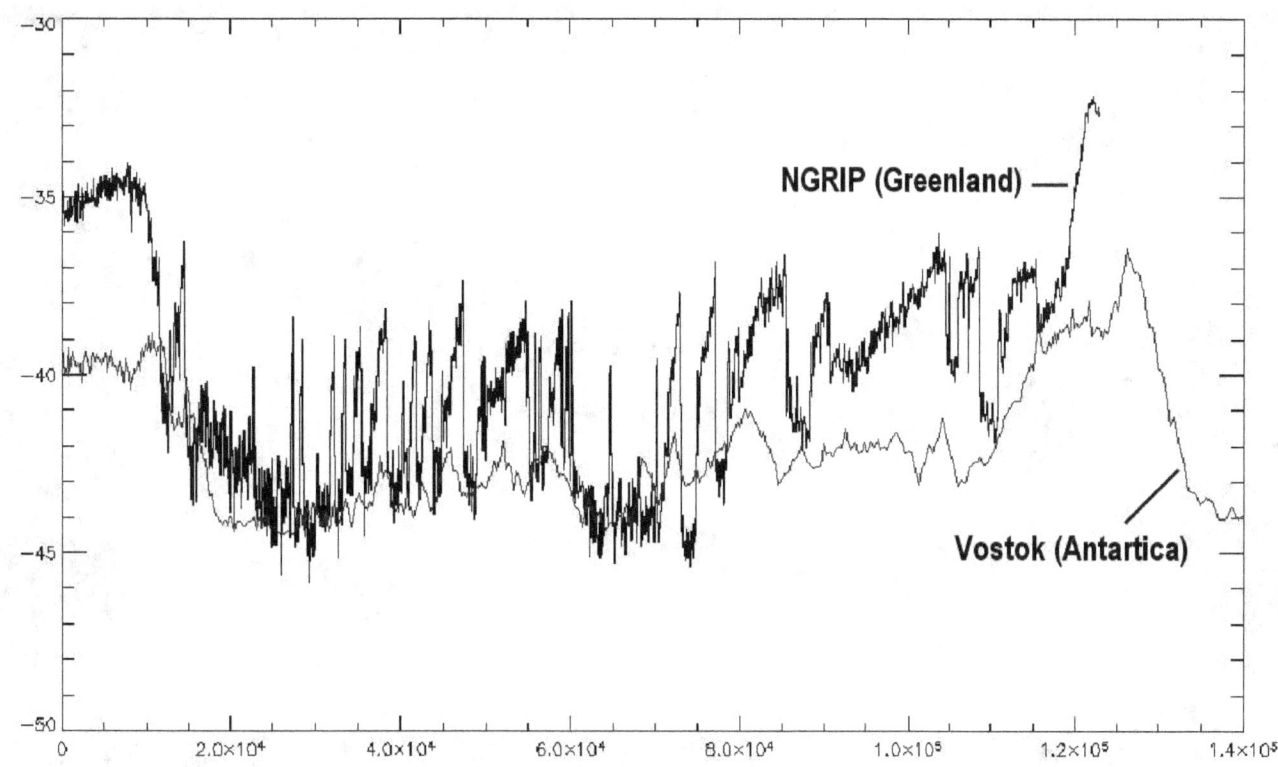

When one overlays the Greenland ice core data, with that of Antarctica, the rapid oscillations appear washed out, so that only the major trends show up in Antarctica. The reason for the difference is that Antarctica is an ice desert with extremely little precipitation whereby the fine details are lost in the coarser resolution.

The coldest, driest, and windiest continent on Earth

"Marie Byrd Land" by Michael Studinger / NASA Goddard Space Flight Center - Flickr: Marie Byrd Land. Licensed under CC BY 2.0 via Commons

It is not widely recognized that Antarctica is the coldest, driest, and windiest continent on Earth, with average 'winter' temperatures below minus 60 degrees Celsius, with the coldest temperature recorded at minus 89 degrees, that's minus 129 degrees Farenheit.

The snowfall there is as minuscule there as the rain in the Sahara. It may be for the lacking depth of resolution that the details in Antarctica are measured in ratios of heavy hydrogen, H-2 isotopes, while in Greenland the measurements are made in ratios of the heavy oxygen-18 isotope..

In comparison with Antarctica, Greenland is a wet place

In comparison with Antarctica, Greenland is a wet place. The same depth of ice that on Greenland covers up to 120,000 years, covers in Antarctica a range of over 450,000 years.

Ice cores go back in time up to 740,000 years

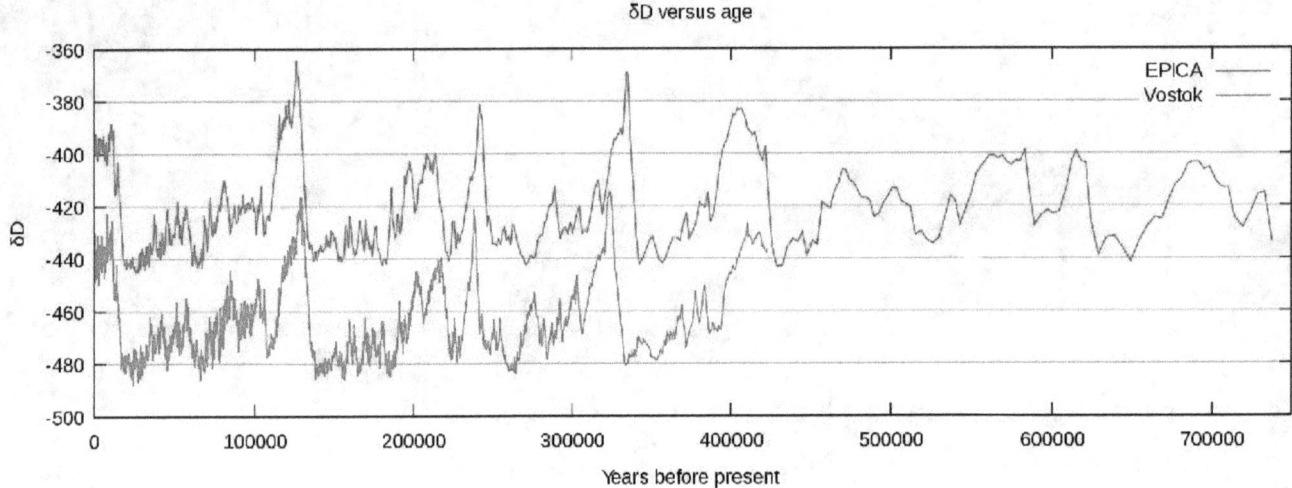

Some ice cores go back in time up to 740,000 years, as at the EPICA drilling site. The European Project for Ice Coring in Antarctica, drilled out an ice core that goes back in time 740,000 years and reveals 8 previous glaciation cycles. The project was completed in December 2004, just slightly over a decade ago.

Differences in the resolution of details

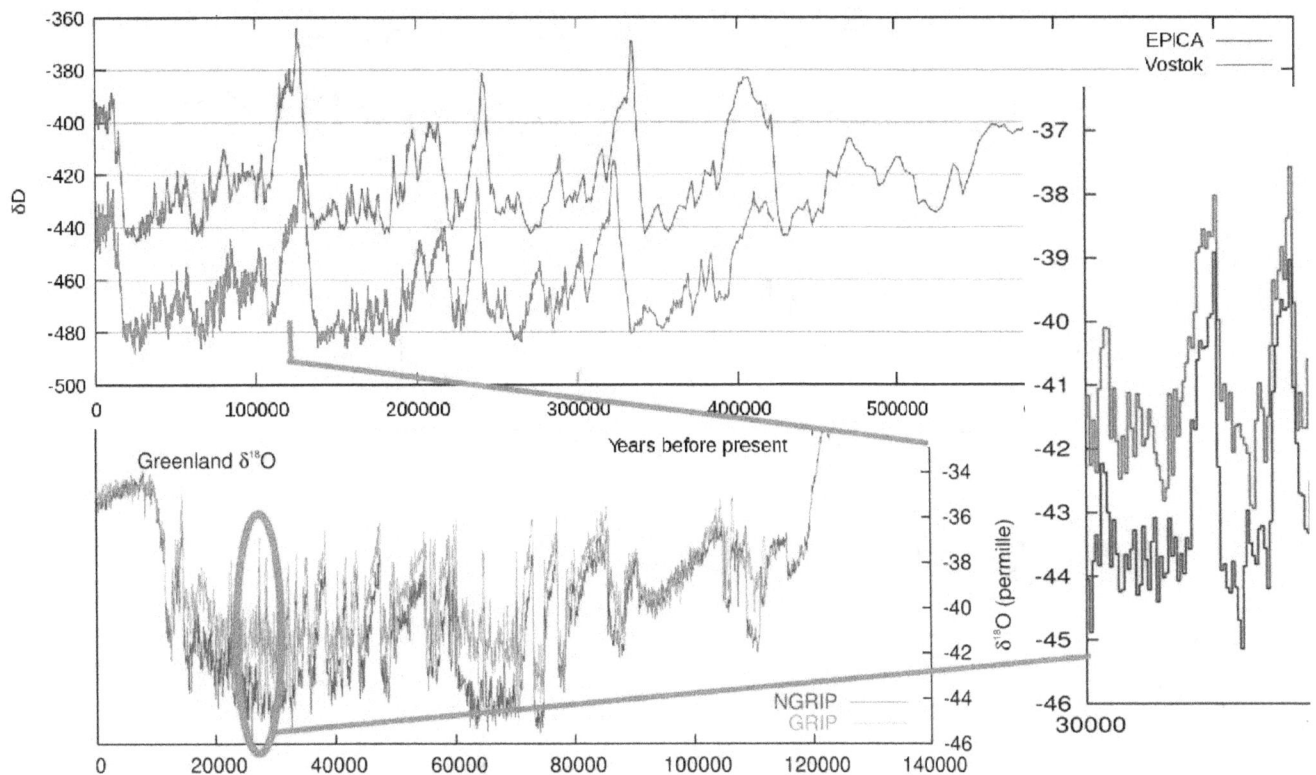

The high compaction of the ice in Antarctica all adds up to corresponding differences in the resolution of details.

The Dansgaard-Oeschger oscillations are very real

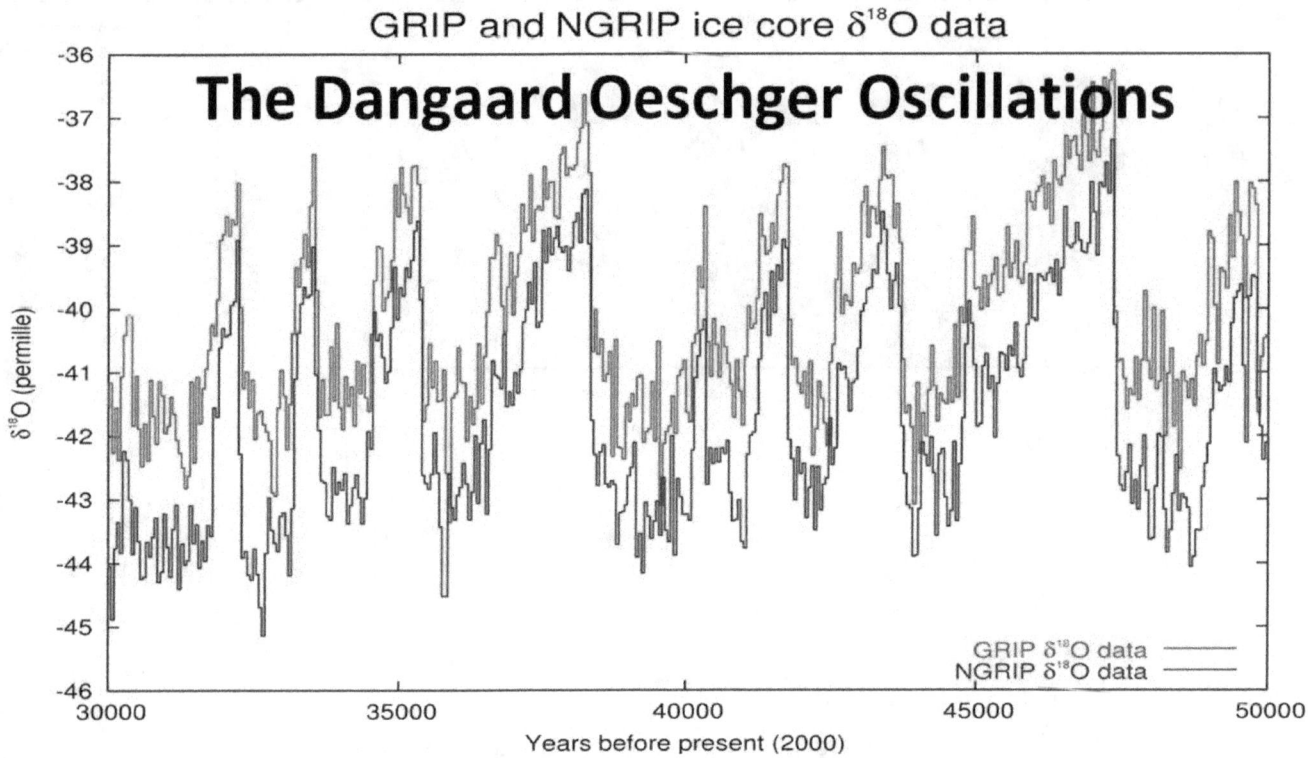

The bottom line is, as the high resolution data from Greenland from multiple drilling sites indicates, the Dansgaard-Oeschger oscillations are very real.

They are real. They are very big. And they are an enigma outside the recognition of the Sun, as being a plasma star that is able to go inactive periodically. In fact, the very existence of the oscillations proves the Sun to be an electric star, because nothing else has the enormous on-off capability that we see reflected in the historic measurements for the climate on Earth.

Spaced 1470 years on average

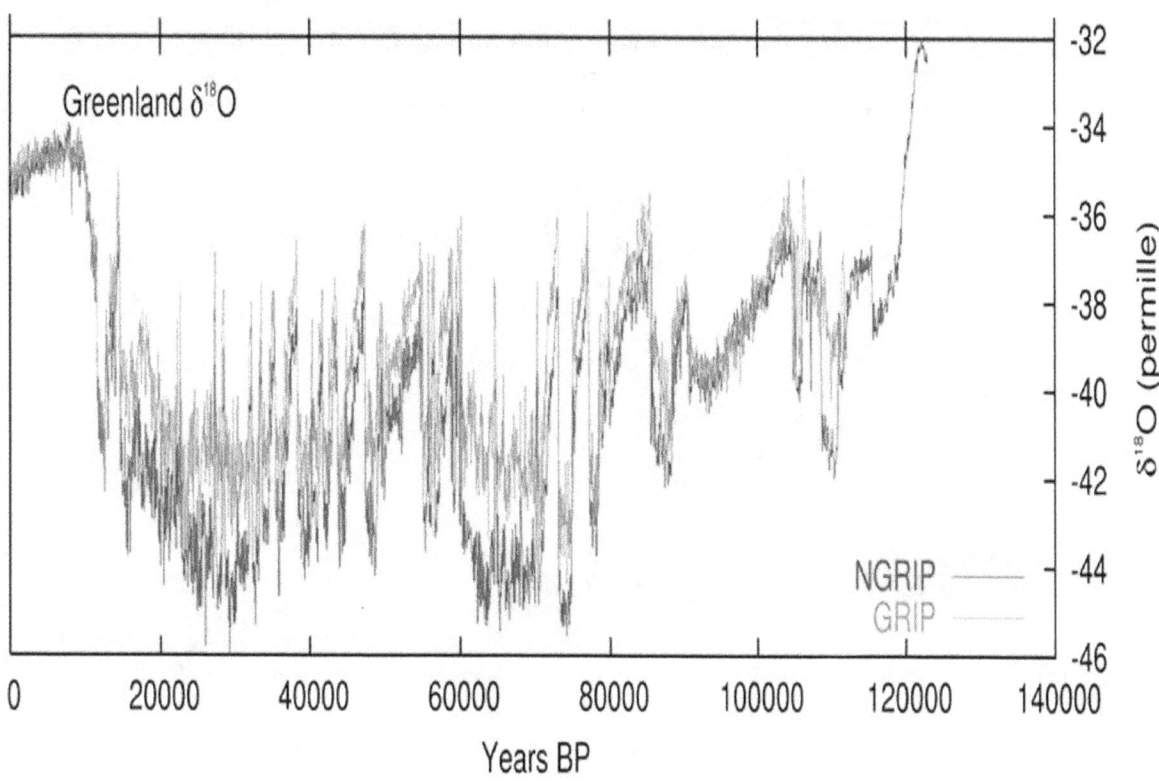

As I said before, researchers have identified 25 major Dansgaard-Oeschger pulses, spaced 1470 years on average. It might be possible that these pulses can become significant as points of reference for determining the next potential cut-off point for the Primer Fields, and with them the powered Sun.

A reasonable determination three events spaced 1470 years apart

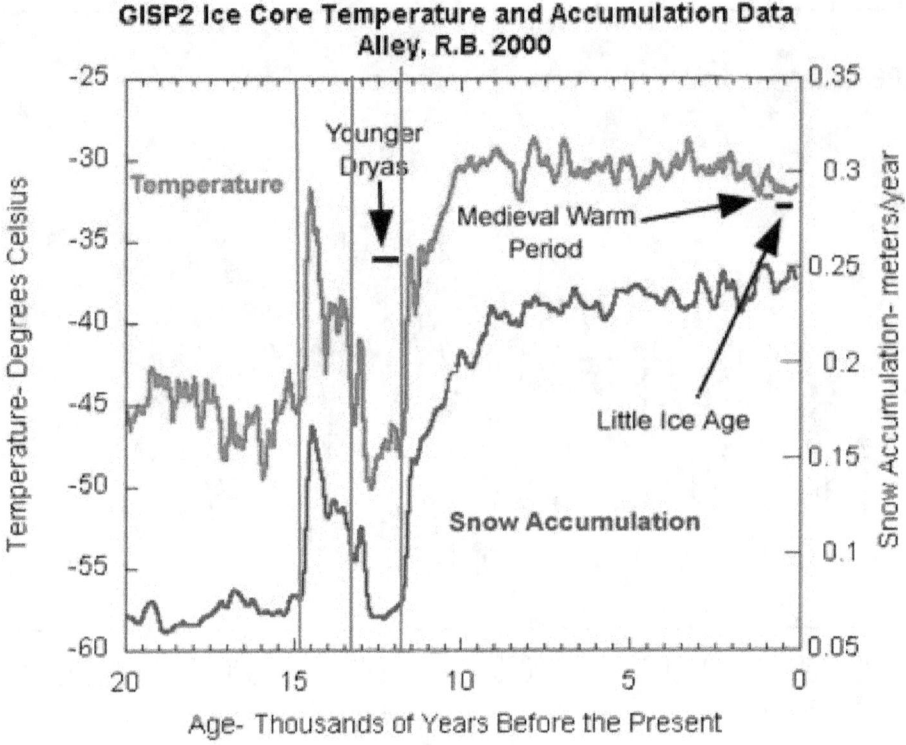

We can make a reasonable determination by looking at the last three of the Dansgaard-Oeschger events that occurred at the end of the most recent Ice Age.

We see evidence of three such events there, spaced 1470 years apart. The first started with a major upswing, followed by a down slope. The second event started on the down slope, but was of short duration. The last upswing was much larger again, and it didn't collapse this time, but unfolded further into the interglacial periode.

The most recent Dansgaard-Oeschger event

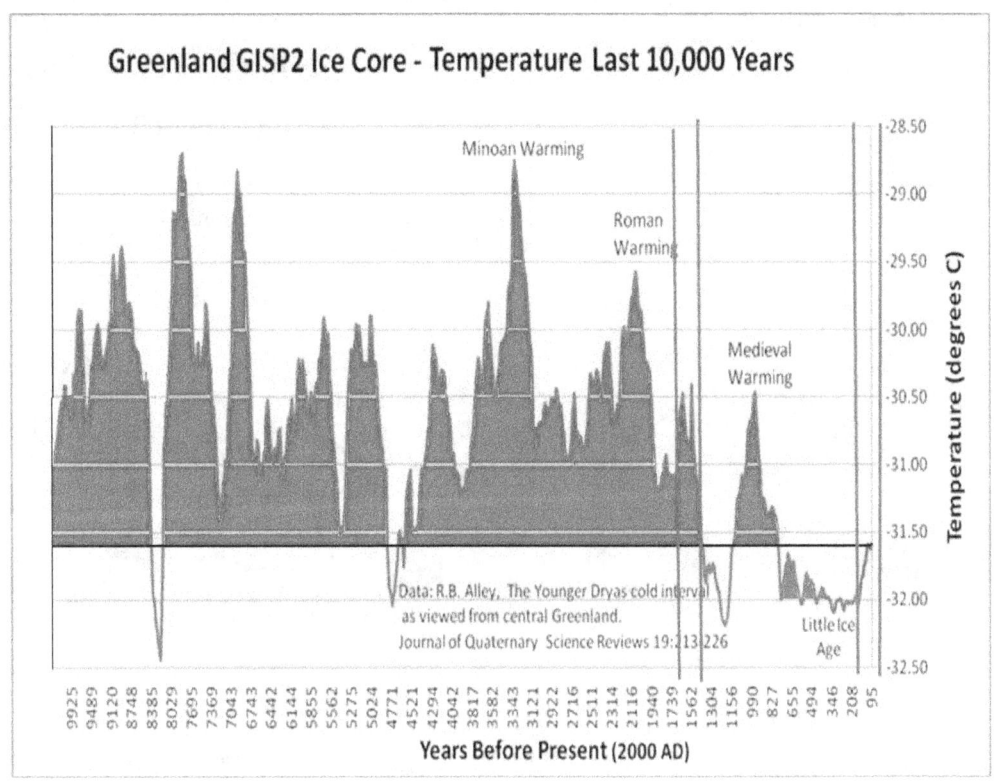

If we run the clock forward from these three steps, in steps of 1470 years, then the most recent Dansgaard-Oeschger event should have occurred roughly two to three hundred years ago. This happens to be the point in time where we see the Little Ice Age ending, and the Great Global Warming beginning.

With this coincidence in mind, a high probability exists therefore that the cosmic invigorating of the Sun and the warming of the world that broke us out of the Little Ice Age, may have been a Dansgaard-Oeschger event.

Preventing the collapse of the Little Ice Age into the next big Ice Age

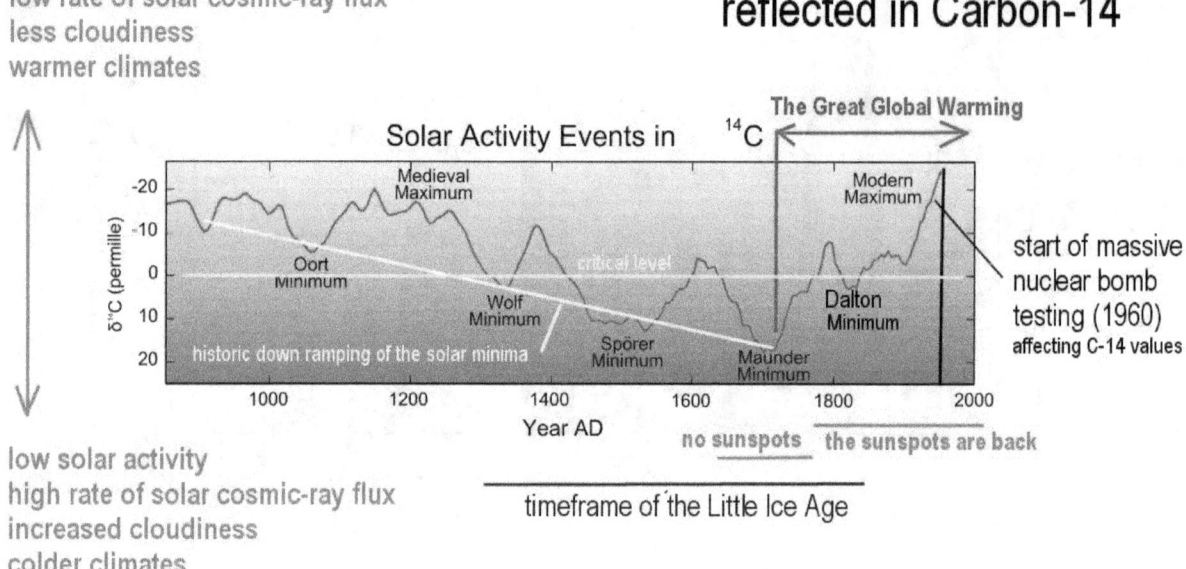

"Carbon14 with activity labels" by Leland McInnes at the English language Wikipedia. Licensed under CC BY-SA 3.0 via Commons

If this is so, the Dansgaard-Oeschger pulse at this time, becoming active in the 1700s, may have saved civilization, and humanity with it. It may have saved us by preventing the collapse of the Little Ice Age into the next big Ice Age. It overshadowed the down-ramping with a major steep up-ramping in solar activity.

The previous Dansgaard-Oeschger event

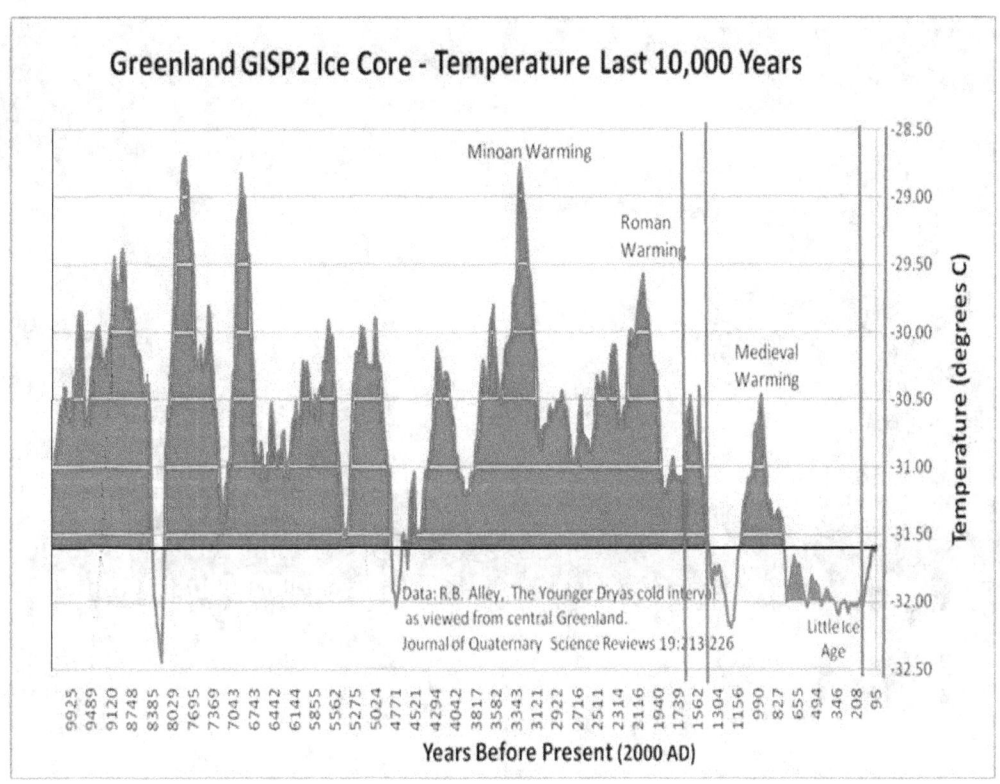

The previous Dansgaard-Oeschger event, which would have occurred 1470 years earlier, takes us back to 1770 years before the present, to the peak time between the Roman Warming, and the Medieval Warming, which had ended steeply, causing the deep low of the Oort Minimum.

If the current Dansgaard-Oeschger pulse ends in the same manner that the previous one had ended, humanity will be in deep trouble, because nothing has been prepared for the kind of world that we will then have to live in. But that's what we see unfolding, resulting from diminishing plasma density from interstellar space that we see reflected in collapsing sunspot numbers, diminishing solar activity, and diminishing solar-wind pressure, and so on.

A two-fold nested system of Primer Fields

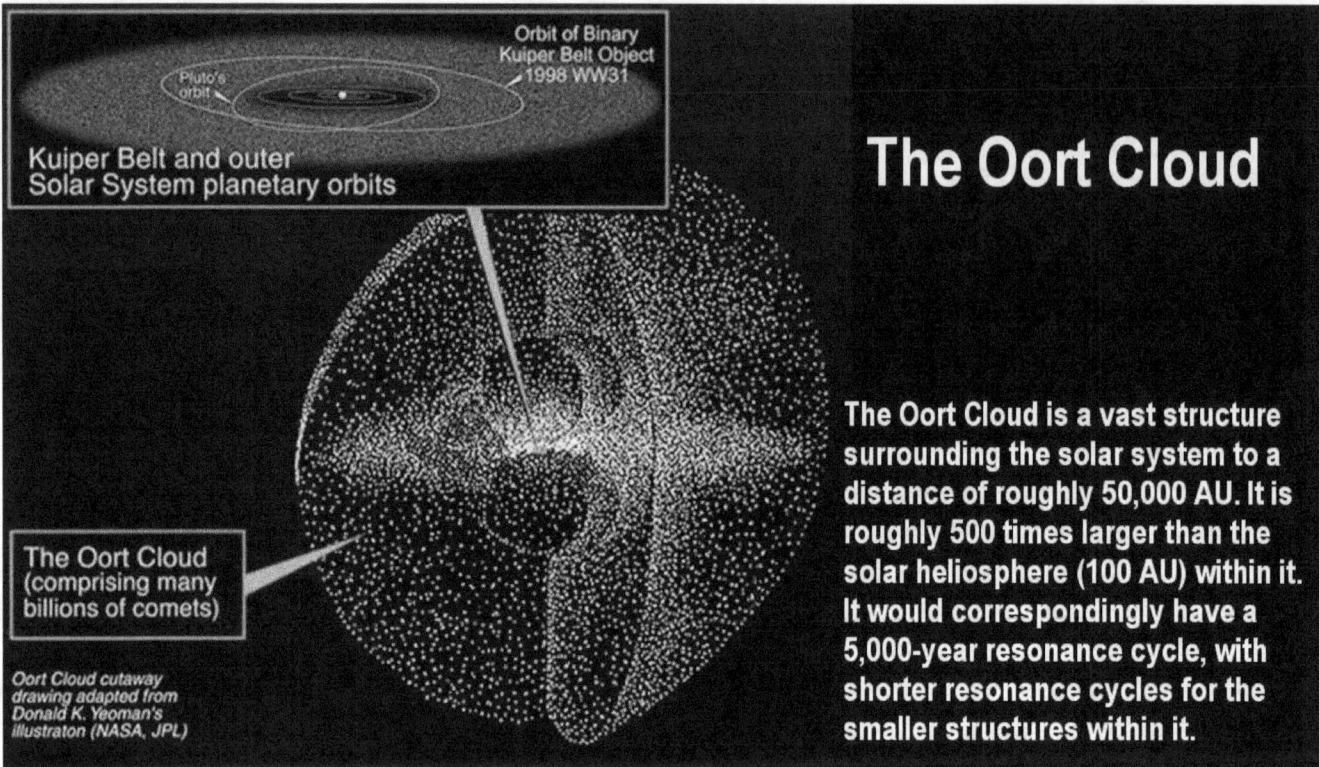

The 1470-years resonance of the Dansgaard-Oeschger oscillations suggests that the electromagnetic system that focuses interstellar plasma onto the Sun is likely a two-fold nested system of Primer Fields, an inner system and an outer system. The inner system with a resonance of 22 years, would comprise the solar system and the heliosphere. The inner system, in turn, would be the focal point of the outer system that would have to be more than 50 times larger and have a 1470-years resonance. This outer system would become the dominant system during the weak time of the glaciation periods. The theorized inner Oort Cloud could meet this pre-staging requirement with a 1470-years resonance cycle. The Dansgaard-Oeschger oscillations, which are very real, may one day be seen as tangible evidence for the existence of the inner Oort Cloud. Actually, it would be surprising if this type of evidence didn't exist.

If the Oort clouds are real

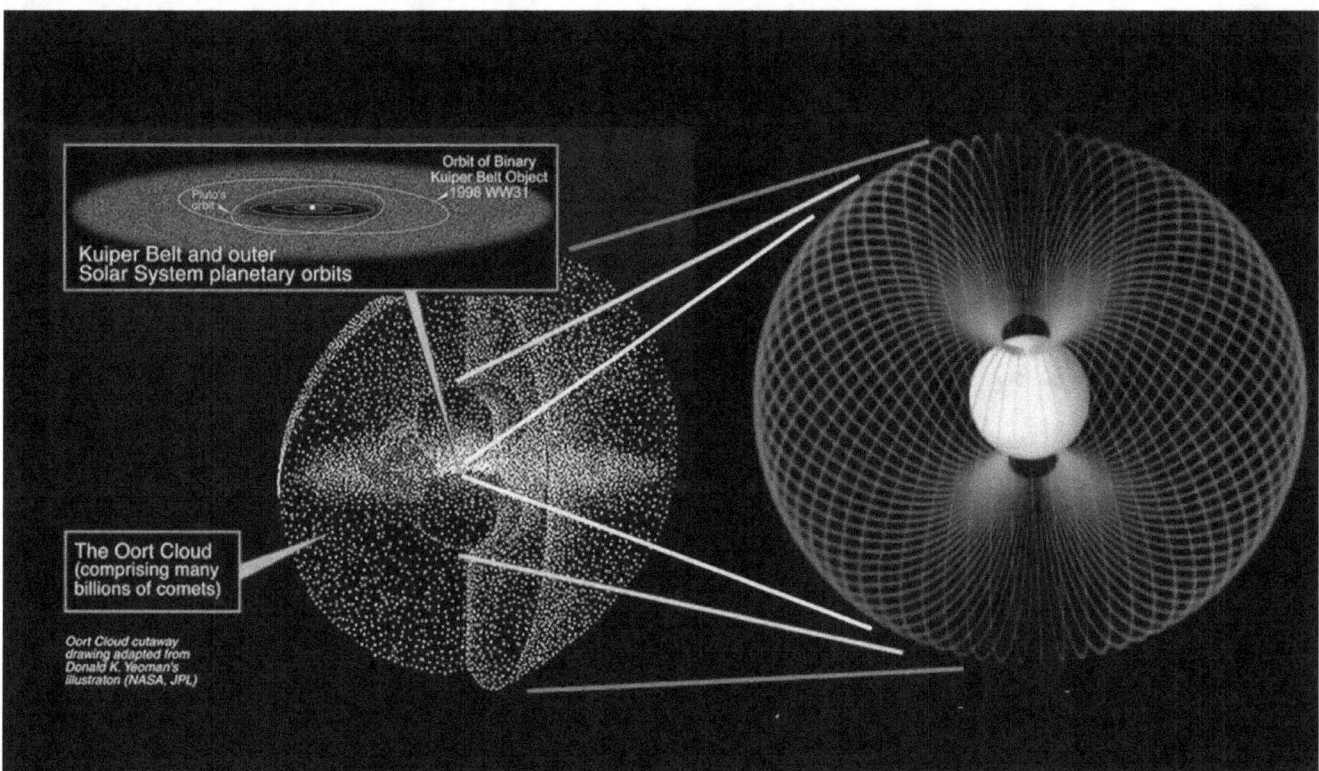

The inner cloud itself may be but the nested focal point of still another larger set of primer fields. This means that the plasma that is focused on our Sun may be supplied by a three-fold nested system of Primer Fields, of which only the innermost is subject to becoming inactive during the glaciation periods, with the resonance of the inner Oort Cloud supplying the recovery pulses that we see in the form of the Dansgaard-Oeschger oscillations in ice core samples, and so on.

If the Oort clouds are real, which they appear to be for numerous reasons, because nothing except the electromagnetic forces of the operation of primer fields would keep the space junk and comets that the clouds are made of in their distant place and centered on the Sun.

Evermore evidence keeps coming to light

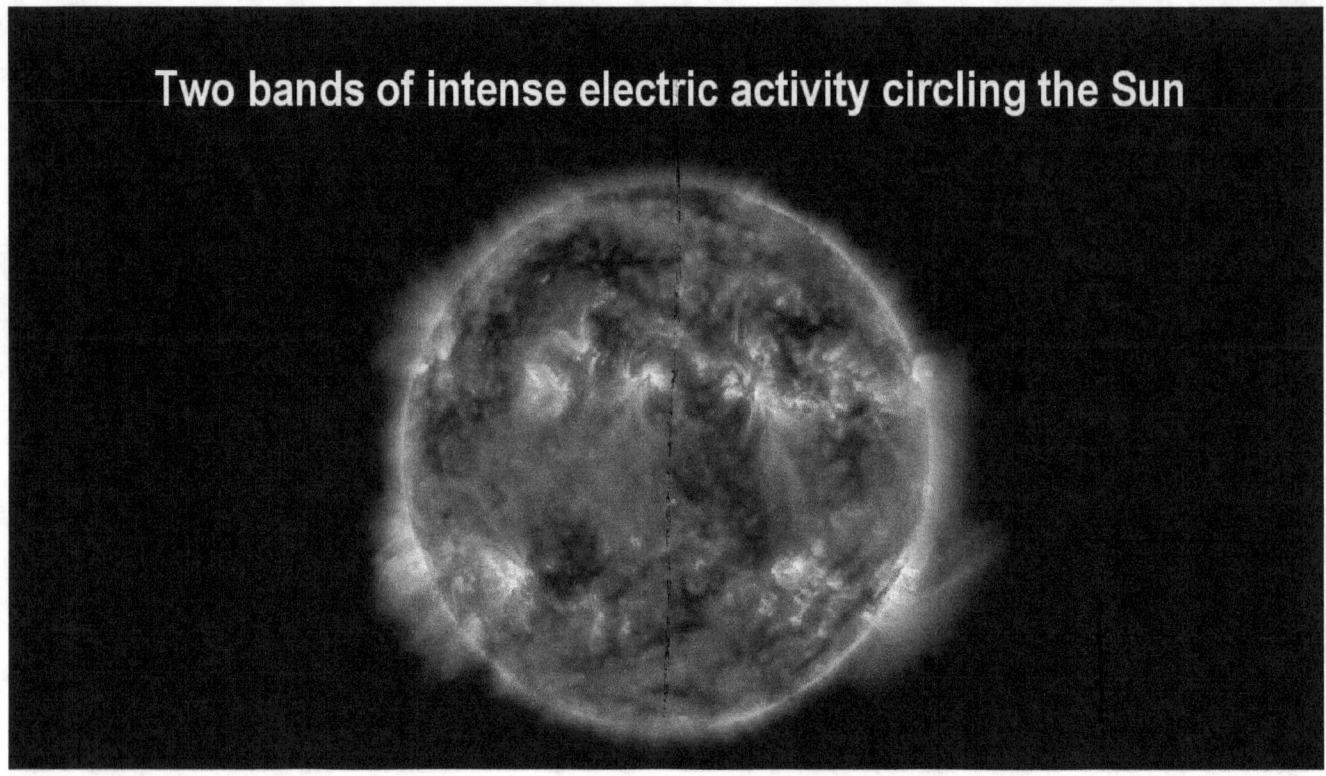

Evermore evidence keeps coming to light for the dynamics of the plasma powered external-fusion Sun. What we find here is ultimately our salvation, especially for the immediate times ahead. the plasma streams that power the Sun, which also extend to the Earth to some degree, may soon be required to power the human economy.

The cosmic plasma streams are not entropic

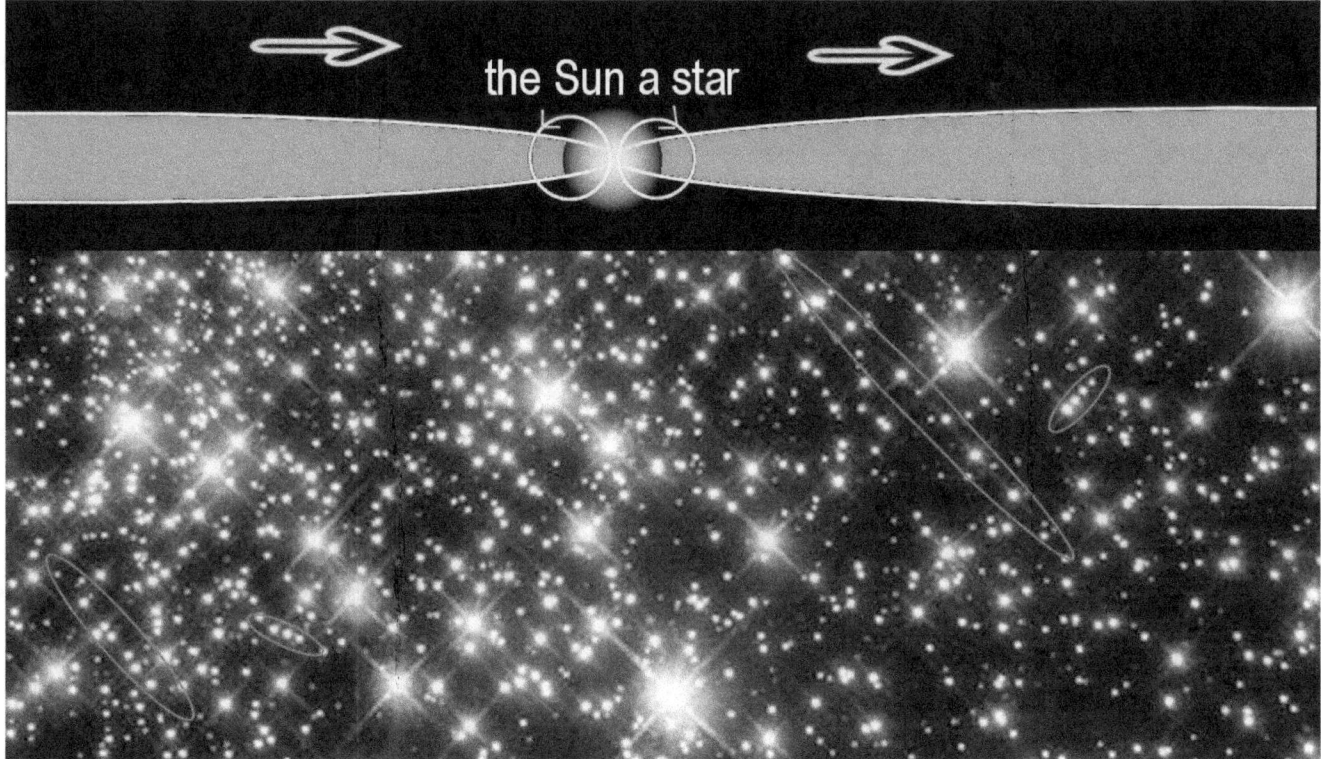

The cosmic plasma streams are not entropic. They are not subject to resource depletion. They are only subject to electric resonance fluctuations. They are inherently self-renewing from the energetic nature of space itself.

The 'Explicate Order' that David Bohm had referred to

The all-pervading plasma streams appear to be unfolding along the line of the 'Explicate Order' that David Bohm had referred to, with his revolutionary concept of universal energy, that all the energies expressed in the universe are but ripples of.

Whom Einstein had once referred to as his successor

David Bohm is the man whom Einstein had once referred to as his successor.

Whatever the case may be, plasma in space is not a fuel that is used up, but is anti-entropic in nature as David Bohm saw the energy background in the universe, which renders energy a dynamic part of the universe itself. The energy background in the universe may be the determining factor that limits the propagation speed of light that nothing supersedes, that even the neutrinos that penetrate everything, including the Earth itself, are bound to.

Finite resources subject to depletion

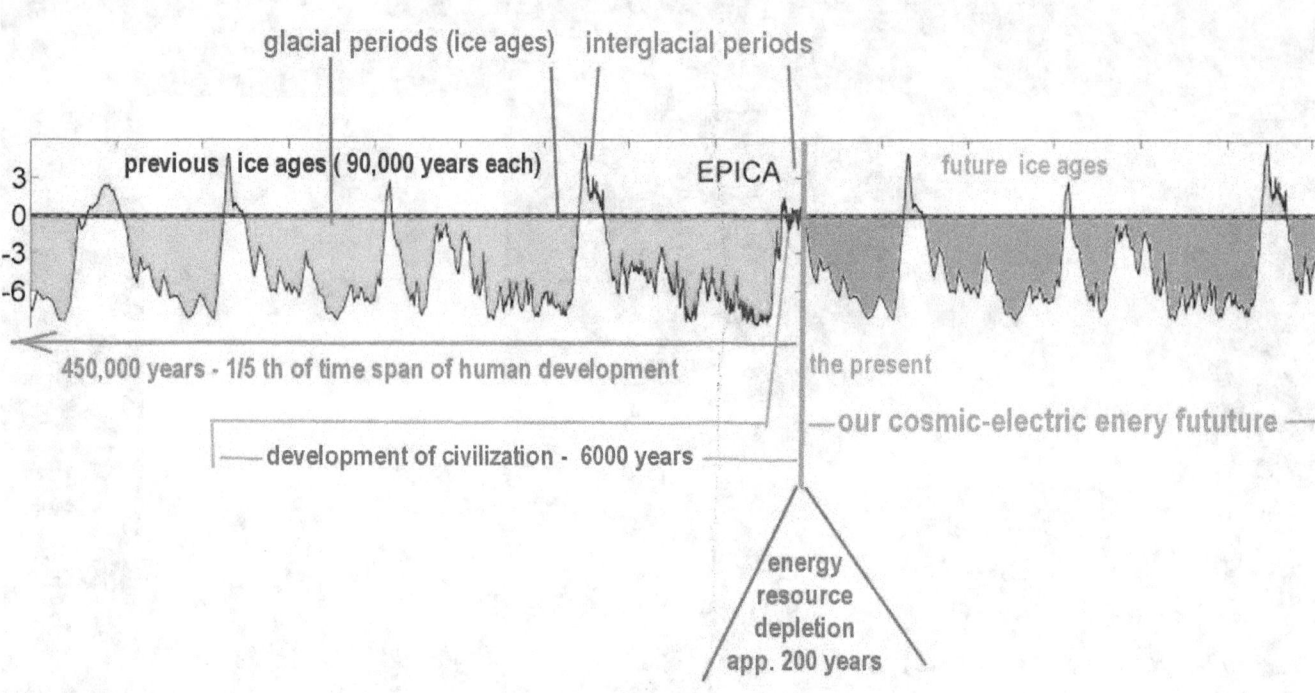

All the energy resources that are presently used to power our economies, are generated with fuels that are fast becoming depleted. Oil, gas, coal, even nuclear power fuels, are all finite resources subject to depletion. A large portion will become inaccessible when the next Ice Age begins, and those that remain accessible will be gone in a few decades, and for some types, in a few hundred years. Windmills and solar cells fall by the wayside then. All that we have remaining then to fall back on, is cosmic electric energy that may be drawn from plasma in space.

Without the utilization of cosmic energy

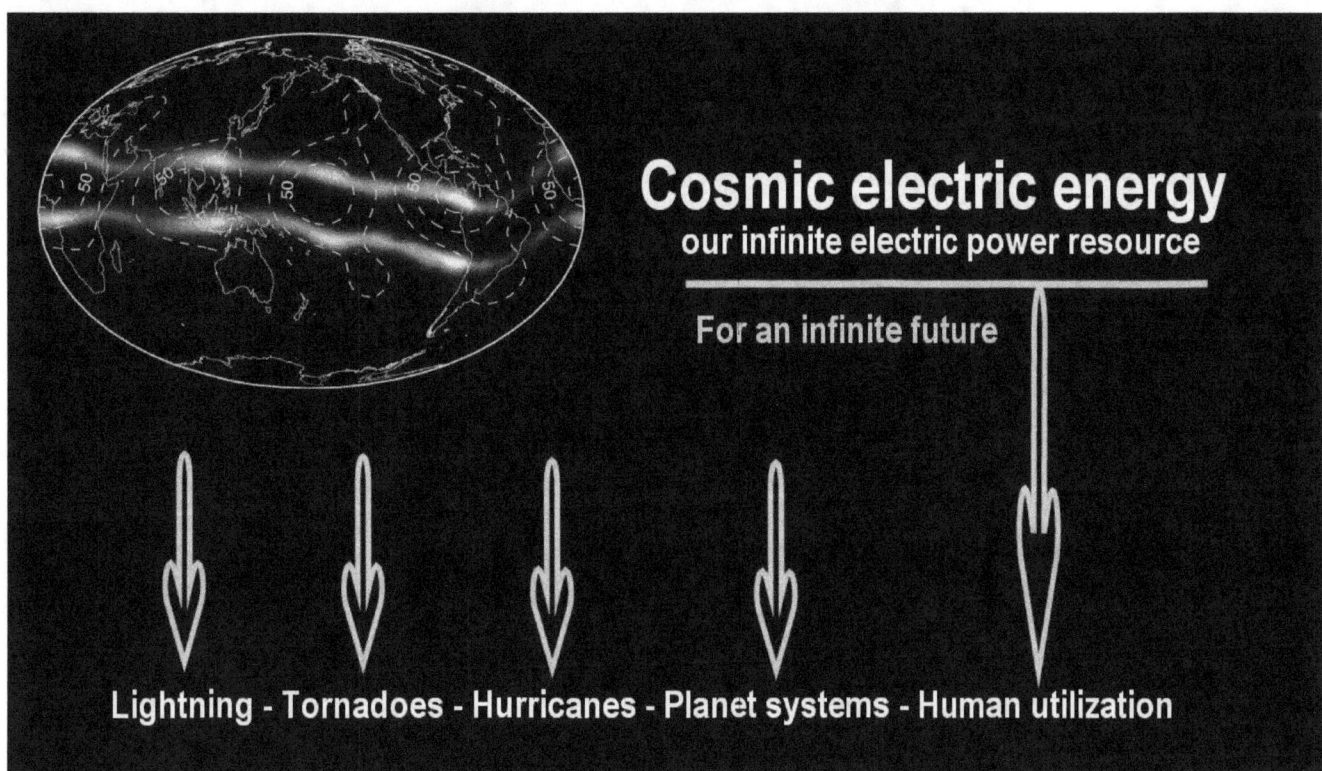

Without the utilization of cosmic energy to drive our economy, we are as good as dead. A large civilization requires large energy resources. The few recourses that we have left in the form of fuels won't last a thousand years, much less for 90,000 years till the end of the coming Ice Age, and for the millions of years thereafter that we expect to exist on the Earth. We really have no choice then, except to master the utilization of the cosmic energy system that nothing can deplete.

An interface to the cosmic power grid

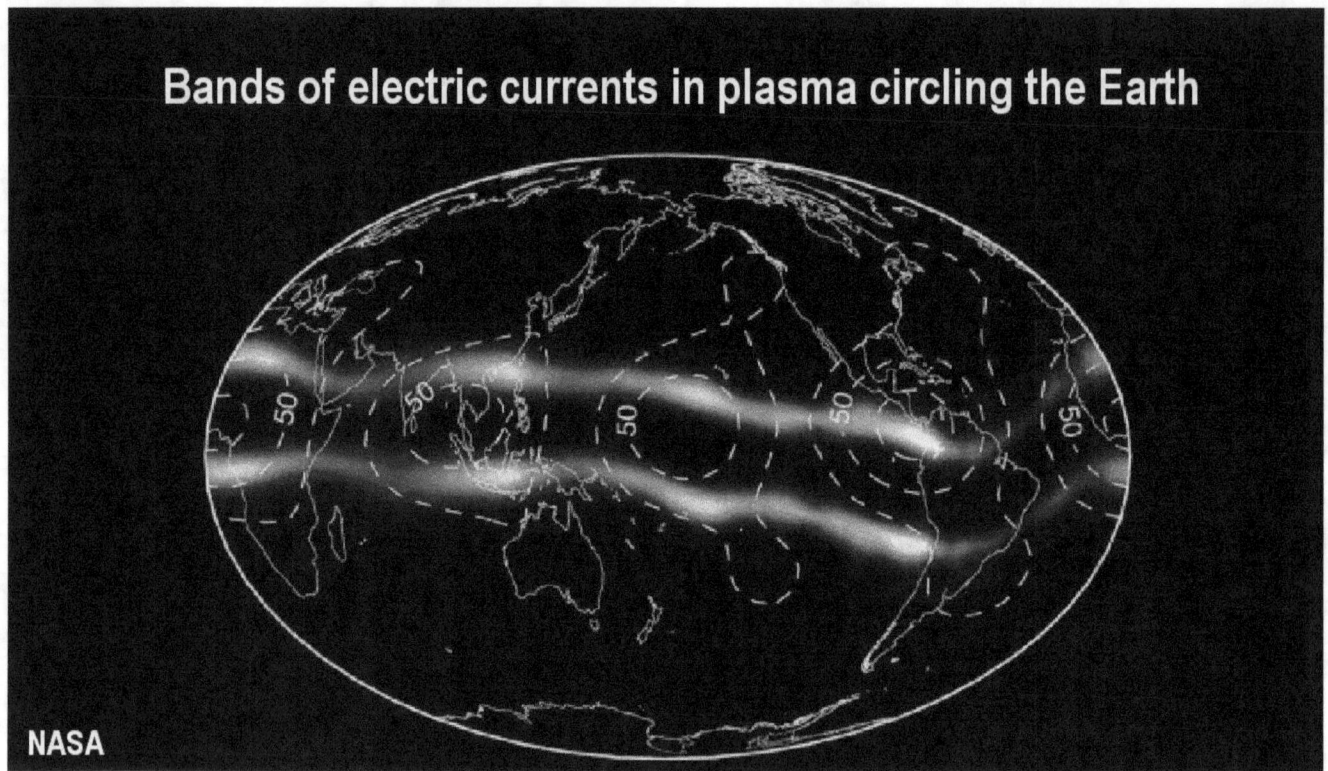

An interface to the cosmic power grid is already been made visible technologically, and exist in the form of two plasma bands encircling the Earth.

Evident on the face of the Sun in UV light

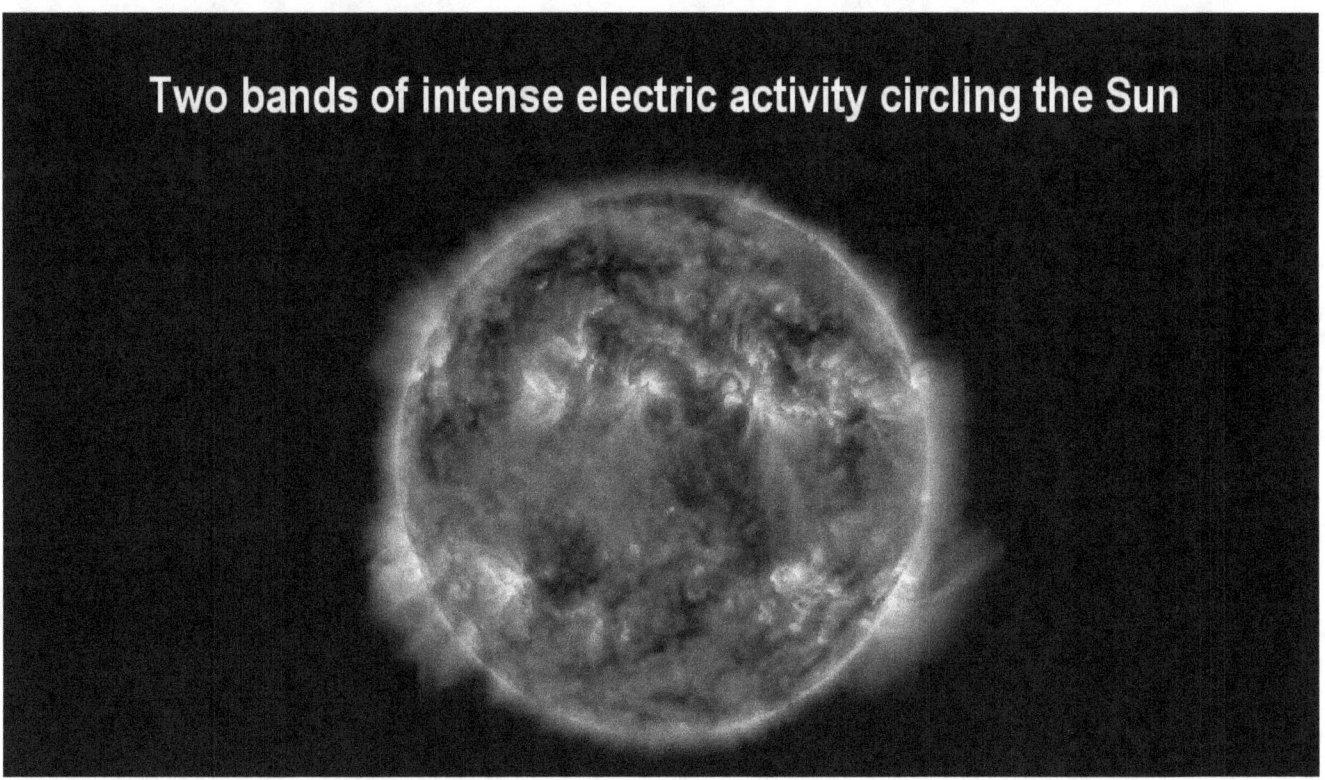

Similar bands of electric activity concentration are evident on the face of the Sun in UV light. the universe is a vast sea of energy utilization. Our galaxy of 400 billion stars is powered by cosmic energy resources, and likewise are the hundredth of billions of galaxies that exist. Energy is the 'trade name' of the universe. The universe is the product of energy. Some day we will learn that we are a part of it and open ourselves up to it. If we fail on this front we will die when the phase shift begins that will dim our Sun.

Let's stop playing those silly and dangerous games

So, let's stop playing those silly and dangerous games that we play, like wars, looting, depopulation, nuclear terror threats, and global warming scares. The universe has called us to attention. The time for the phase shift is getting nearer.

Soil temperatures at the Solar Terrestrial Institute in Irkutsk

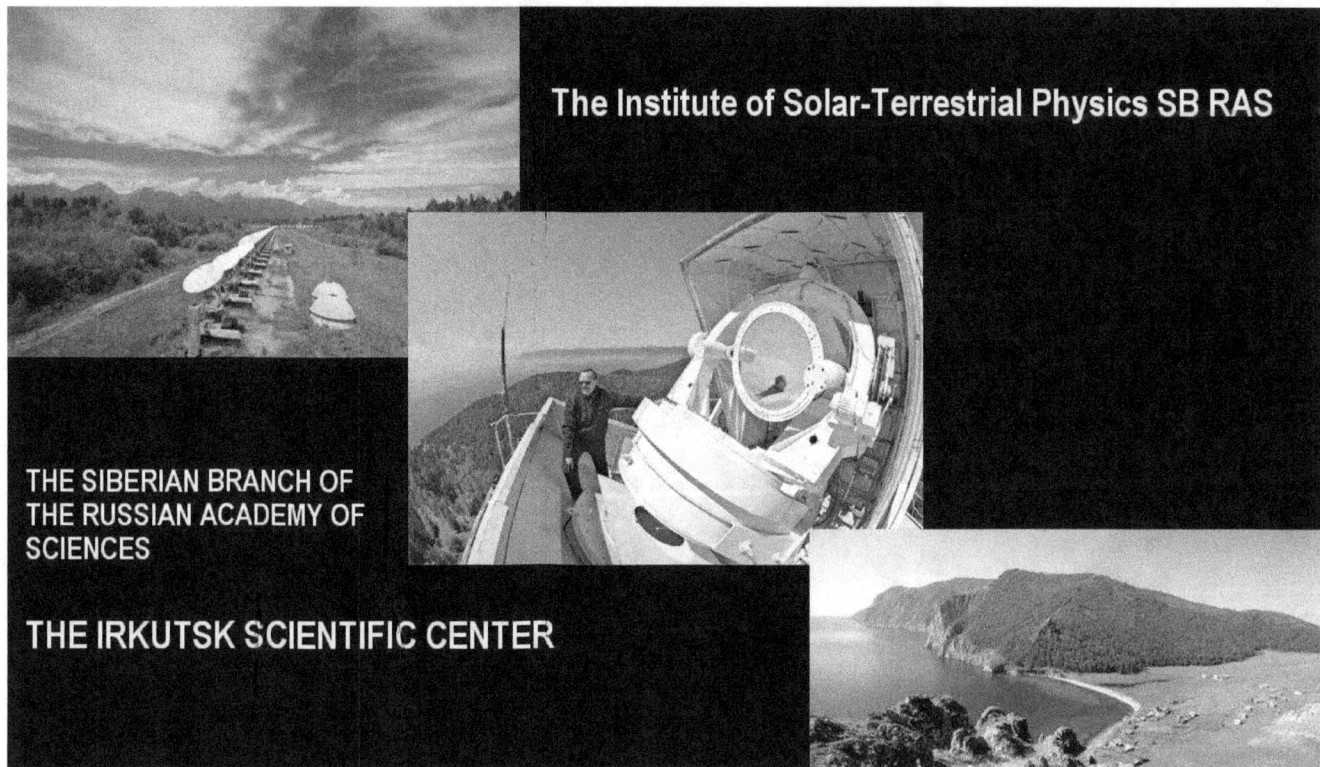

The movement towards the phase shift appears to have begun in earnest in the late 1990s. On-the-ground measurements of soil temperatures at the Solar Terrestrial Institute in Irkutsk in southern Siberia, saw a steep decline in annual average temperatures beginning in 1998 of almost two degrees over the span of 4 years. That's huge.

NASA's Ulysses spacecraft

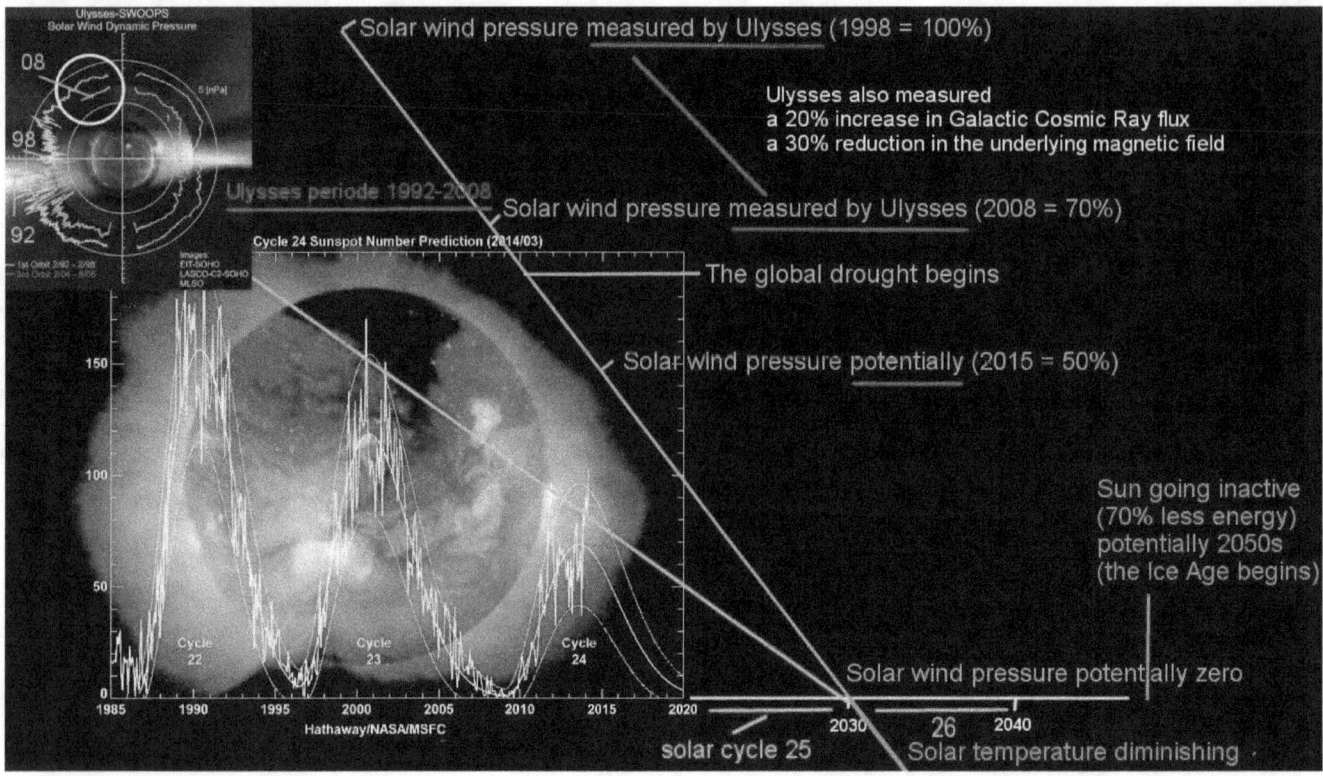

NASA's Ulysses spacecraft saw the solar-wind pressure diminishing by 30% over 10 years, likewise from 1998 on. This too, is huge for such a short time frame.

Similarly did we see the start of a steep decline of the sunspot cycles after the 1990s, which reflect the diminishing intensity of solar activity.

Something big is in the making, and it isn't pretty.

The phase shit to the next glaciation cycle

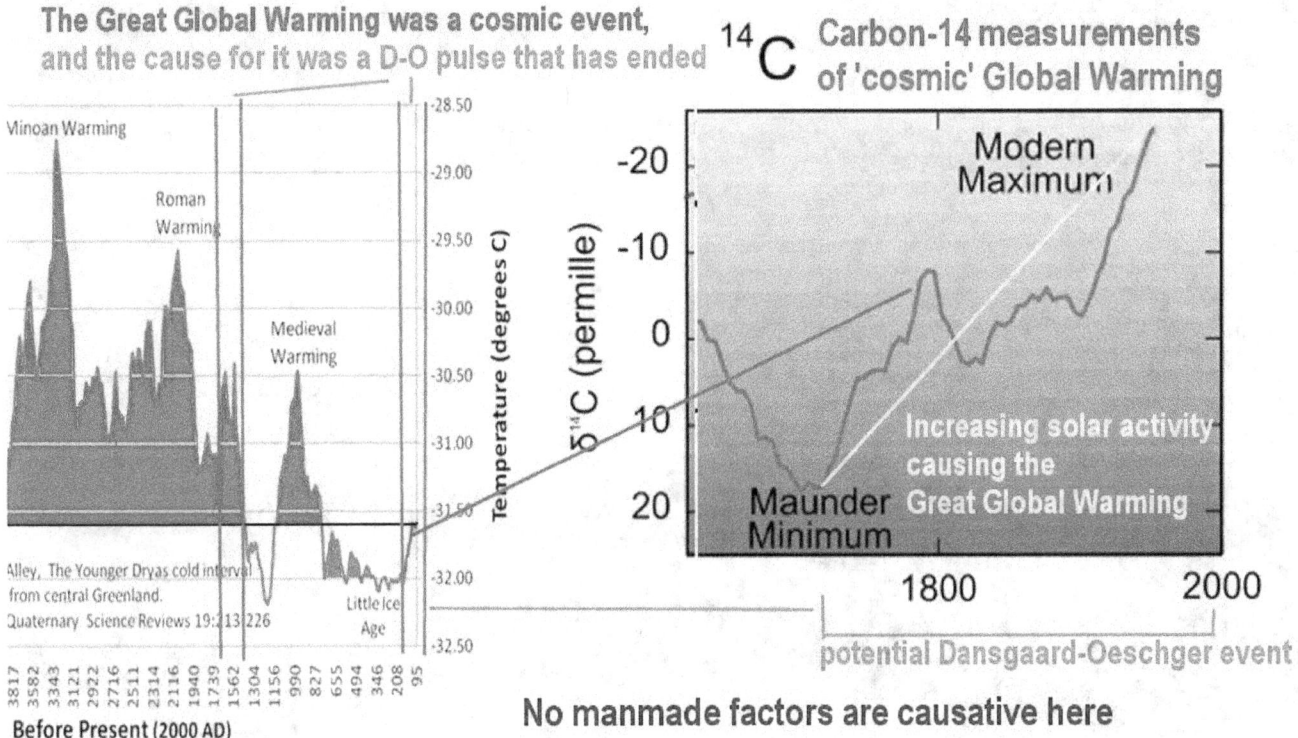

The phase shit to the next glaciation cycle even matches in time the potential end of what may have been the end of the most recent Dansgaard-Oeschger event that had pulled us out of the Little Ice Age and gave us the Great Global Warming afterwards.

And even this event is already being reversed again as the pulse that apparently caused it, is ending.

We are heading towards a new Ice Age in 30 years

All the big indicators tell us that we are heading towards a new Ice Age in 30 years, with enormous changes unfolding for humanity.

Society likes to play fantasy games

Poster of the Climate Conference. Licensed under Fair use via Wikipedia

COP 21: Heads of delegations by GUSTAVO-CAMACHO-GONZALEZ - Licensed under CC BY 2.0 via Commons by Presidencia de la República Mexicana -delegates

Unfortunately society likes to play fantasy games, such as the global warming game, instead of being concerned with what is really going on. It is betting its life, even the very existence of humanity as a whole, with just a few exception, that the massively indicated phase shift to the next Ice Age with a dimmer Sun will not happen, so that by this betting against all odd, nothing is being done to protect humanity's existence at its most critical stage in history. Of course, this is a bet that humanity will likely loose. The certainty stands against it.

The global warming hoopla

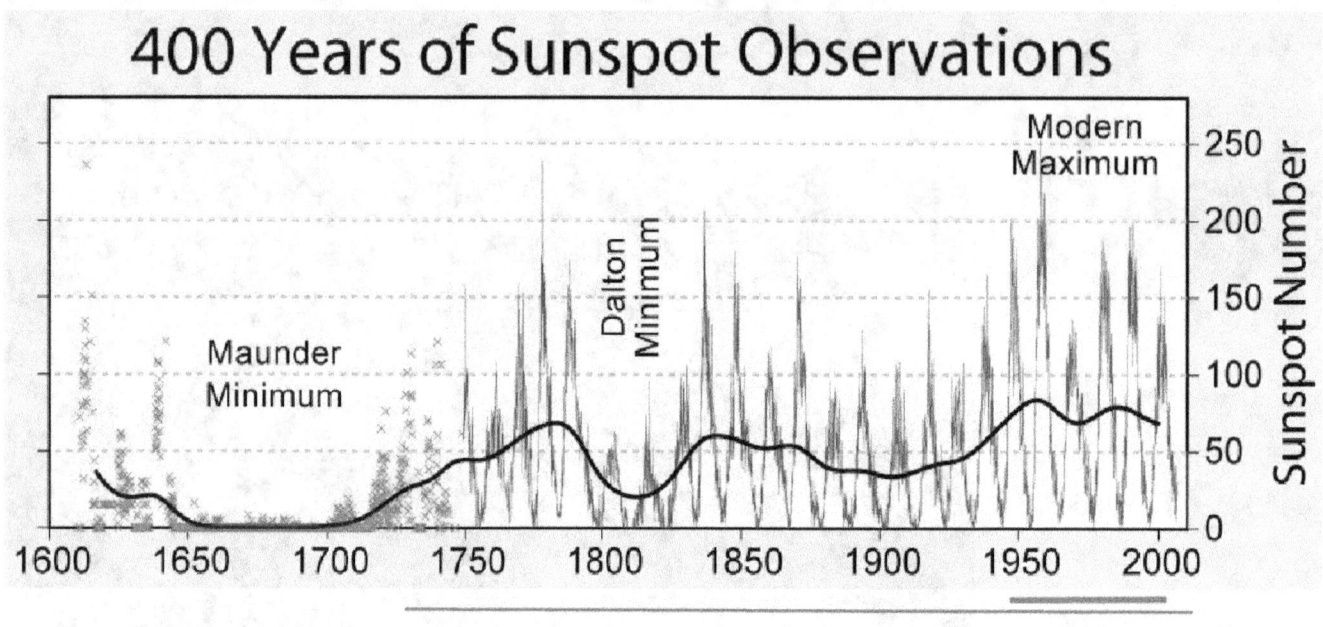

This means, of course, that all the global warming hoopla that was draged onto the world scene in 1974, was essentially a fairy tale spun around the effects of what was essentially an astrophysical Dansgaard-Oeschger event. Of course, the existence of the Dansgaard-Oeschger pulses hadn't been discovered at the time, in 1974, and would not be solidly recognized until almost 20 years later.

In earlier ice cores the Dansgaard-Oeschger events were noted, roughly around 1985, as violent oscillations and were then simply dismissed and attributed to climatic anomalies. It wasn't fully apparent until the big NGRIP project was completed that these oscillations are not local anomalies, but are basic global climate oscillations.

The link of the Dansgaard-Oeschger events with the global recovery from the Little Ice Age, however, cannot be measurably confirmed. No evidence for such a link exists, other than the evidence in timing, and the fact that the recovery from the Little Ice Age with the Great Global Warming has happened and that it was a cosmic event. This part is history. This should be enough, however, to move forward with, which unfortunately isn't happening. Instead, everything is stalled, and apparently intentionally so.

Part 5 - Dansgaard Oeschger oscillations

Give it more time to prepare

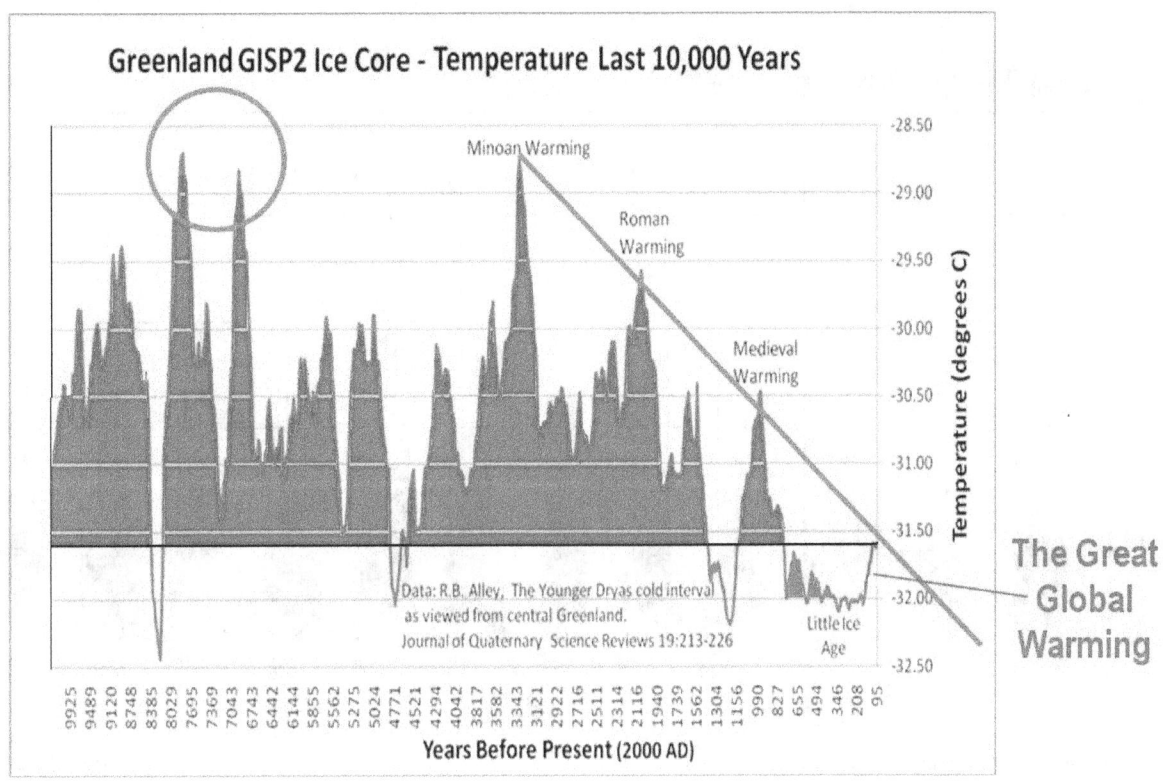

Let's hope that the currently weakening plasma environment will remain strong enough a bit longer, long enough that it will hold back the cut-off point for the Sun by a few more decades, while society learns to open its eyes. The few extra decades would give it more time to prepare the world for the dim and cold glacial environment that awaits us beyond the cut-off point of the powered Sun. Of course, this kind of pure dreaming won't get us anywhere, but in the grave.

The 25th solar cycle

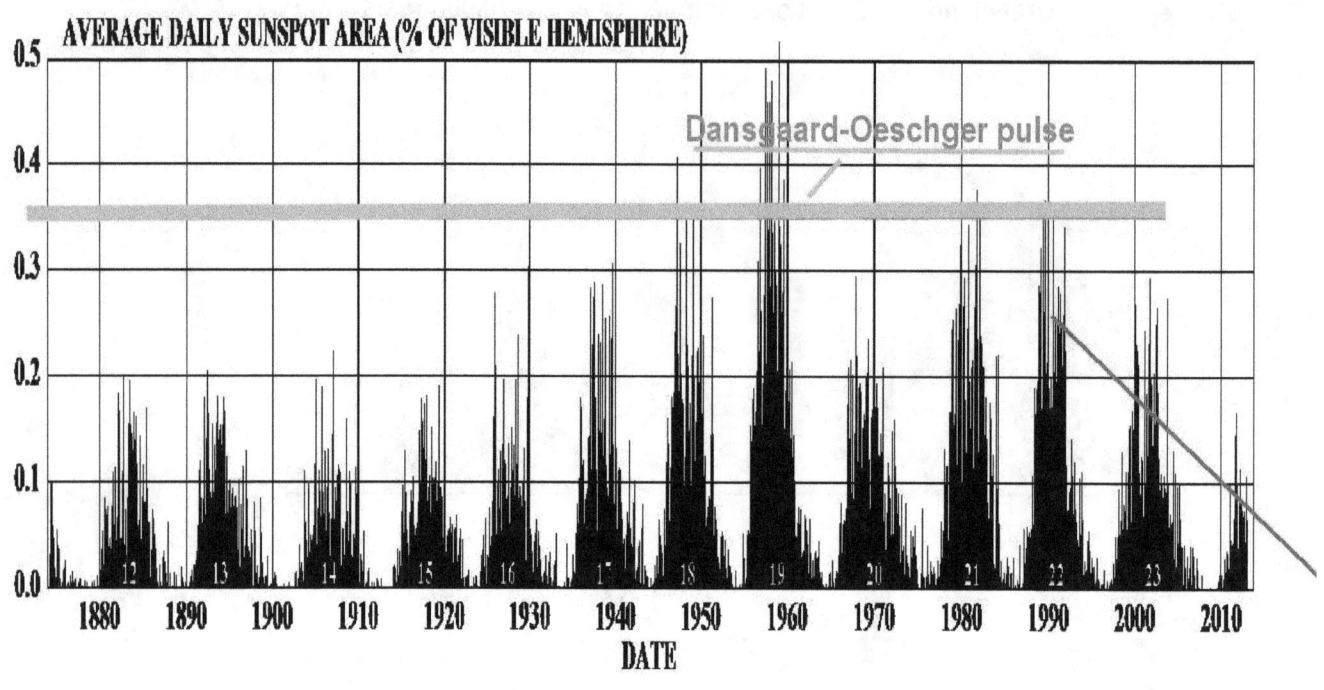

The sunspot cycles are definitely getting weaker. By the current trend, the 25th solar cycle may not have any sunspots at all. The drop-off that we see happening here, may be the ending of the current Dansgaard-Oeschger pulse that had its beginning during the Little Ice Age. It appears to have peaked from the 1950s into the 1990s, and is now dropping off fast.

The Sun is already on the track of a large weakening trend

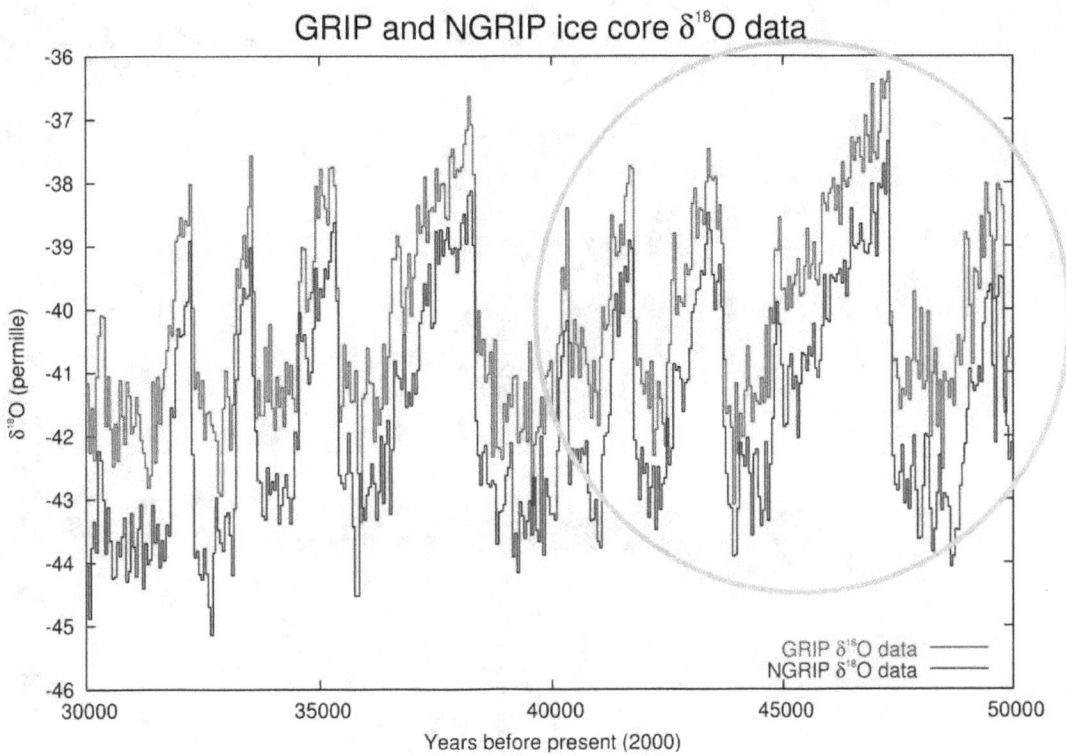

The great danger of our time is that, with the last Dansgaard-Oeschger pulse having now run its course, there is nothing much left of the interglacial plasma density to sustain the Sun, so that a free-fall collapse with the big phase shift at the end, may occur sooner instead of later and overwhelm us before we are ready for it.

The large Dansgaard-Oeschger warming events during the last glaciation period, were historically of short duration.

How close we are to the current Dansgaard-Oeschger pulse ending, which appears to be reflected in the current climate, cannot be determined. The remaining time will likely be in the range of decades, seeing that the Sun is already on the track of a large weakening trend.

What our response should be is rather obvious

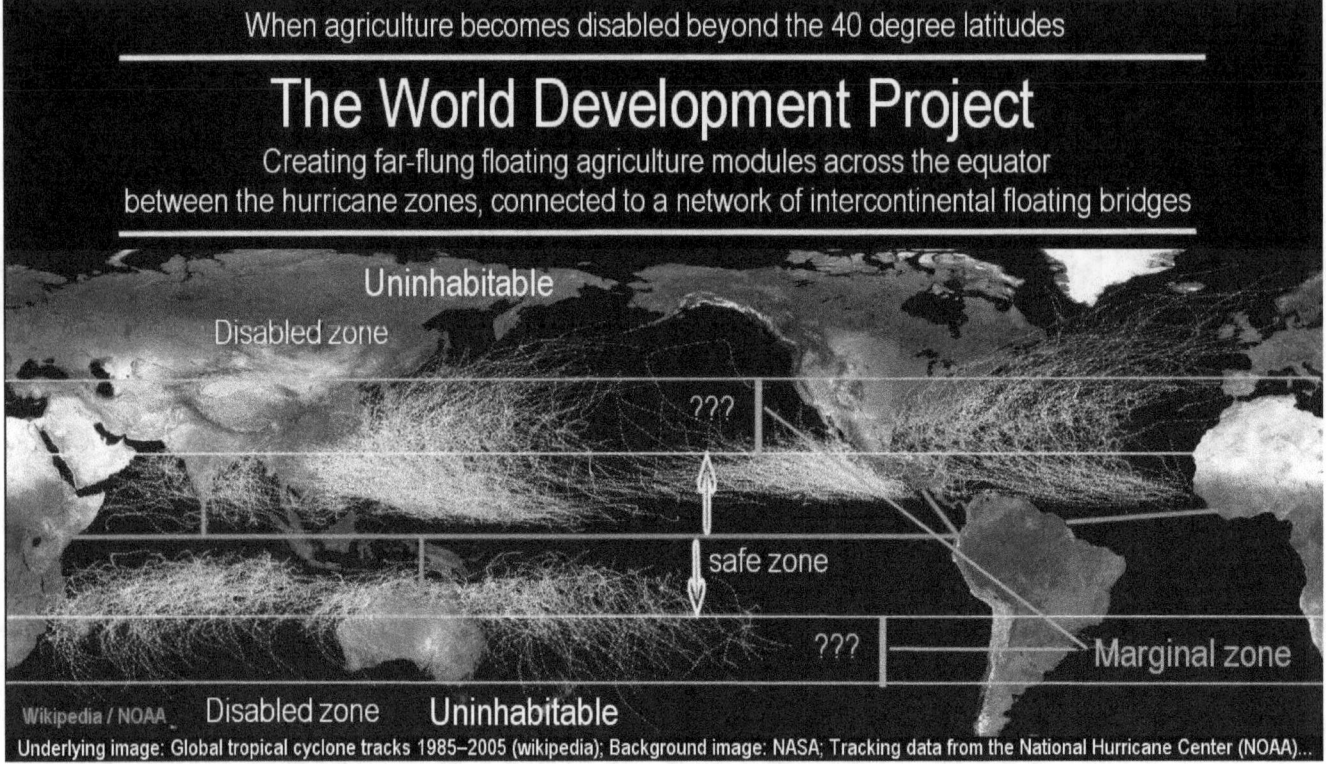

What our response should be is rather obvious. While we are committed to taking the politically correct path and do nothing, singing the global warming song that lulls us to sleep and prepares us to die in the coming Ice Age starvation, or the in the much nearer thermonuclear war that is already fully prepared for, we still have time remaining and the power on hand to secure our living with the greatest economic, cultural, and scientific development of all times that demonstrates what a human being is and is able to create with love for one-another. We have the power to choose this for our future.

Part 6 - The UFO phenomenon

The UFO phenomenon

Rapid on-off conditions are natural occurrences

The Dansgaard_Oeschger oscillations are only one type of example in which rapid on-off conditions are natural occurrences for electric plasma phenomena. The so-called UFO phenomenon is another example of the same type.

Many cases of UFO 'sightings' have been reported. Many have been photographed. But note, what has been photographed conforms perfectly to the shape of the electromagnetic bowl-type structure that forms the Primer Fields, which in turn form a sphere of concentrated plasma at their focal point. This happens to be the typical shape that has been captured in UFO photographs.

We see the bowl-type electromagnetic structure, and the plasma sphere at the focal point of it. Both are clearly evident in this photograph. Reports say that the UFOs suddenly appear, that they can stand still, and that they can move quickly. They are seen existing in space, and are even hovering above the moon. They can accelerate immensely fast in their movement, and make extremely sharp sudden turns. They can do all this, because the so-called UFOs are not space ships, but are electromagnetic phenomena that have no mass that would impede such movements. They are merely shapes created by electromagnetic interaction of plasma flowing in the atmosphere, and in space, which can become visible when the conditions are right. Some are seen as moving points of light, even pulsating light, some even appearing in groups.

By being electric plasma concentrations, the UFO phenomenon also reflects radio waves, whereby it can be seen on radar. However, when aircraft are launched to intercept, the electric field around the aircraft affects the field that forms the UFO, which thereby changes its position, or causes it to simply vanish. All this makes interception quite impossible.

No UFO craft does actually exist

NASA - Constellation: trans-lunar injection

Of course interception is also impossible for the simple reason that no UFO craft does actually exist. Extraterrestrial visitors travelling in a physical spacecraft to our planet would have to fly against all the known laws of the physical universe and time. It would take a spacecraft 12,000 years to reach the earth from our closest star, flying at a speed of 360,000 km per hour, which is roughly 30 times the speed of the Apollo 11 flight to the moon, which might be the limit for spacecraft velocities, moving against the plasma background in space. Exotic theories have been invented in attempts to side-step the physical limits and barriers, in attempts to support the UFO theory, while the obvious reality is simply being ignored.

The UFO as an example

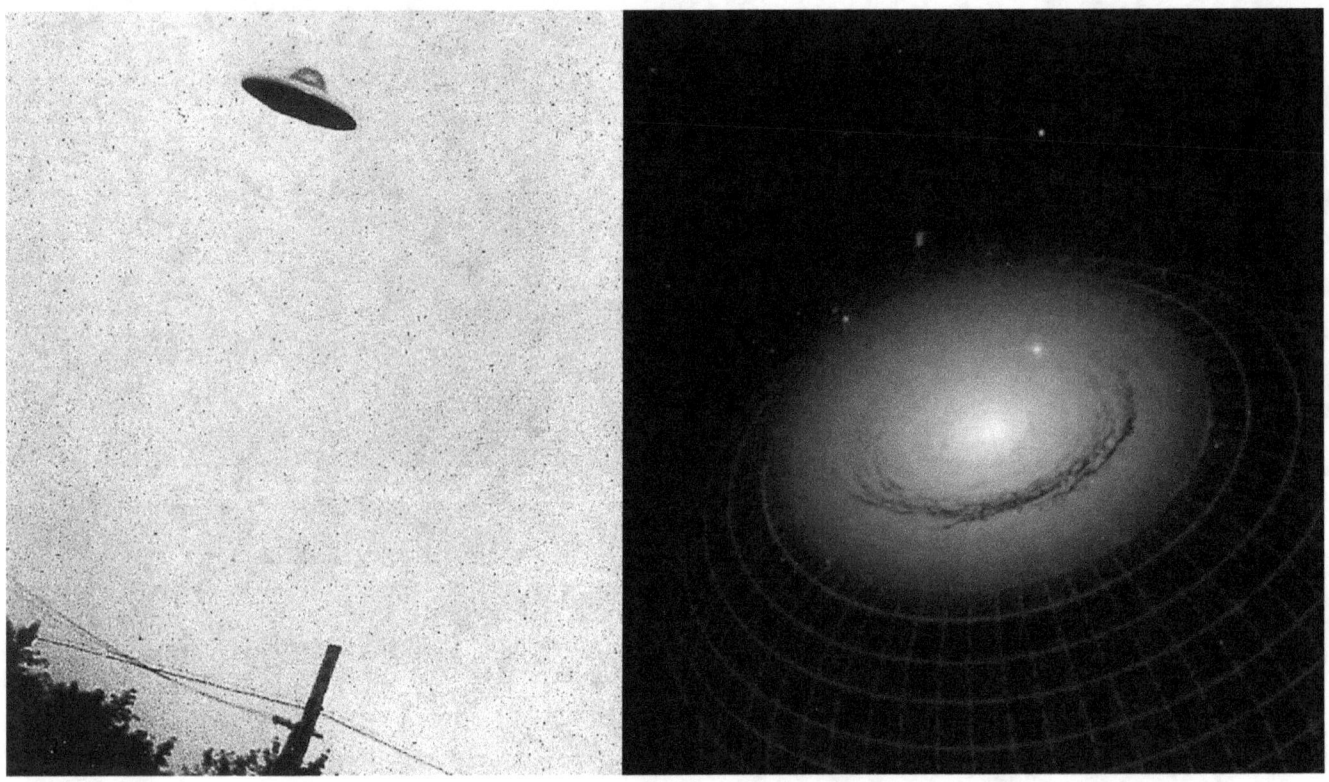

In real terms the UFO phenomenon stands as an example of electromagnetic structures that form in space with gigantic effects and in the atmosphere with minute effects, all formed by Primer Fields that are dynamically created in electric plasma streams.

UFOs comparable to the sprites

The UFOs, thus, become comparable to the sprites that appear in the upper atmosphere. The resulting effects, of course, depend on the coincidence of conditions that enable them, which are inherently rare and fragile in nature, and are specific for the type of atmosphere in which they occur. When the conditions are satisfied in the lower atmosphere, they often enable a number of small plasma events simultaneously, which then become regarded as multiple UFO sightings.

Fragility of the sprites in the stratosphere, and UFO events on the lower atmosphere

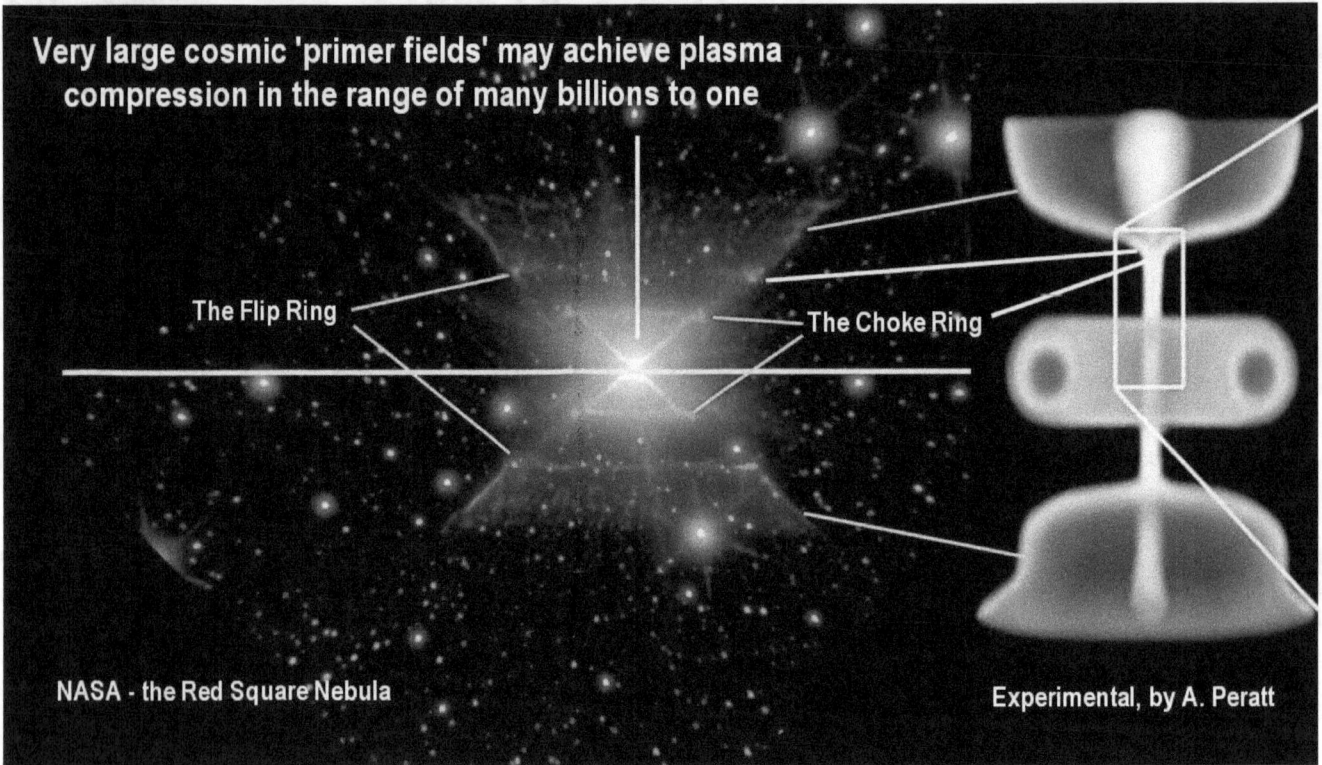

This fragility of the sprites in the stratosphere, and UFO events on the lower atmosphere evidently also applies to the complex structures and magnetic fields that create the conditions for our Sun to be electrically powered. When the conditions are right, the fields form, and when the conditions no longer exist, the formed fields vanish.

UFO phenomenon fragile

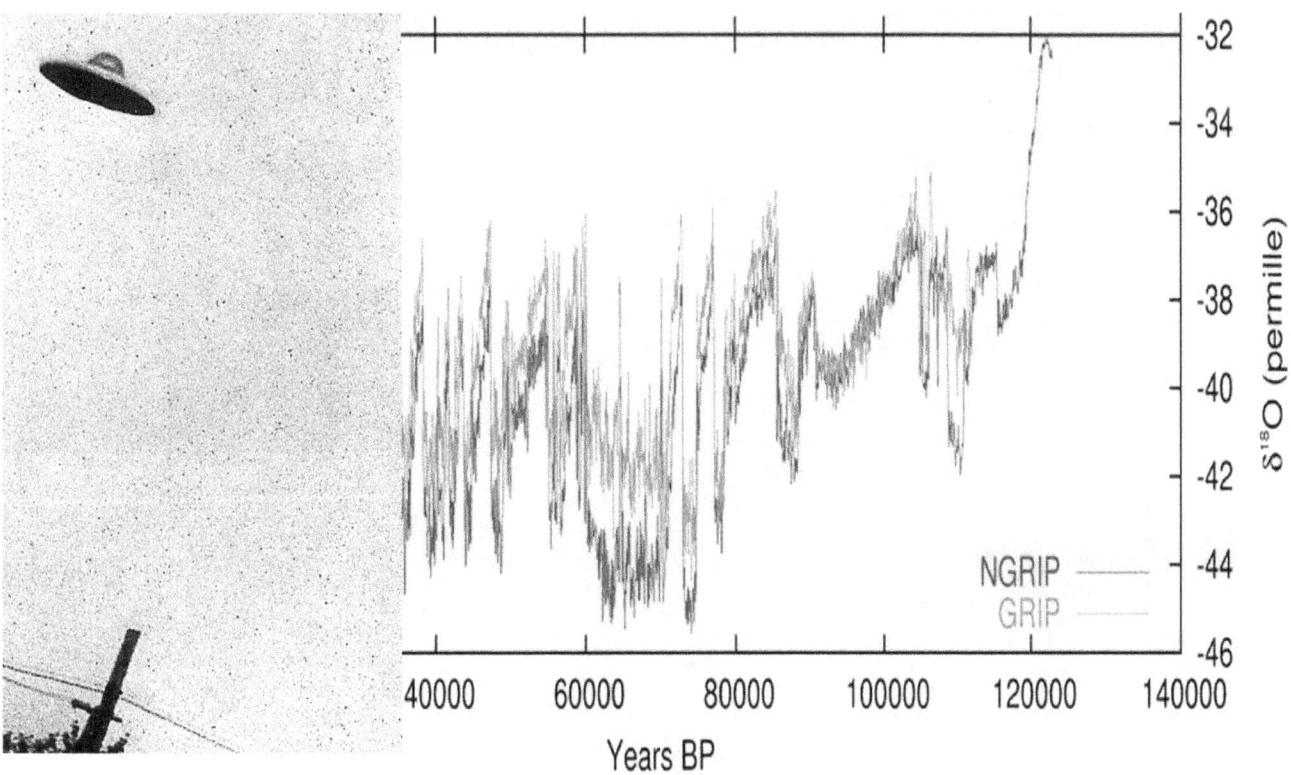

Both the UFO phenomenon and the Dansgaard_Oeschger oscillations illustrate to some degree how fragile the conditions inherently are that affect the powered state of a star, especially that of a small star as our sun.

Powered state of the Sun rare

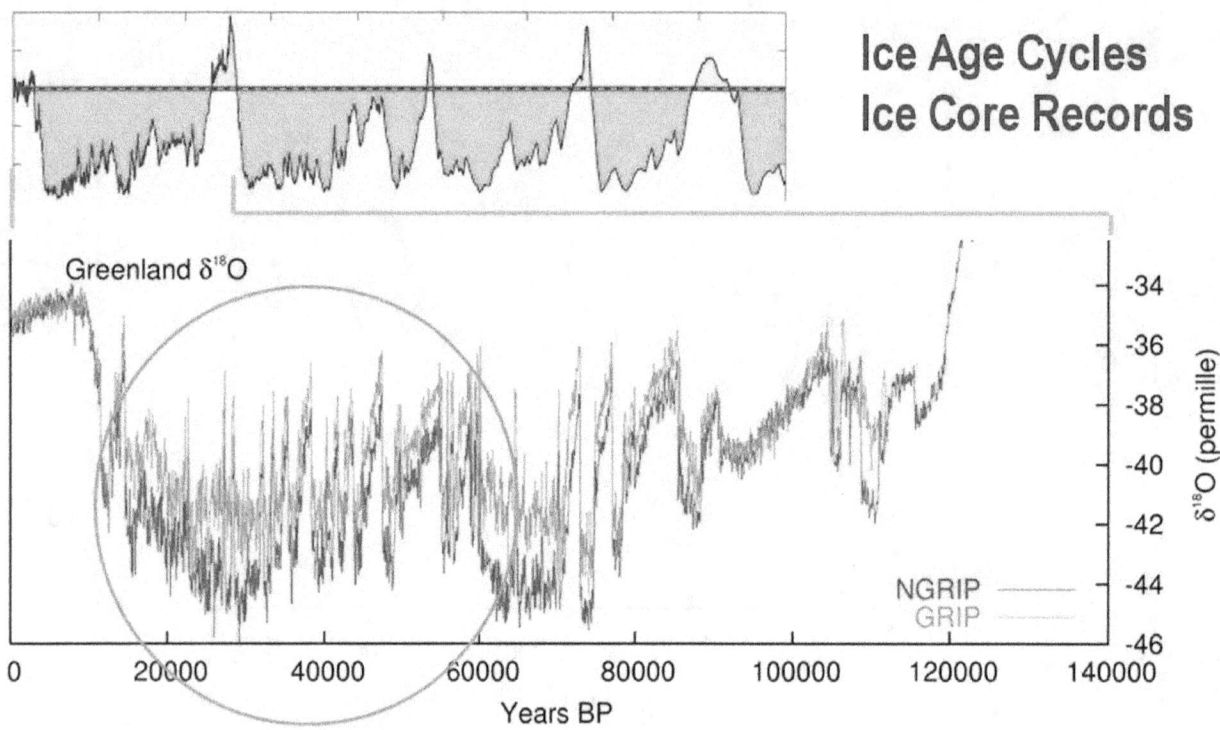

The powered state of the Sun has actually been a rather rare occurrence in the last two million years of the great ice ages. The few long periods that have the Sun constantly active, called the interglacial periods, merely interrupt the 'normal' dark and cold world of the long ice ages that only few people have come through alive.

UFO sightings from the 1950s on

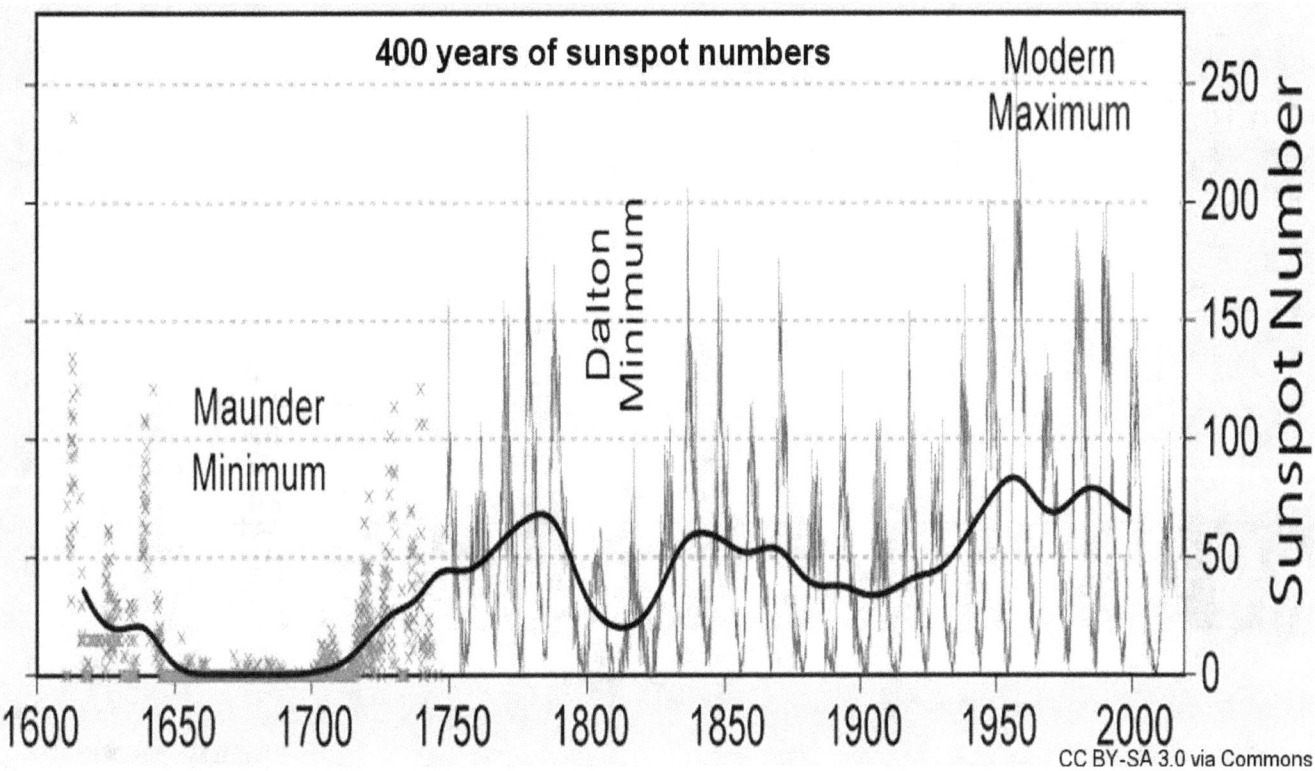

The same fragile nature that applies to our solar system, also applies to the UFO phenomenon. It is not coincidental, therefore, that the UFO sightings became most frequent from the 1950s on, through to the 1990s, when the current Dansgaard_Oeschger pulse peaked, when the solar activity was the strongest in recent time, in which the electric conditions on the Earth were likewise the strongest.

The solar system is on a path to the vanishing point

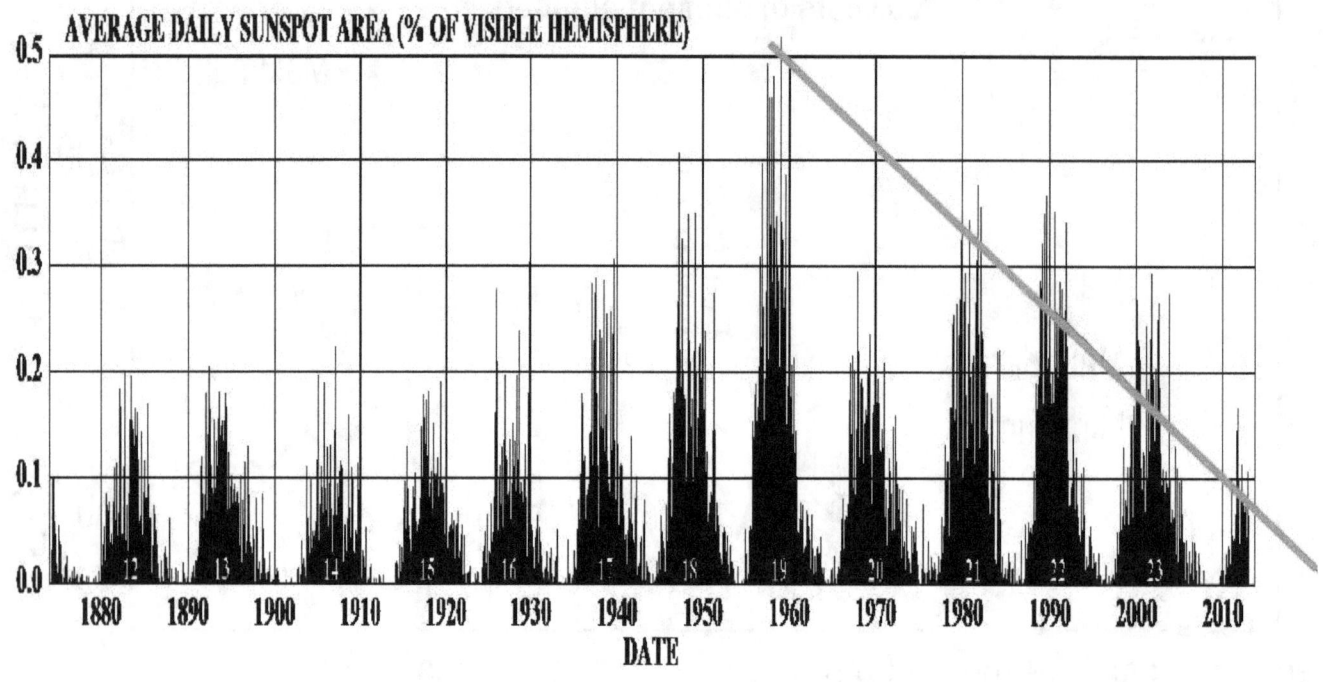

With the electric environment around our Sun now getting rapidly weaker again, according to the down-ramping that the Ulysses satellite has reported, we appear to have entered a transition zone of extremely fragile conditions. The solar system, as an UFO, is on a path to the vanishing point.

The critical warning

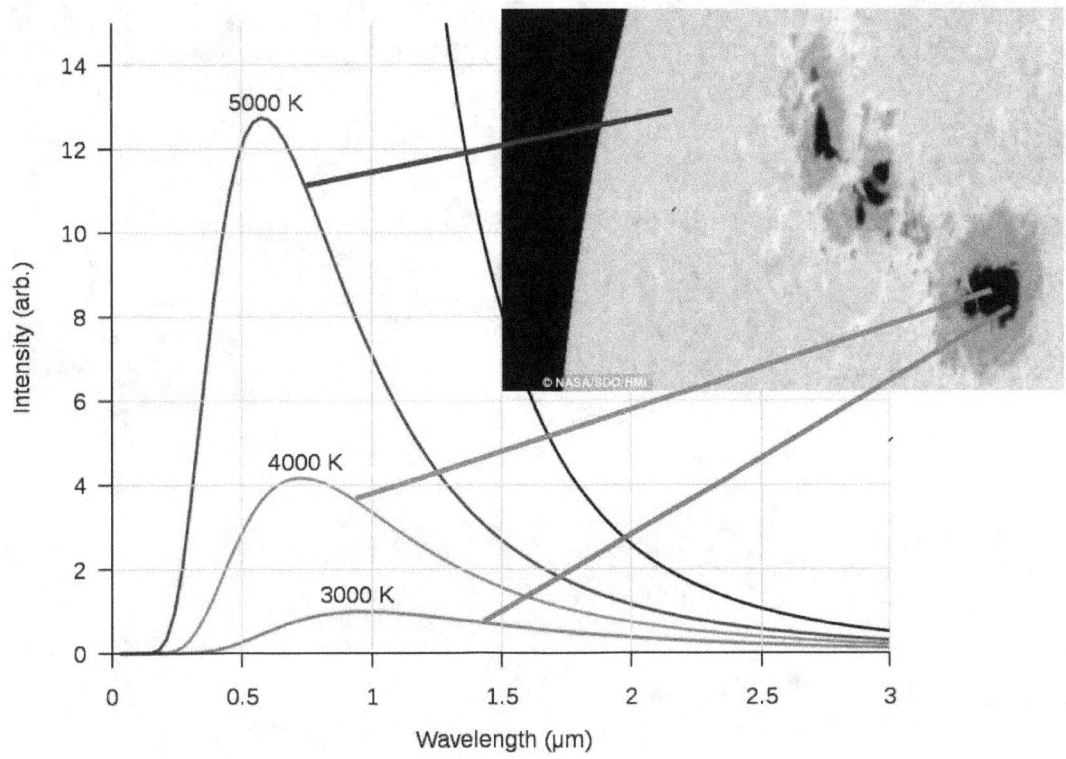

The critical warning that science affords us on this basis, unfortunately, is not precise to the day, nor in exact details. Still, the down-ramping that one sees in many areas happening simultaneously presents strong points of scientific correlation of the numerous elements that the astrophysical arena can provide, that points towards a big phase shift in the making.

The exact day for the Sun reverting to its unpowered state, cannot be determined. But do we need to know the day of the event when the principle is known and the dynamics of its manifestation? The evidence that we have before us, presents a strong case for us all to make extreme efforts towards building the vast new infrastructures that our continued existence on this planet WILL most certainly depend on.

The entire world is affected

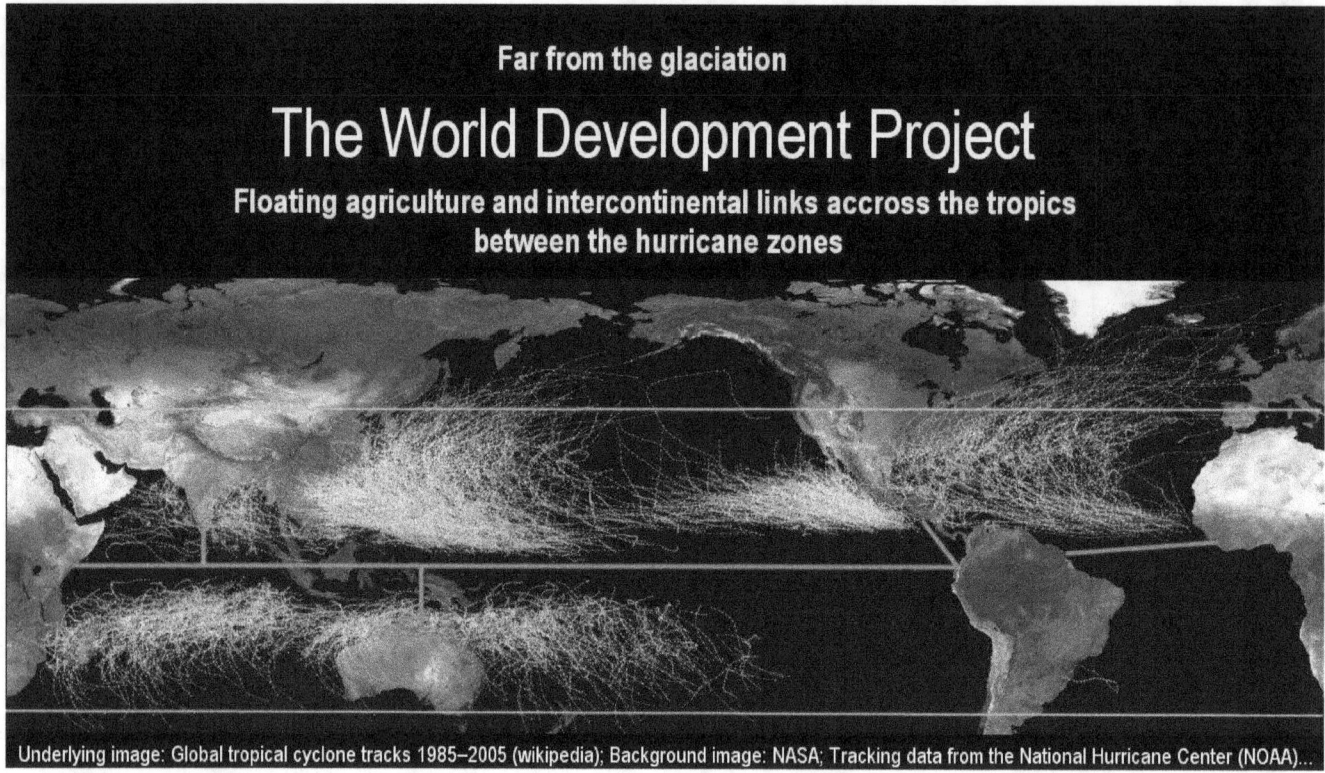

It is tempting to assume that only the far northern countries and regions, like Canada, Europe, Russia, and China, are affected by the coming ice age transition in which the Sun becomes inactive. This is a foolish assumption. When the Sun turns dim and the food supply is collapsing, the entire world is affected. Thus it becomes the task of humanity as a whole to build the infrastructures to protect human existence on this planet. Not to act decisively on this front amounts to committing universal suicide. The changing astrophysical environment that affects our world will affect everything. It affects even the magnitude of earthquakes. It is highly likely on this front too, that we haven't seen anything yet in terms of consequences, which of course we can avoid by placing our civilization afloat onto the sea.

Part 7 - Orbit dynamics of the planets

Orbit dynamics of the planets

The Sun will

Evidence exists that the Sun will likely loose a portion of its mass without the action of the Primer Fields, when its inactive state begins. The external plasma pressure is thereby removed.

At a mere 71% of the escape velocity

Planetary data

Location	1. orbital period	2 mass re. Earth	3 distance (AU)	4 orbit velocity (km/sec)	5 escape velocity (km/sec)
on Sun (equator)		332,000	0		617.5
Mercury,	88 days	0.382	0.387	47.9	67.7
Venus,	225 days	0.949	0.723	35.0	49.5
Earth,	365 days	1.000	1.000	29.8	42.1
Mars,	1.88 yrs	0.533	1.520	24.1	34.1
Jupiter,	11.9 yrs	11.200	5.200	13.1	18.5
Saturn,	29.5 yrs	9.450	9.540	9.46	13.6
Uranus,	84.0 yrs	4.100	19,200	6.81	9.6
Neptune,	165.0 yrs	3.880	30.100	5.43	7.7

At the present time the planets orbit at a mere 71% of the velocity that a stable orbit requires. However, for the long glaciation periods with a lighter Sun, the present velocity would be just about right.

This means that under present conditions the Sun's gravitational attraction is 29% stronger than the centrifugal force of the planets in orbit. The resulting large difference is evidently compensated for by the electromagnetic effects of the Primer Fields that maintain the orbits electrically against the stronger gravity. Evidence suggests that the orbits of the planets are electromagnetically ordered.

Non-magnetic steel balls magnetically self-spacing

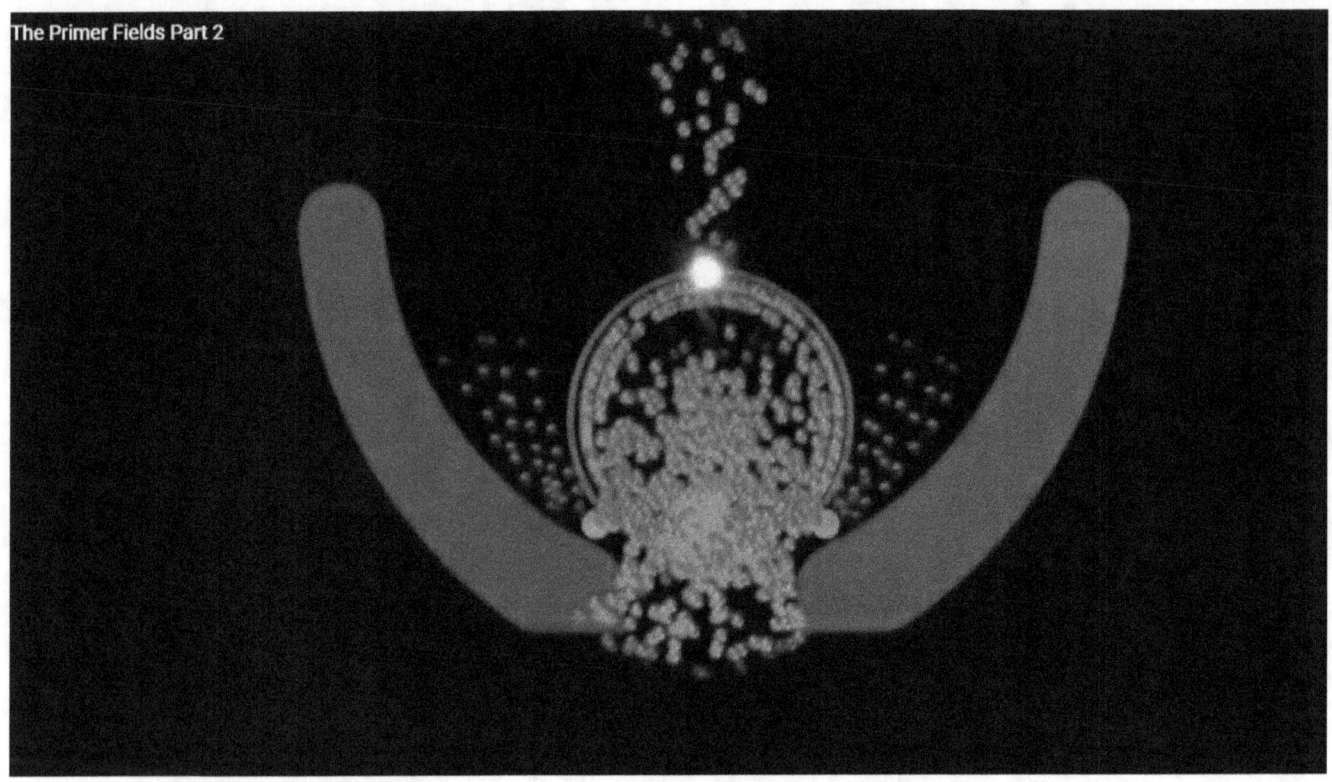

In a lab experiment, non-magnetic steel balls were laid on a sheet of glass. They were drawn into a specific orbit around the center of two bowl-shaped magnets. The steel balls seen here are magnetically self-spacing. If they are disturbed and let go, they return to their magnetically determined positions.

The experiment illustrates to some degree how the orbits of the planets are likely magnetically assisted, so that the Sun's gravitational variation is not a big factor.

However, during the long solar off-times in the glacial period, when the Primer Fields no longer exist, or exist only during the short periods of the Dansgaard_Oeschger warming events, the Sun's gravity would be the sole factor for keeping the orbits intact, especially in the time when the Sun's gravity is changing. While this won't likely affect the orbits of the planets significantly, it appears to have a major effect on orbiting asteroids. that are affected by cosmic drag, because of their larger surface to mass ratio.

Large dust accumulations in the ice of Antarctica

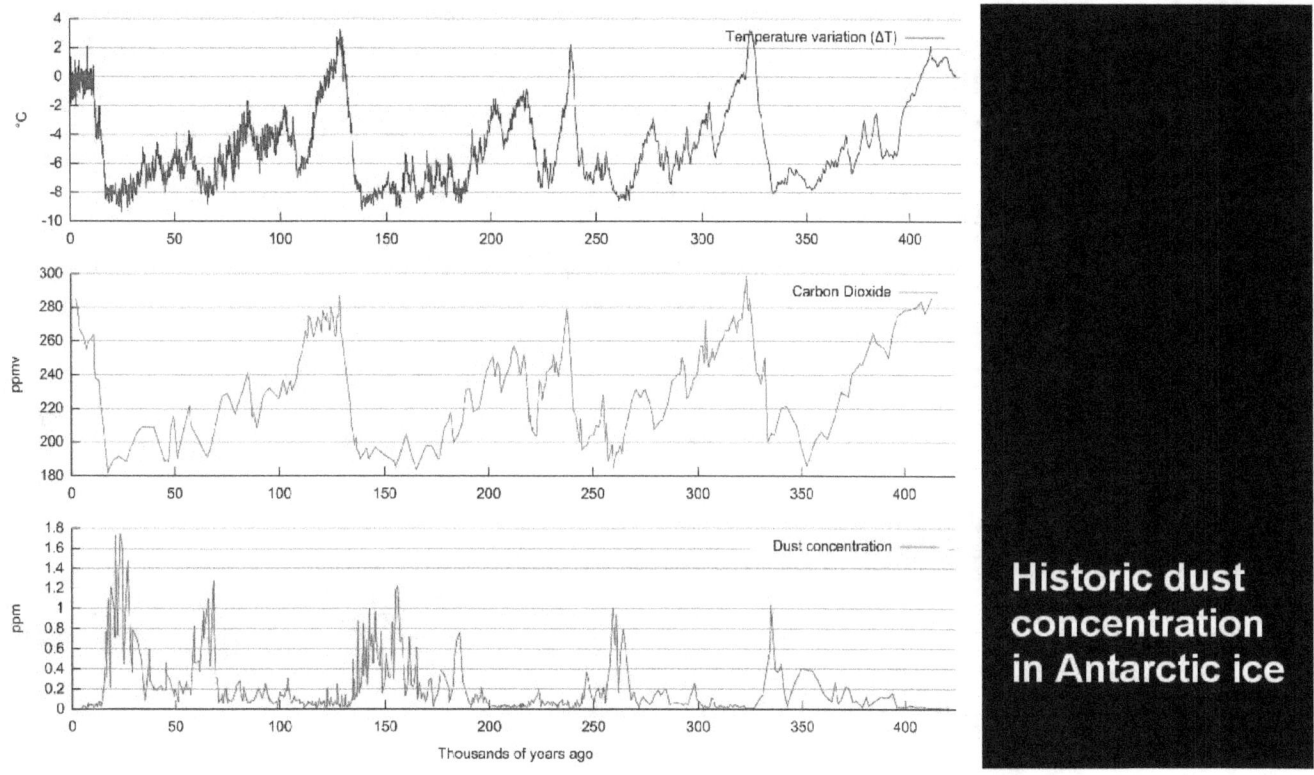

The ice core data tells us that the loss of the primer fields has a far greater effect than we care to imagine. Near the end of each glacial period large dust accumulations are found in the ice of Antarctica, which appear to have resulted from increasing impacts of asteroids, or asteroids disintegrating in the atmosphere.

The dust always stops

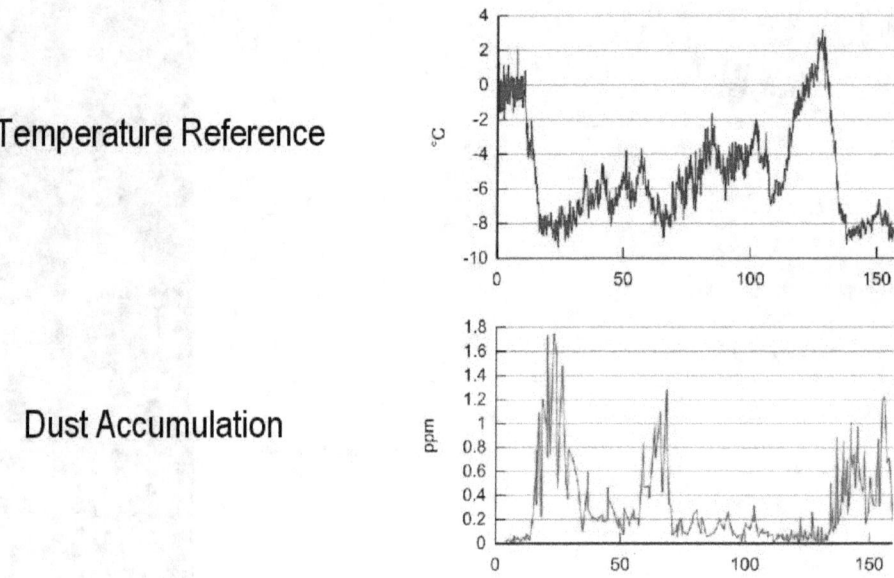

The dust always stops when the interglacial period begins in which the primer fields are active again.

A wider field of phenomena stands as evidence for the Primer Fields

The operation of the primer fields is evidently also responsible for the planets orbiting in a tightly maintained ecliptic plain. No purely mechanistic cause would organize the planets into an ecliptic plain. No mechanistic principle would prevent the planets from orbiting the Sun in a random pattern, or force them to orbit at all. This means that the Primer Fields have a much larger effect and purpose than merely focusing plasma onto the Sun. A wider field of phenomena stands as evidence for the Primer Fields.

The near-geometrically expanding spacing

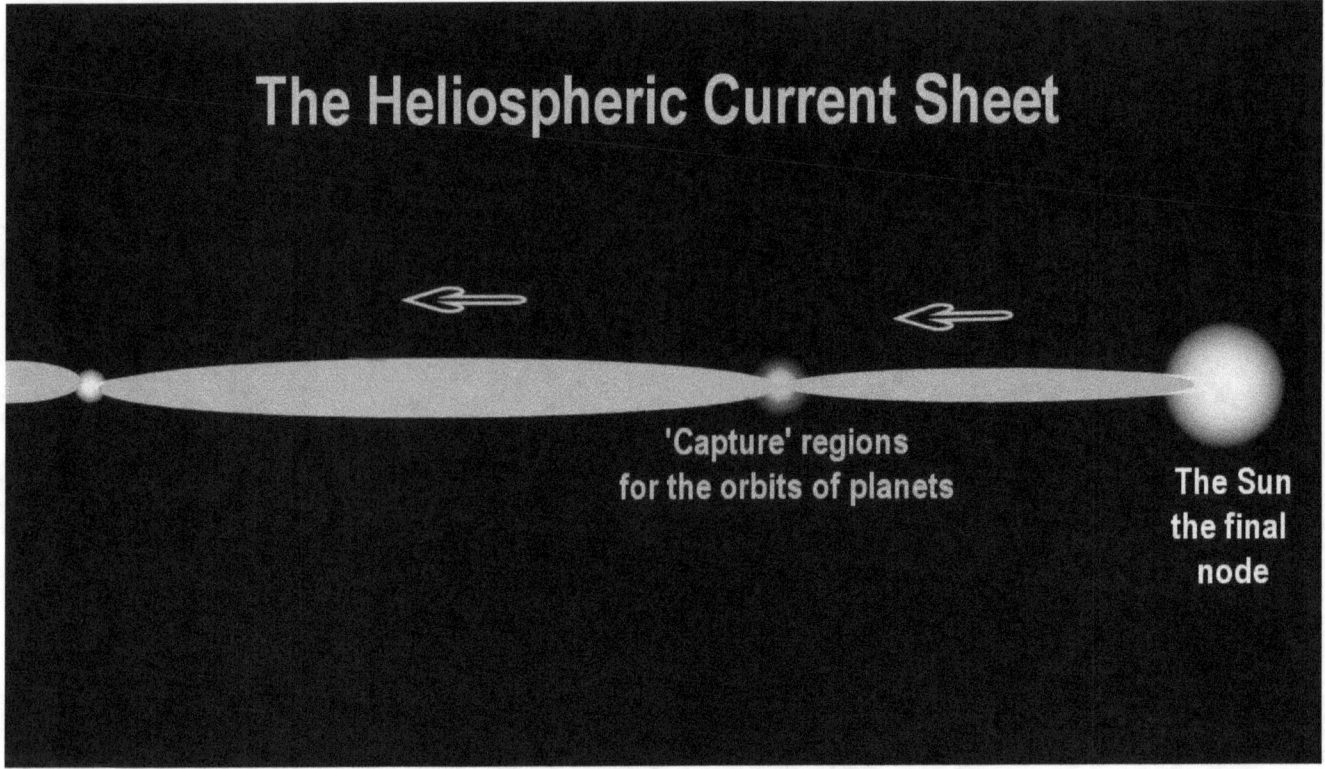

The near-geometrically expanding spacing of the orbits of the planets reflects the progressive spacing that one would expect to see between the node point in the heliospheric current sheet that becomes diffused by its expanding in geometrically widening space.

The geometrically expanding spacing of the orbits

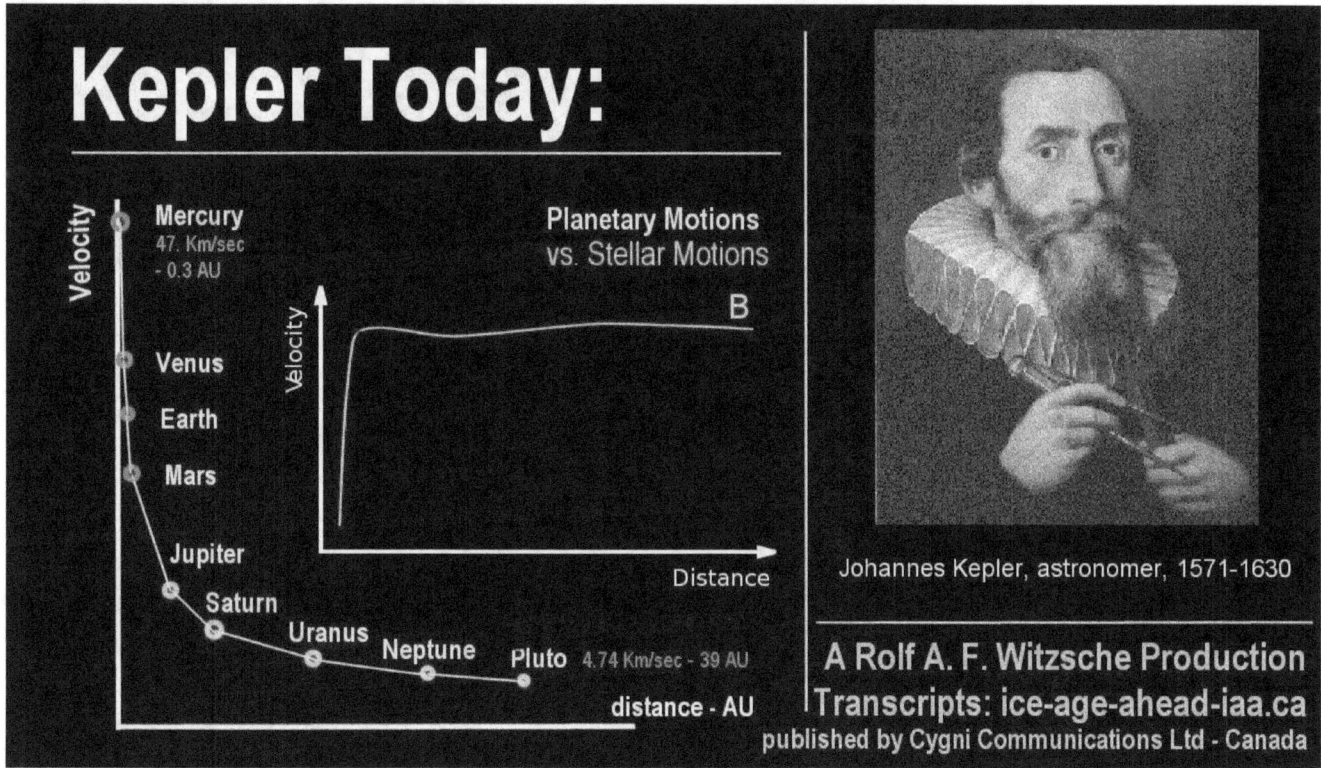

The geometrically expanding spacing of the orbits of the planets, too, adds another item of evidence for the operation, and for the far-reaching effect, of the Primer Fields that power our Sun.

The spin-rotation of the Earth itself is evidently electrically powered

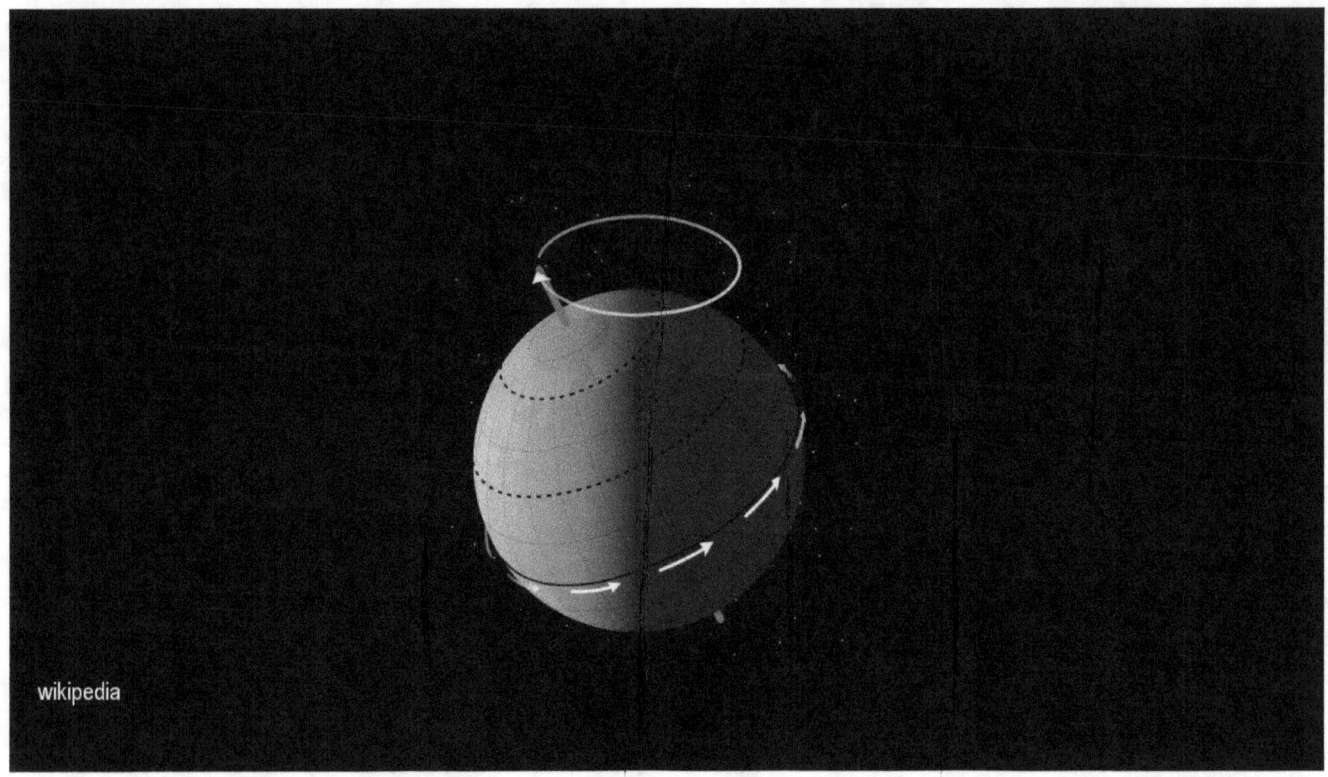

Actually, we don't have to look quite as far for evidence. We have mechanical evidence for the operation of the primer fields right here on the Earth.

The spin-rotation of the Earth itself is evidently electrically powered by the effects of the primer fields. If it wasn't for that, cosmic drag would have stopped the spinning long ago.

Faster than the rotation of the Earth itself

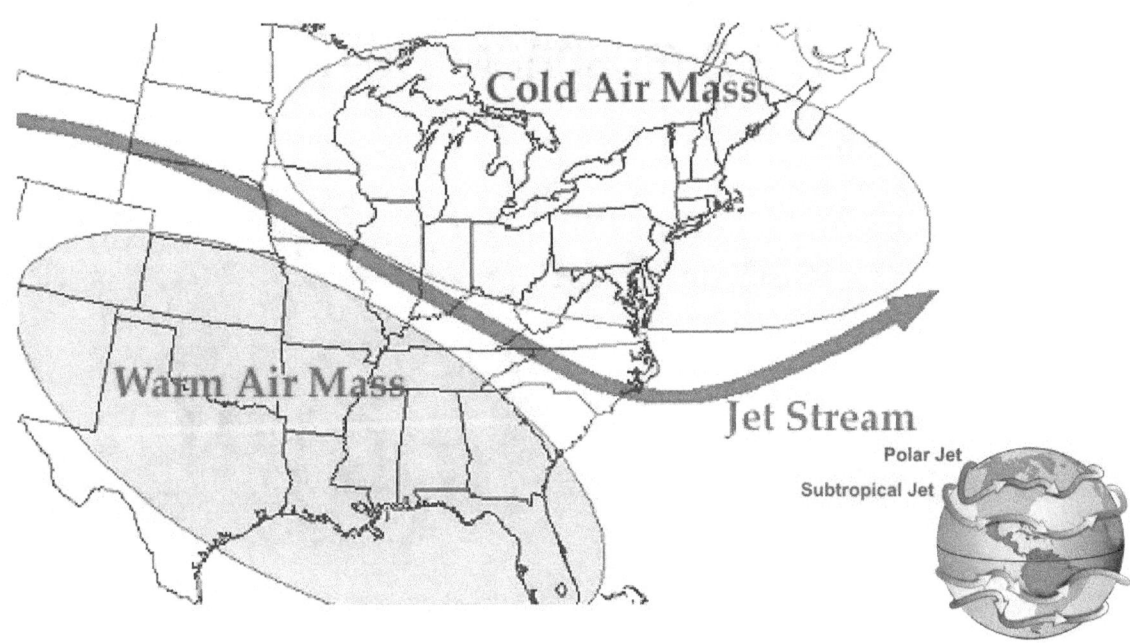

Image/Text/Data from the University of Illinois WW2010 Project - wikipedia http://ww2010.atmos.uiuc.edu/%28Gh%29/guides/mtr/cyc/upa/jet.rxml

A further item of evidence that supports this recognition is found in the jet-streams in the atmosphere that are extremely fast moving air currents that flow in the direction of the the Earth's rotation, but move significantly faster than the rotation of the Earth itself.

Believed that the jet streams are powered by the Coriolis effect

The atmospheric Jet Streams

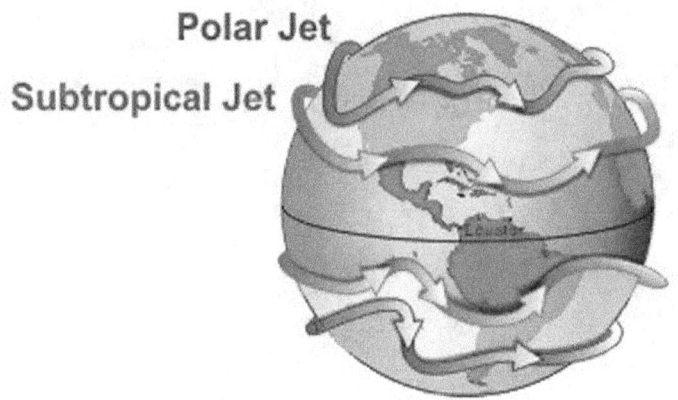

It has been long believed that the jet streams are powered by the Coriolis effect of air masses from the polar regions, reversing direction in the warmer regions. However, in this case the northern and southern jet stream would flow in opposite directions, which they don't.

Another phenomenon where rotational movements occur faster

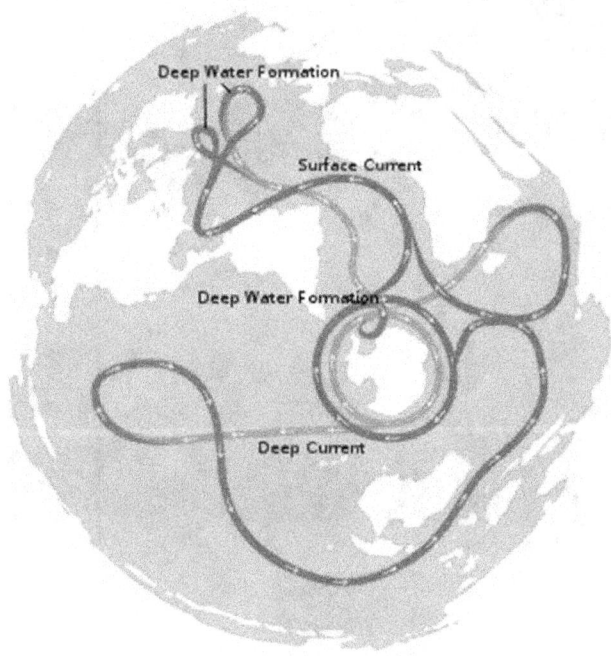

The ocean currents conveyor belt centered on the deep cold waters encircling Antarctica

"Conveyor belt" by Avsa - under CC BY-SA 3.0 via Commons -

Another phenomenon on Earth where rotational movements occur faster than the rotation of the Earth, are the massive ocean currents that flow in a circle around Antarctica.

The currents shown here are a part of a worldwide cold-water recycling system.

It appears to be thermally powered. But it also has the potential to be electrically powered, which is probably more likely.

Cold waters in the polar regions The Arctic pool has an outflow

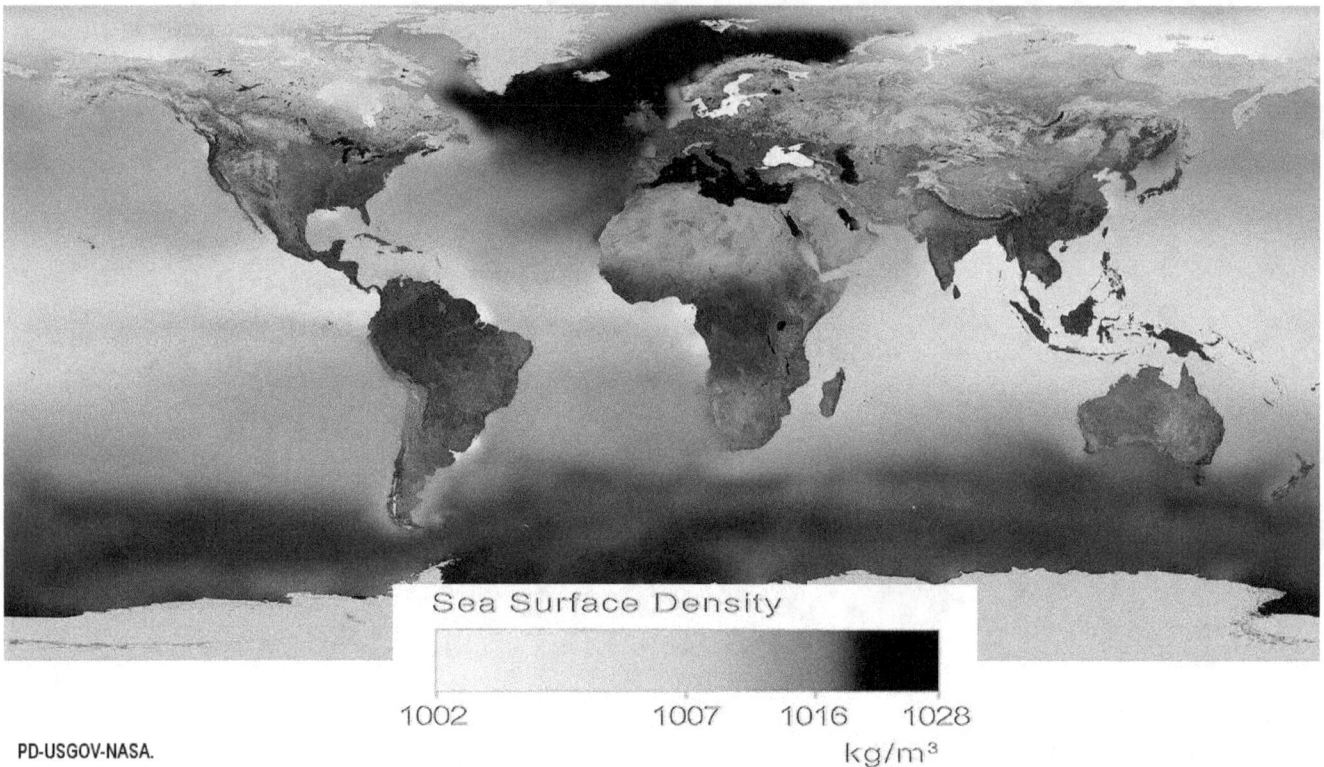

PD-USGOV-NASA.

Cold waters in the polar regions are so dense by contraction and salination, that they sink into deep pools.

~ocean_currents_belt_big

The Arctic pool has an outflow that creates a deep current that flows all the way to Antarctica where it joins the deep pool there, that encircles the continent.

Two currents 'spin' off from the rotating pool

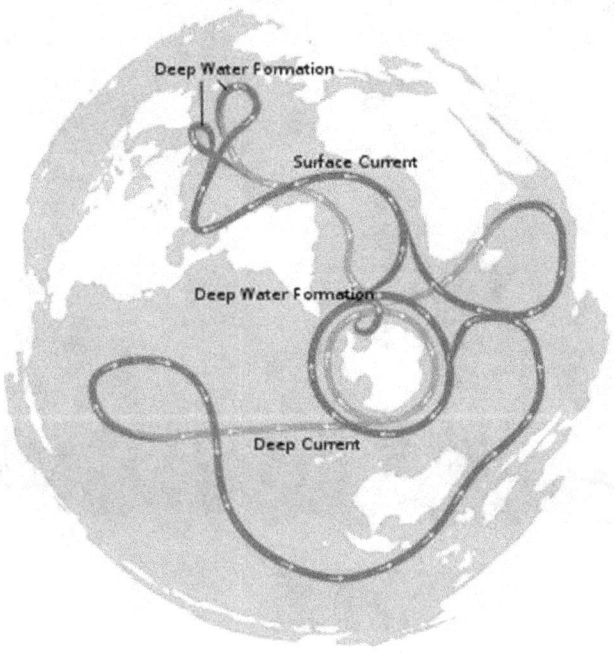

The ocean currents conveyor belt centered on the deep cold waters encircling Antarctica

"Conveyor belt" by Avsa - under CC BY-SA 3.0 via Commons -

With the encircling deep pool flowing faster than the rotation of the Earth, two currents 'spin' off from the rotating pool. One flows to the coast of Africa and into the Indian Ocean where the cold waters warm up and resurface. The second branch flows into the central Pacific, where it too warms up and resurfaces.

The cold deep waters originating in the polar regions

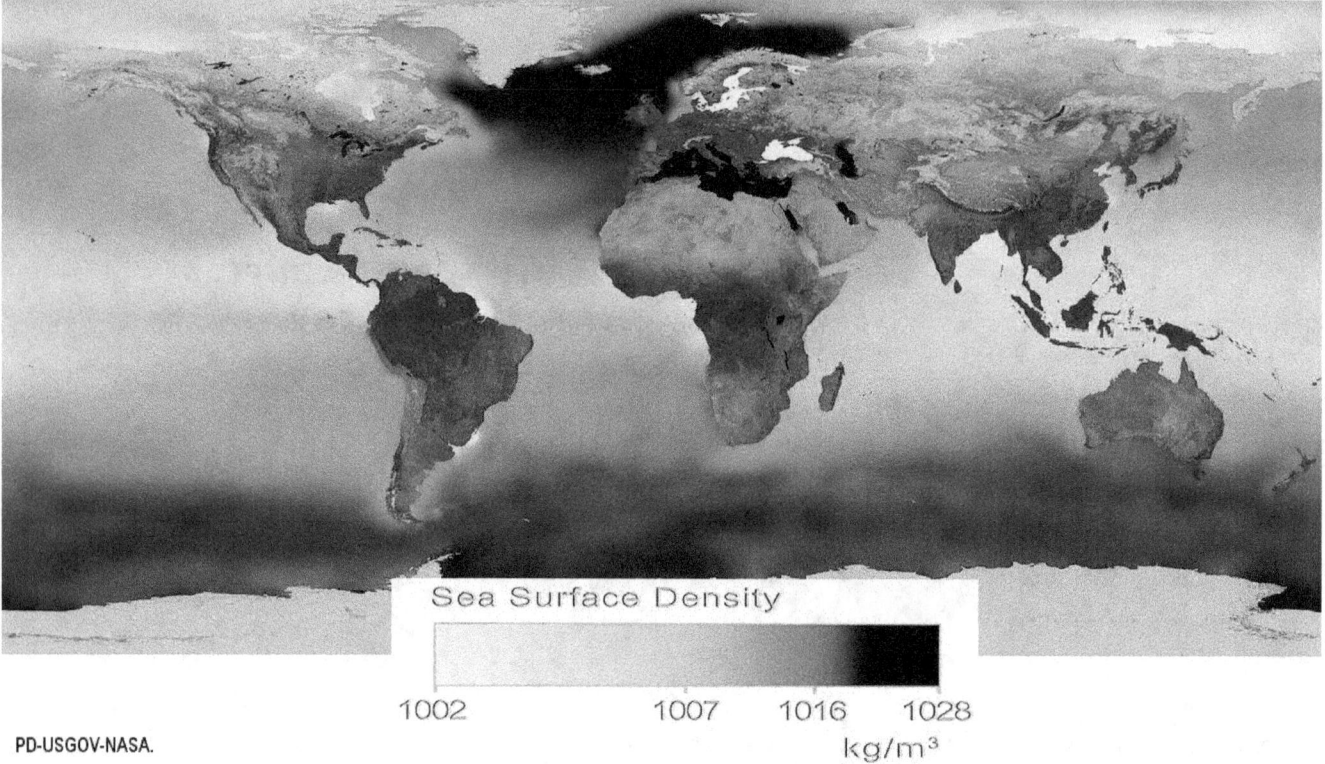

PD-USGOV-NASA.

With the cold deep waters originating in the polar regions, they carry with them large volumes of dissolved CO2. Cold waters can dissolve CO2, to up to 10 times greater density than the density than the atmosphere presently holds. In this manner, dissolved CO2 takes a ride in a very slow moving recycling system with a transit time of 350 years to a thousand years or more, which is evidently electrically powered by rotational actions motivated by the primer fields.

Recycling system emits CO2 dissolved 350 years ago

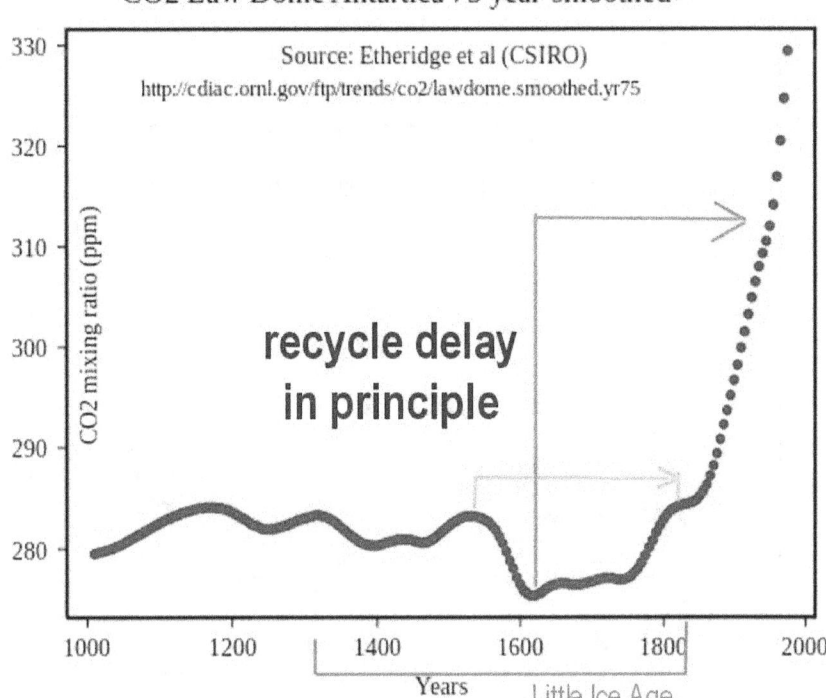

This means that the actively-powered recycling system emits CO2 into the atmosphere today that had been dissolved into the oceans during the extreme cold time of the Little Ice Age. This also means that the large CO2 increase in modern time originated at the Little Ice Age 350 years ago, or in other cold periods in more distant times, rather than being the product of human activity.

The recycling system incorporates three different transit times

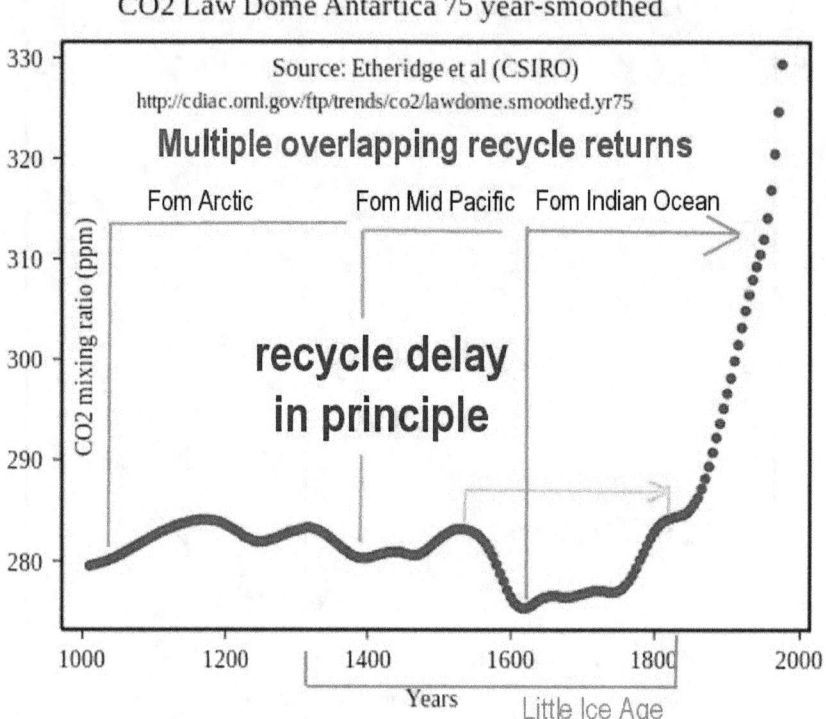

If one considers that the recycling system incorporates three different transit times, it is possible for CO2 to be released by the recycling system from 3 different historic times simultaneously.

The transit time to the Indian Ocean

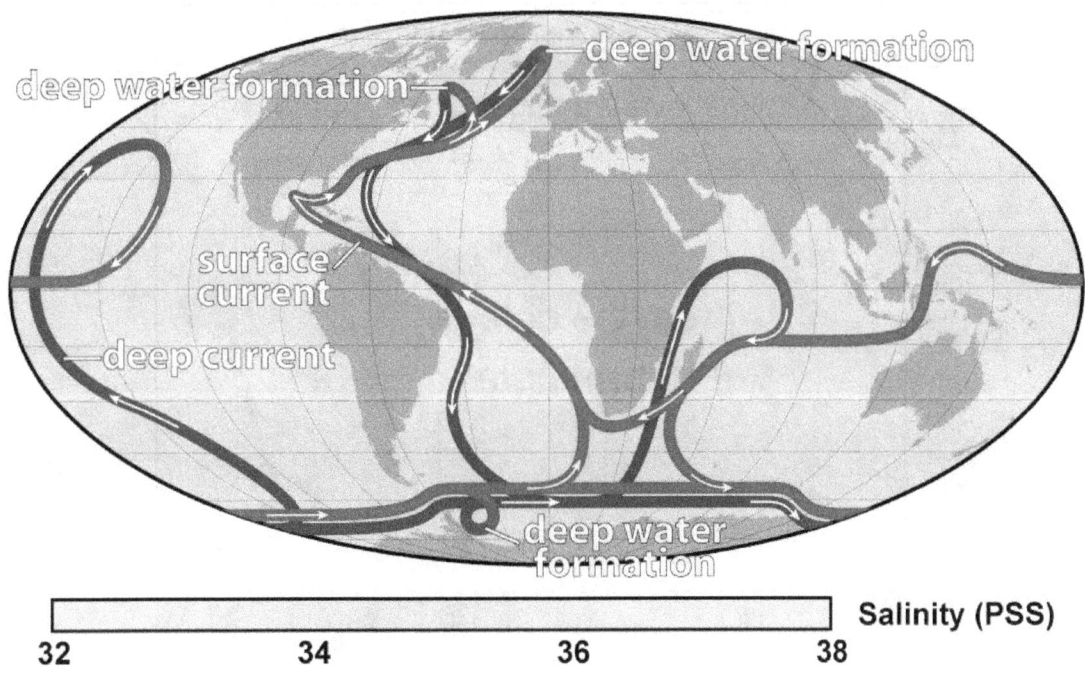

The transit time to the Indian Ocean is the shortest, possibly in the range of 350 years, with the island of Madagascar in the way. The transit time to the Mid Pacific appears to be significantly longer. And for the CO2-rich waters from the Arctic in the North, the total transit time, probably exceeds a thousand years. The exact times are not known.

Different transit times resurface CO2 from three different eras

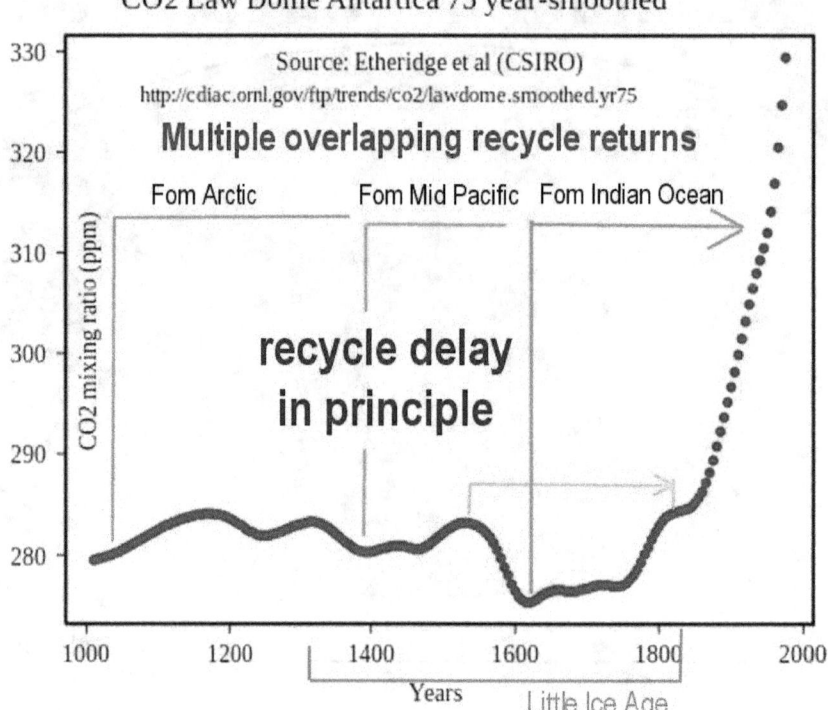

Since the different transit times resurface CO2 from three different eras simultaneously, the large increase in CO2 that has been measured in recent times appears to be the result of overlapping recycling returns from three of the big cold periods of the last twelve hundred years.

The large increase in CO2 that has occurred

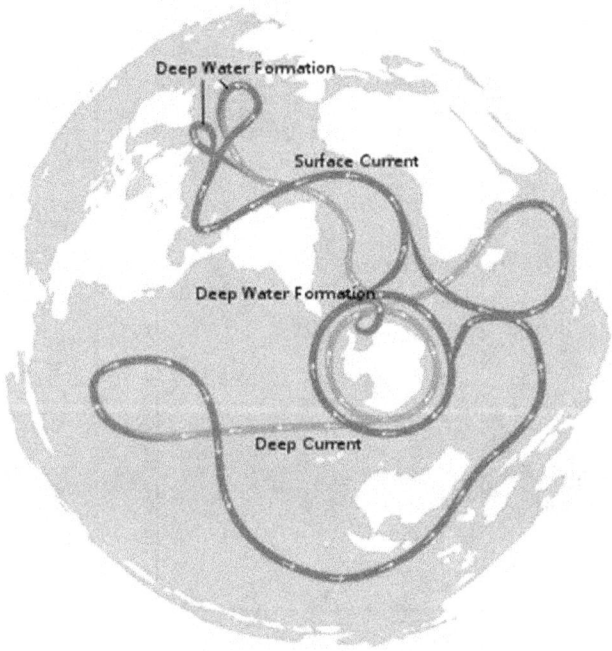

The ocean currents conveyor belt centered on the deep cold waters encircling Antarctica

"Conveyor belt" by Avsa - under CC BY-SA 3.0 via Commons -

This means that the large increase in CO2 that has occurred, which has invigorated the biological system, is a benefit provided for humanity by the electromagnetic effects of the Primer Fields that evidently power the recycling system by the rotation of ocean currents around Antarctica at a speed greater than the rotation of the Earth. The increased CO2 is a benefit for us, for the simple reason that CO2 is critical for plant growth and thereby increases harvests.

The Great Global Warming resulted from increase in solar activity

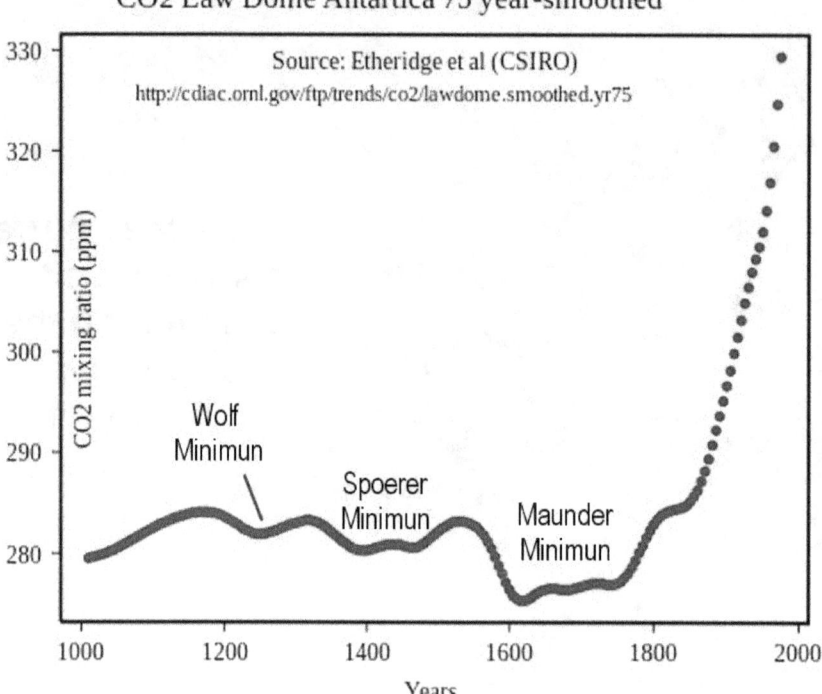

Of course other factors also enter the consideration of the CO_2 increase that the measurements tell us of. One factor is, that the increase in CO_2 that is shown here, has occurred during the timeframe of the Great Global Warming, which has measurably resulted from the sharp increase in solar activity that has occurred during this timeframe from the Maunder Minimum on.

The Great Global Warming that has recovered the global climate

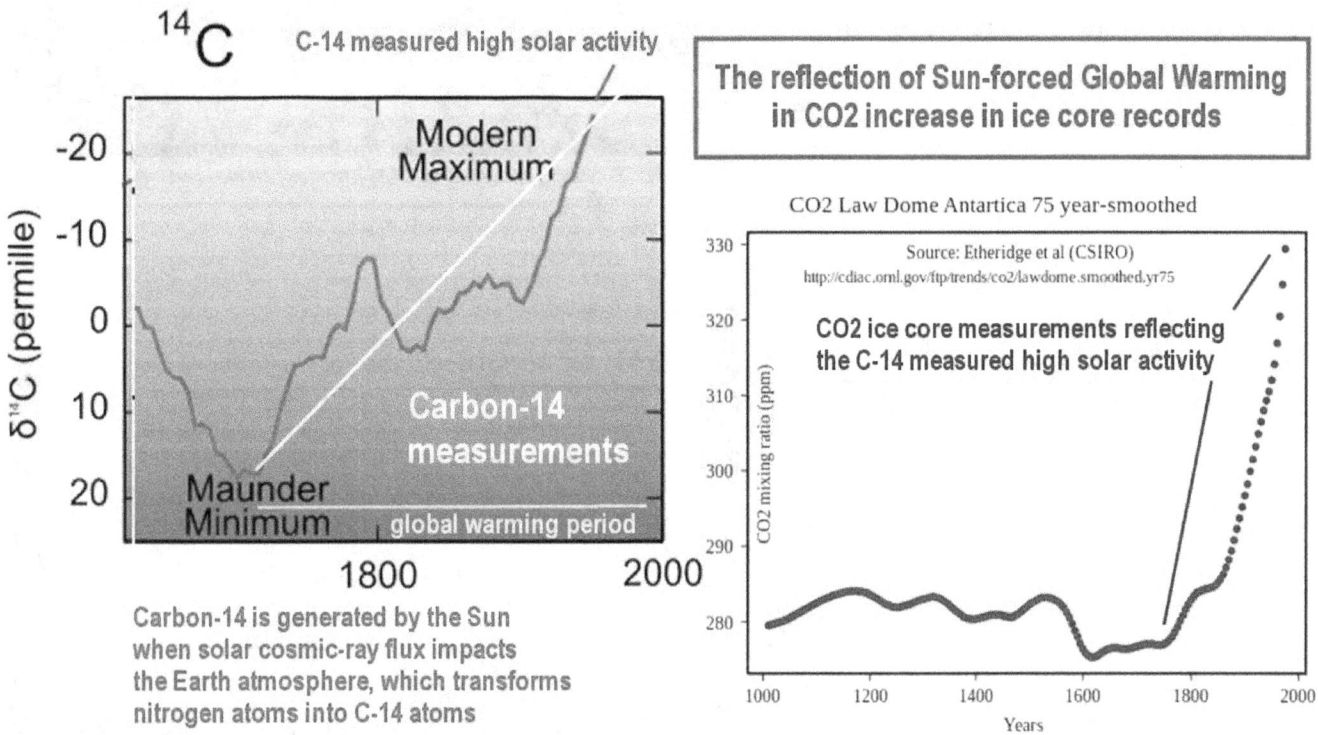

The Great Global Warming that has recovered the global climate from the Little Ice Age, was clearly the result of the sharp increase in solar activity that occurred after the Maunder Minimum. The sharp increase in solar activity has been recorded in increased sunspot cycles, and also in Carbon-14 measurements, which measure solar cosmic-ray flux. The resulting warm climate from increased solar activity would naturally increase the atmospheric CO2 density, as less CO2 is being dissolved into the oceans during the warm periods.

Topics to think about:

* The reflection of Sun-forced Global Warming in CO2 increase in ice core records

* Carbon-14 is generated by the Sun when solar cosmic-ray flux impacts the Earth atmosphere, which transforms nitrogen atoms into C-14 atoms.

* The C-14 measured, high solar activity

*CO2 ice core measurements reflecting the C-14 measured high solar activity

* Solar-forced global warming period

The measured 15% increase in CO2 from the 1950s on

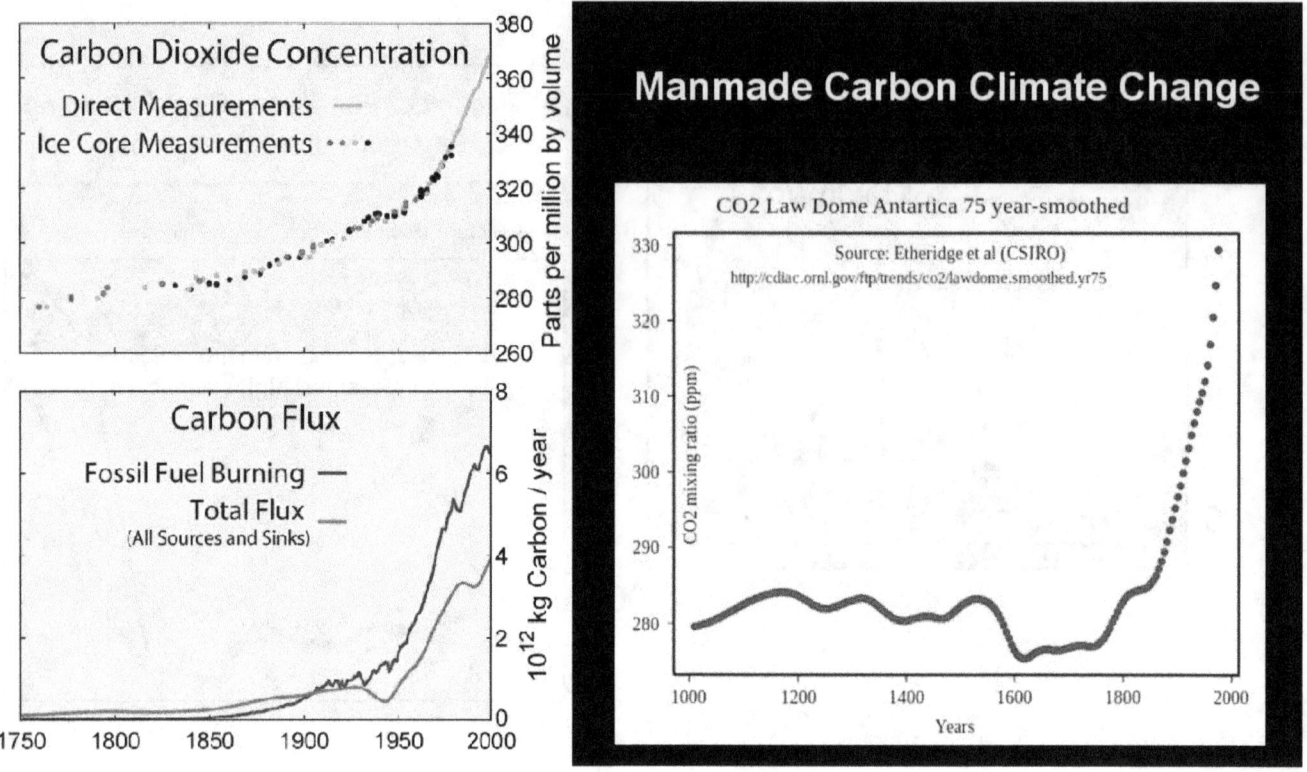

It is interesting to note that the measured 15% increase in CO2 from the 1950s on, couldn't have originated from manmade contribution that, at the peak of the increase, amounted in total a mere 1% of the global CO2 per year.

A quarter of the global atmospheric CO2 gets recycled annually

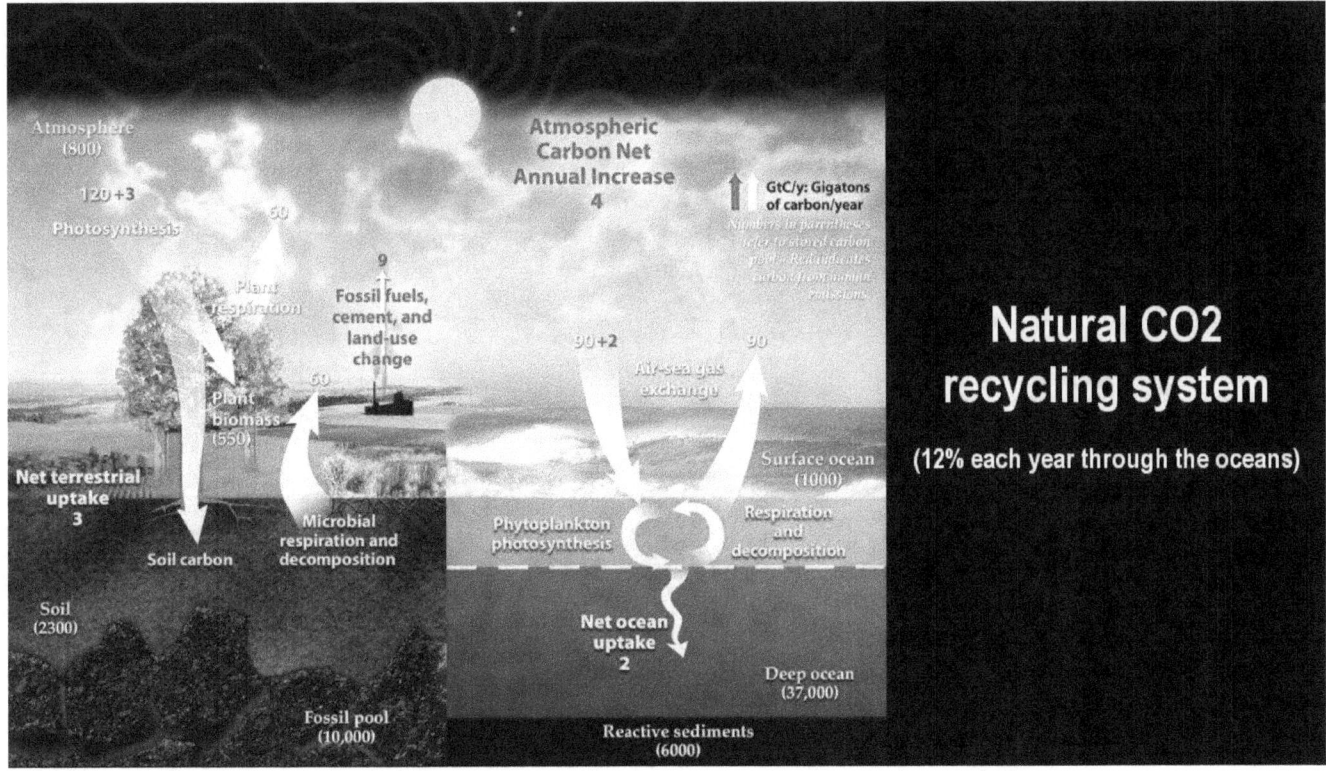

While a quarter of the global atmospheric CO2 gets recycled annually, with slightly half of that flowing through the ocean recycling system.

It is further interesting to note

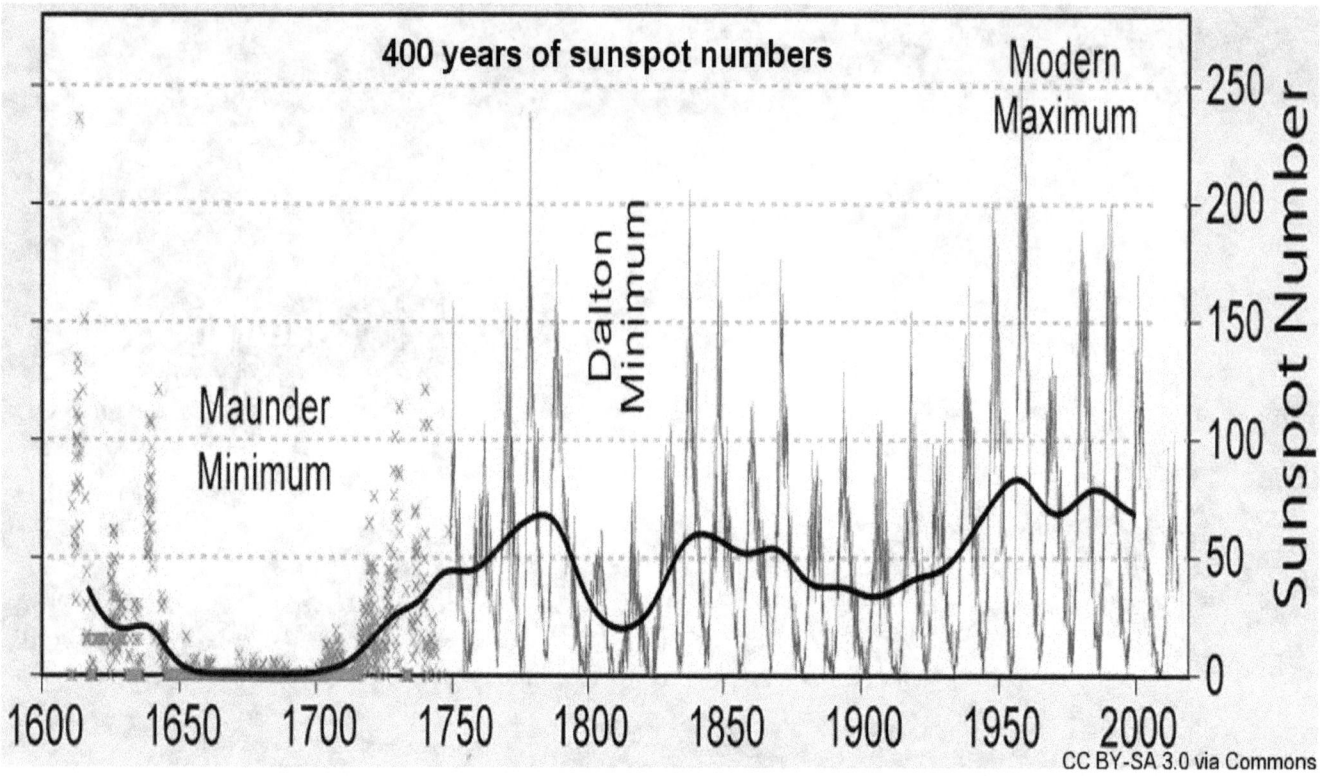

It is further interesting to note that the period from the 1950s on to slightly before 2000 was a peak period in solar activity, and the warmest in hundredth of years, which should therefore coincide with a high rate of CO2 increase due to the warming.

The sharp CO2 increase that the ice core data indicates

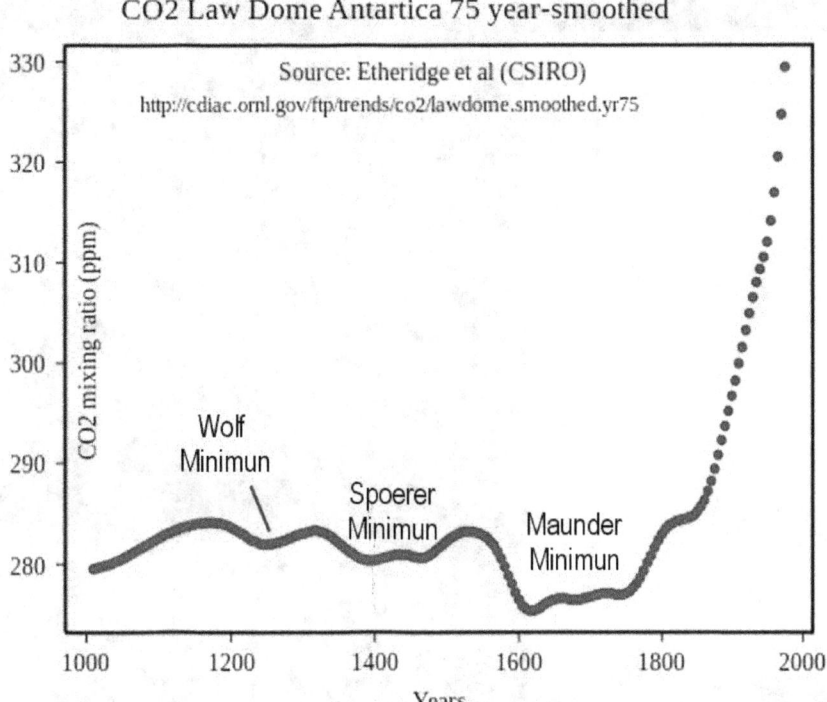

In addition, the sharp CO2 increase that the ice core data indicates, has been disputed by a high-level expert in the field.

Dr. Zbigniew Jaworowski,

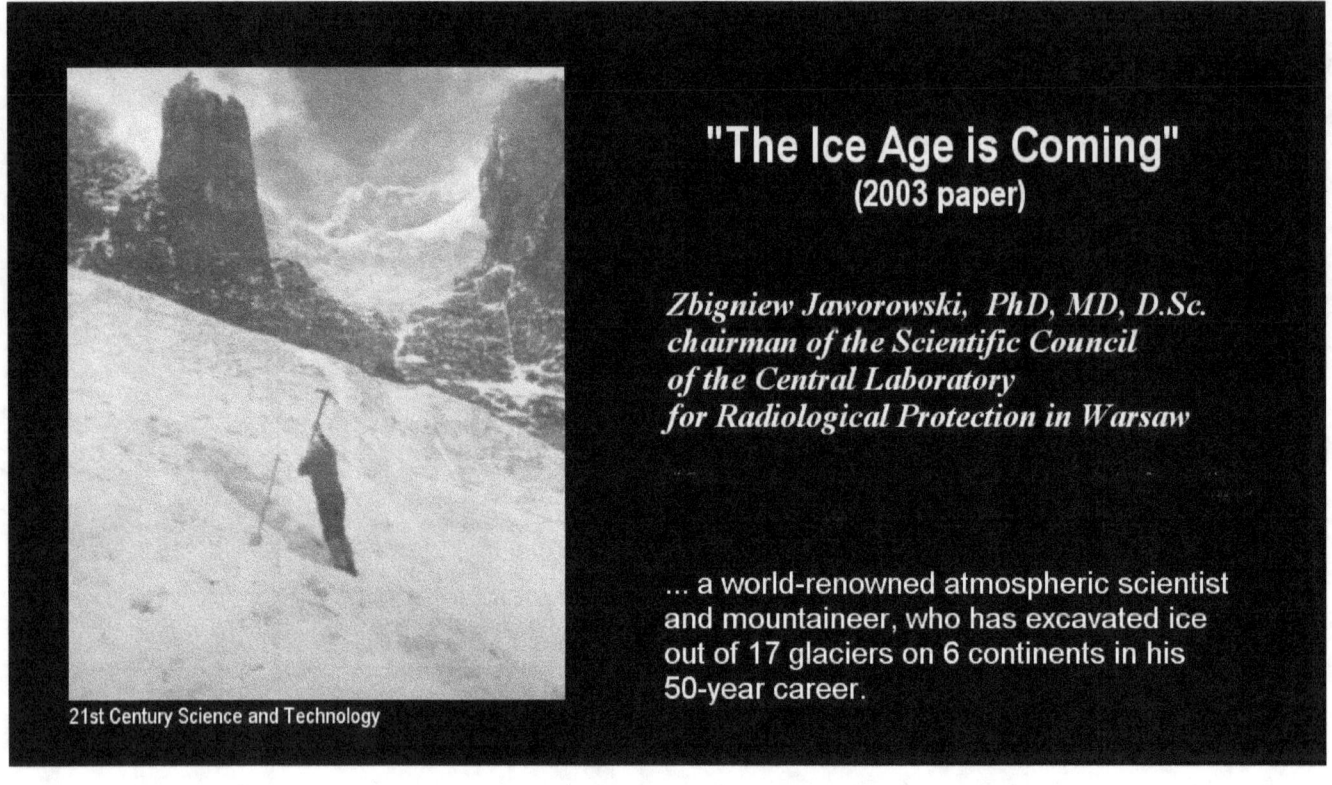

Dr. Zbigniew Jaworowski, the one-time chairman of the Scientific Council of the Central Laboratory for Radiological Protection in Warsaw, has pointed out that ice core measurements have an inherent problem.

With CO2 being highly soluble in water

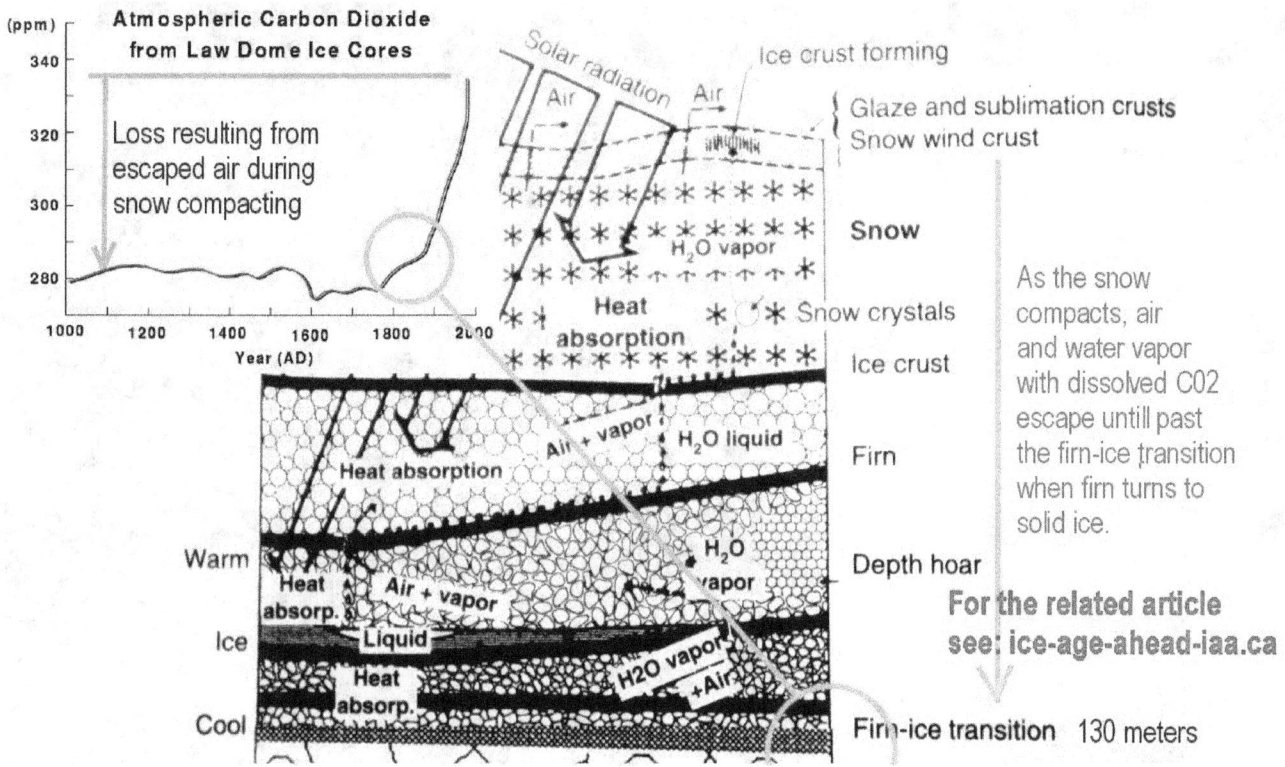

His experience has been, resulting from a 50-year career of ice explorations on 6 continents, that large losses of CO2 result in the period of snow becoming compacted, first into firn, then into deep hoar, and so on, until it becomes compressed into solid ice. With CO2 being highly soluble in water, a large portion of CO2 in the original snow, becomes lost with the loss of water vapor that is expelled by the compaction process. His take is that the CO2 portion that is finally compressed into ice, is significantly less than what was originally present.

Dr. Jaworowski compared the CO2 concentration in ice

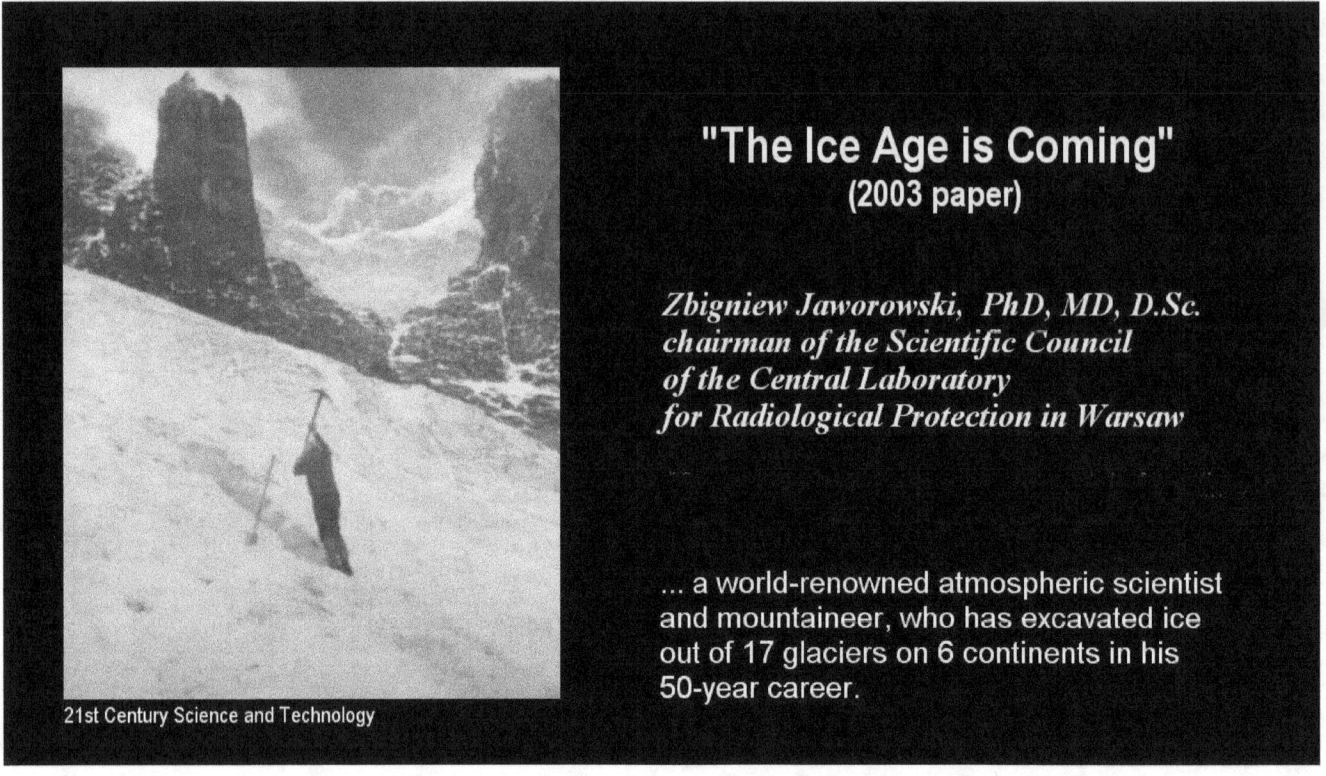

Dr. Jaworowski compared the CO_2 concentration in ice from the stage just past the firn-to-ice transition, with CO_2 concentrations of the same time in deep sediments, and reports that the CO_2 level in the sediments hadn't significantly changed in this period.

Dr. Jaworowski notes that the hokey-stick phenomenon

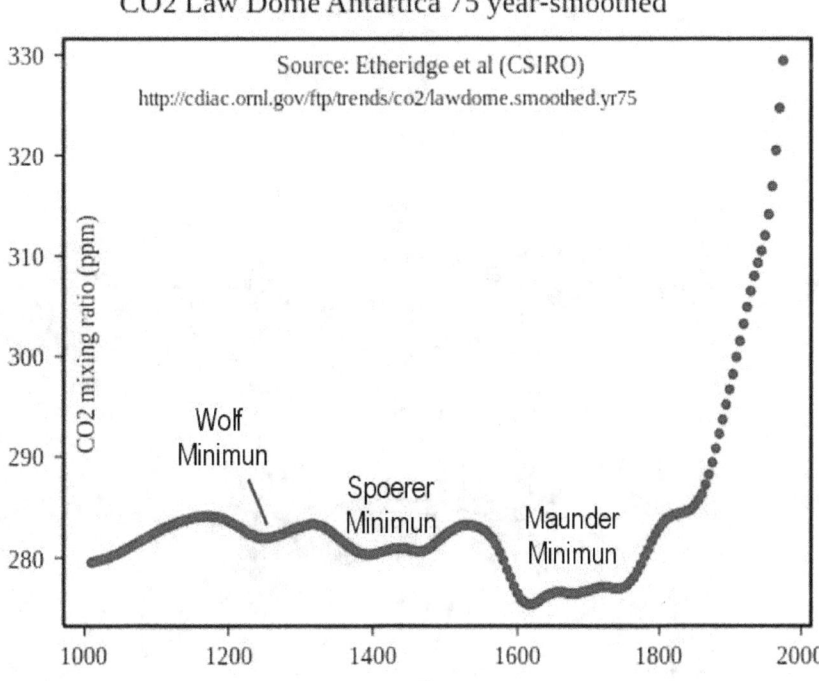

Dr. Jaworowski notes that the hokey-stick phenomenon that has been deployed to scare humanity into economic suicide measures and depopulation, is not actually true, but is simply a characteristic of the CO_2 loss in the ice compaction process. He notes that the characteristic is known, but appears to be intentionally abused for political purposes.

In this sense the Primer fields even affect politics

Poster of the Climate Conference.
Licensed under Fair use via Wikipedia

COP 21: Heads of delegations by GUSTAVO-CAMACHO-GONZALEZ - Licensed under CC BY 2.0 via Commons by Presidencia de la República Mexicana -delegates

In this sense the Primer fields even affect politics. Under political dogmas the imagined CO2 increase in modern time is regarded as man-made, and is deemed to be the sole cause for the Great Global Warming that occurred after the Little Ice Age when industrial activity begun, which, as it is said, must be prevented with political will, at all cost, from further increasing. Not a word was ever spoken in these arenas that the Great Global Warming was demonstrably caused by the Sun, and was impelled by cosmic processes and resonating astrophysical conditions, while CO2 is not a climate factor at all in comparison.

Nor was it ever noted in these politically-driven operations that the hockey-stick CO2-increase is an illusion caused by the false interpretation of the ice core data. Nor was it ever said that the actually, physically measured increase in CO2 levels, which only goes back 50 years, coincides with the greatest peak in solar caused global warming, and coincides in addition with overlapping CO2 emissions from historic high CO2 concentrations via the slow moving global recycling system.

Nor was it ever said that CO2 isn't a climate factor anyway, as it contributes only a millionth part to the global greenhouse effect, which itself isn't the prime factor either, since the absolute prime factor is the rate of cloudiness that is affected by solar activity, which in turn is affected by astrophysical factors that flow through the Primer Fields.

Since all these factors are known, the political will is falsely placed and evidently with intention, especially considering that it is known that CO2, which the manmade global warming dogma is centered on, is not a climate factor at all, never was, and never can be as the claimed effect is physically impossible in the overall context.

The Great Global Warming that pulled us out of the Little Ice Age

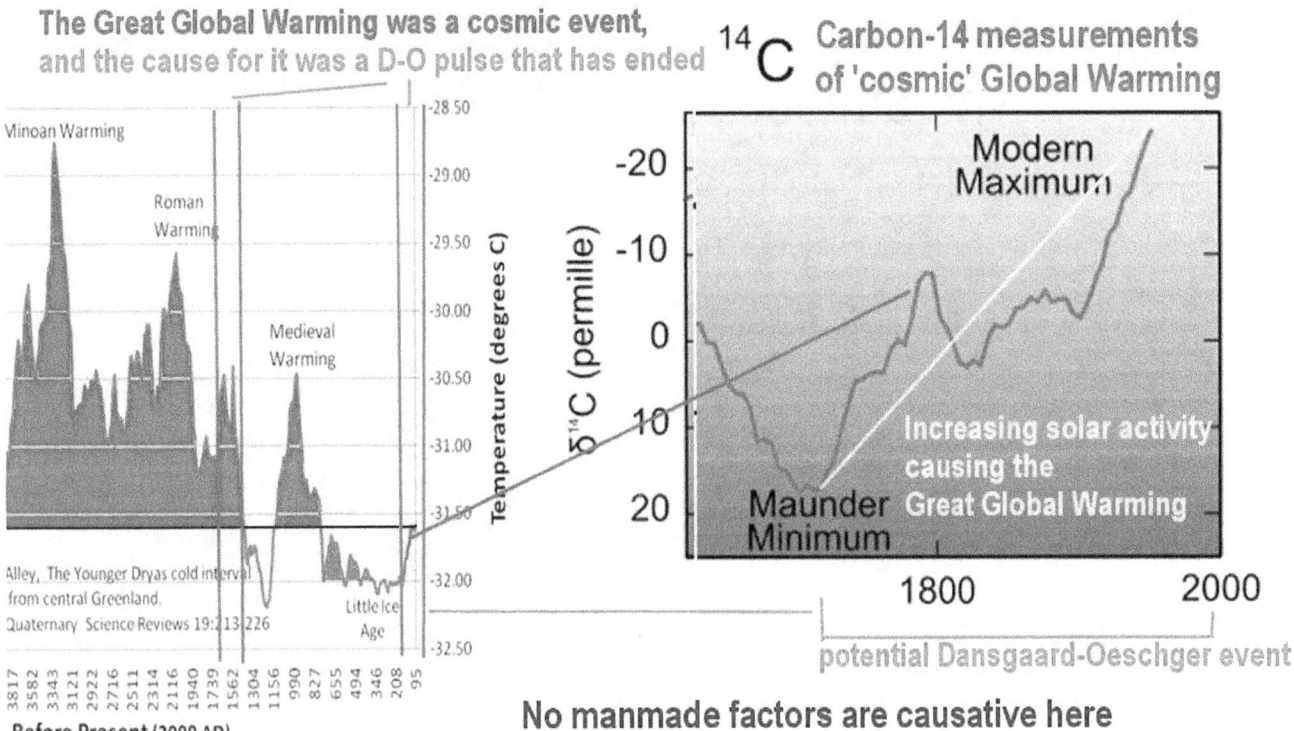

Nor was it ever said in the political brainwashing projects that include to a large degree also the media, that the Great Global Warming that pulled us out of the Little Ice Age was demonstrably and measurably a solar-forced cosmic phenomenon, instead of being manmade in any way.

** the exploration continues

The CO_2 subject is far bigger than a mere academic concern

The atmospheric Jet Streams

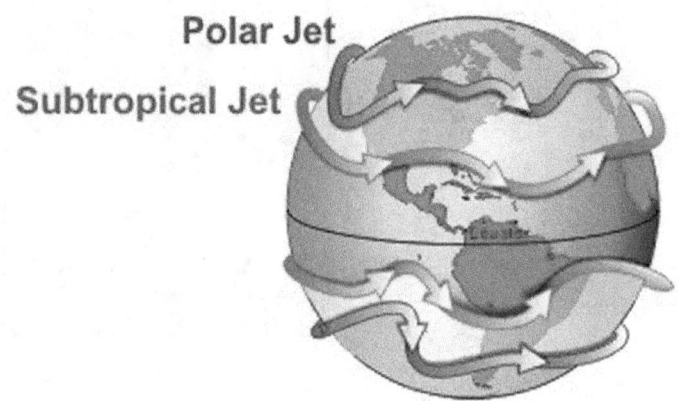

I have diverged into the CO_2 subject, because it is a part of the global recycling system, which is powered by astrophysical principles that cause air and sea movements on Earth that flow faster than the rotation of the Earth. The CO_2 subject, of course, is far bigger than a mere academic concern. It affects your dinner table with increasingly high food prices, and raises the fuel prices for transportation and for heating your home.

The mandated mass-burning of food for biofuels production

Under the banner of "CO2-forced Manmade Global Warming", the mandated mass-burning of food for biofuels production, has diverted huge agricultural resources to be burned, that would have normally nourished 400 million people. In a world that has a billion people living in chronic starvation, the food burning holocaust is claiming more than 100 million victims each year, of death by starvation.

The CO2 issue far out of mere academic concerns

Poster of the Climate Conference. Licensed under Fair use via Wikipedia

This takes the CO2 issue far out of the realm of mere academic concerns, into the political arena with the often stated objective of mass-depopulation, that is to weed out the 'useless eaters,' as the victims were once called, to shrink world population to less than a billion people. The academic issues, if they were widely understood, would block the politically-forced tragedies for which no physical cause actually exists.

Getting back to the phenomena of fluid movements that are faster

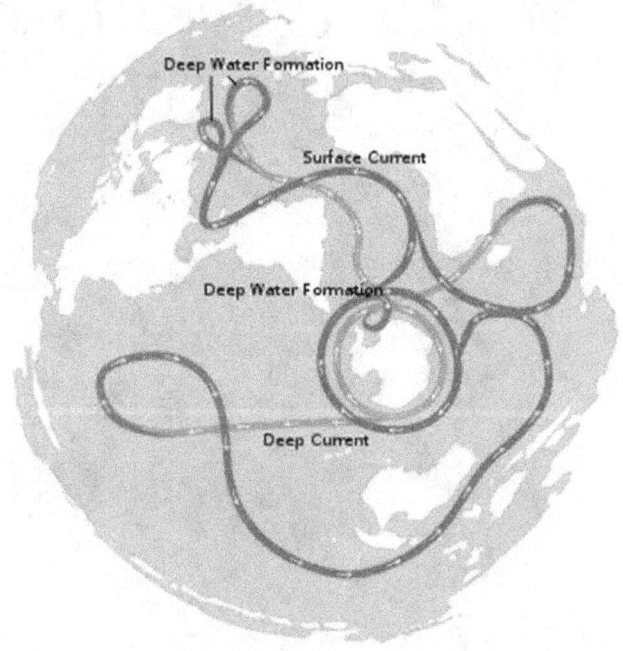

The ocean currents conveyor belt centered on the deep cold waters encircling Antarctica

"Conveyor belt" by Avsa - under CC BY-SA 3.0 via Commons -

In getting back to the phenomena of fluid movements that are faster and than the rotational speed of an object under the Primer Fields, as we see it reflected in the Jet Streams and the ocean-current movements around Antarctica, one finds of course, that the Earth is only a small part of the solar system, which means that it is not alone in being affected by the Primer Fields that cause fluid movements exceeding the spin-axis' rotational speed.

The Sun rotates significantly faster at the equator

The Sun in visible light as seen through a dark glass

The phenomenon is highly prominent on the Sun, though it is barely visible there. The phenomenon is discernable by the differential movements of its feature, such as sunspots. The Sun rotates significantly faster at the equator than at the poles. It takes the Sun 35 days for a single rotation at the poles, but only 25 days for a rotation at the equator.

The faster equatorial rotation stands as evidence that the Sun is actively rotated by an external electric force, which in addition, acts most strongly on its equatorial region from where its heliospheric current sheet extends.

The Sun's two high-activity bands spaced centered off the equator

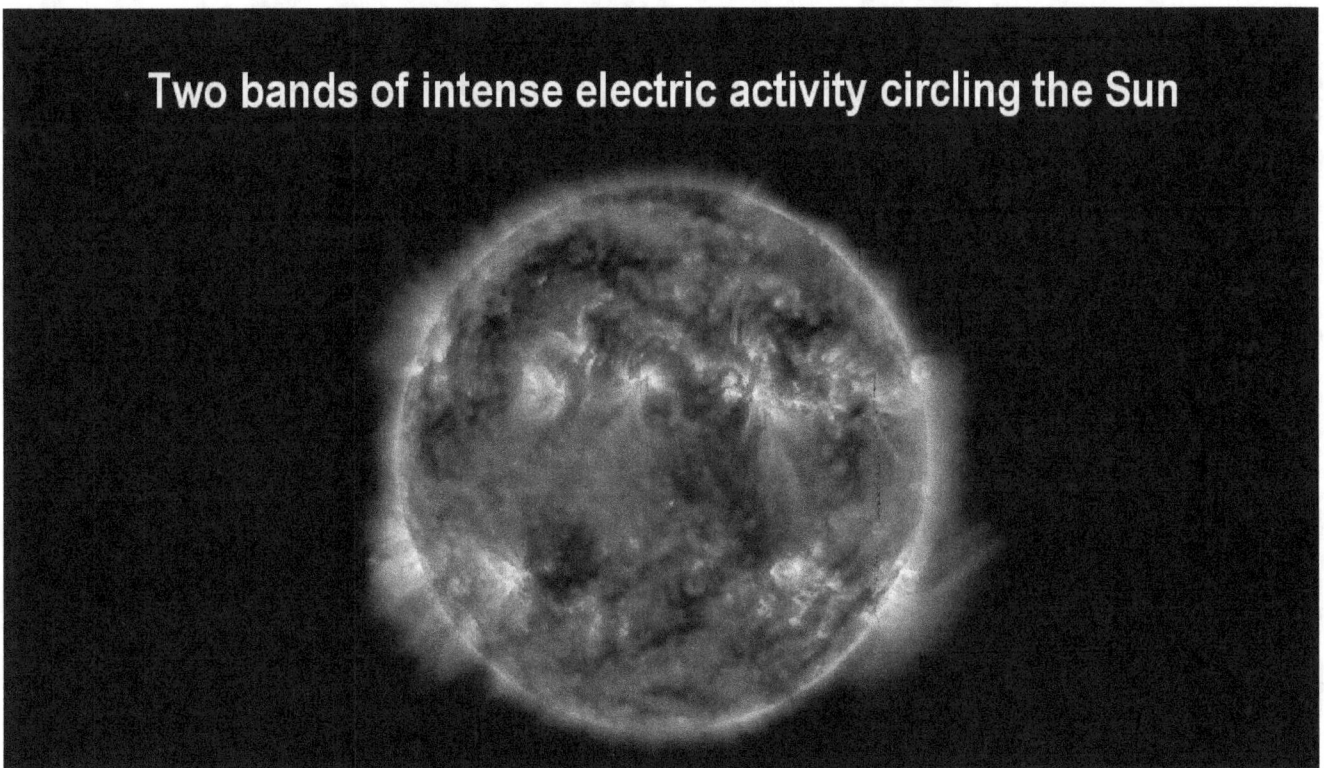

The Sun's two high-activity bands spaced centered off the equator, actually reflect in the large what became evident in principle in plasma experiments.

The evidence illustrates the functioning of the Primer Fields

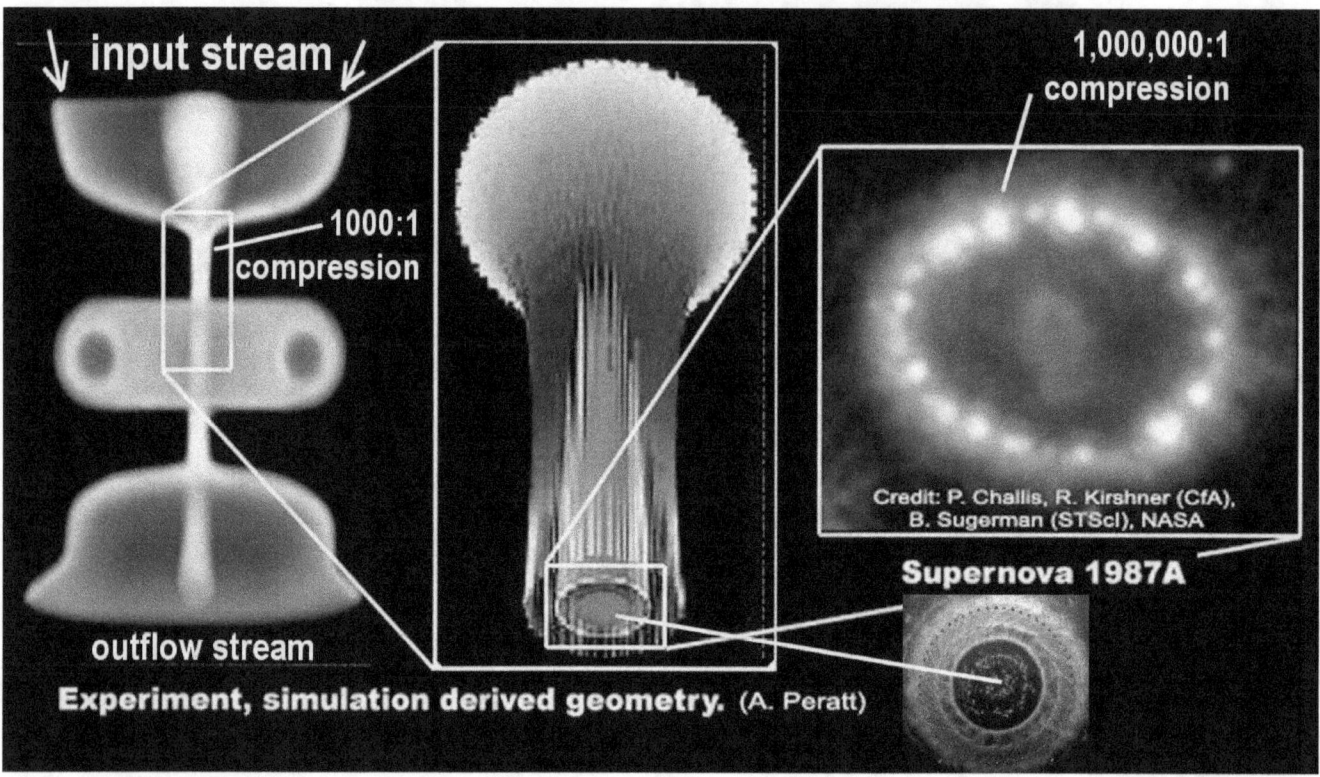

The evidence illustrates the functioning of the Primer Fields, and their effects that we see evident in the form of an ecliptic ring of induced plasma currents, which we see in principle reflected on the Sun in the form of the equatorial high-activity zones.

The ecliptic principle is also apparent

The ecliptic principle is also apparent to some degree in David LaPoint's static exploration of the primer fields.

The resulting shape apparent in the ecliptic shape of galaxies

He points out that the resulting shape in his experiments is dramatically apparent in the ecliptic shape of galaxies, which suggests a commonality of principle in both cases.

A solar system, that is in all major aspects electrically powered

The evidence that one sees here is of a solar system, that is in all major aspects electrically powered and electrically motivated, and shares this principle with the galactic systems.

This means that galaxies are not stable entities

Galaxies at node points

This means that galaxies are not stable entities, but are subject to plasma-density resonances in intergalactic space, and intergalactic plasma streams that are focused by large primer fields onto the galaxies.

For our galaxy, the Milky Way Galaxy

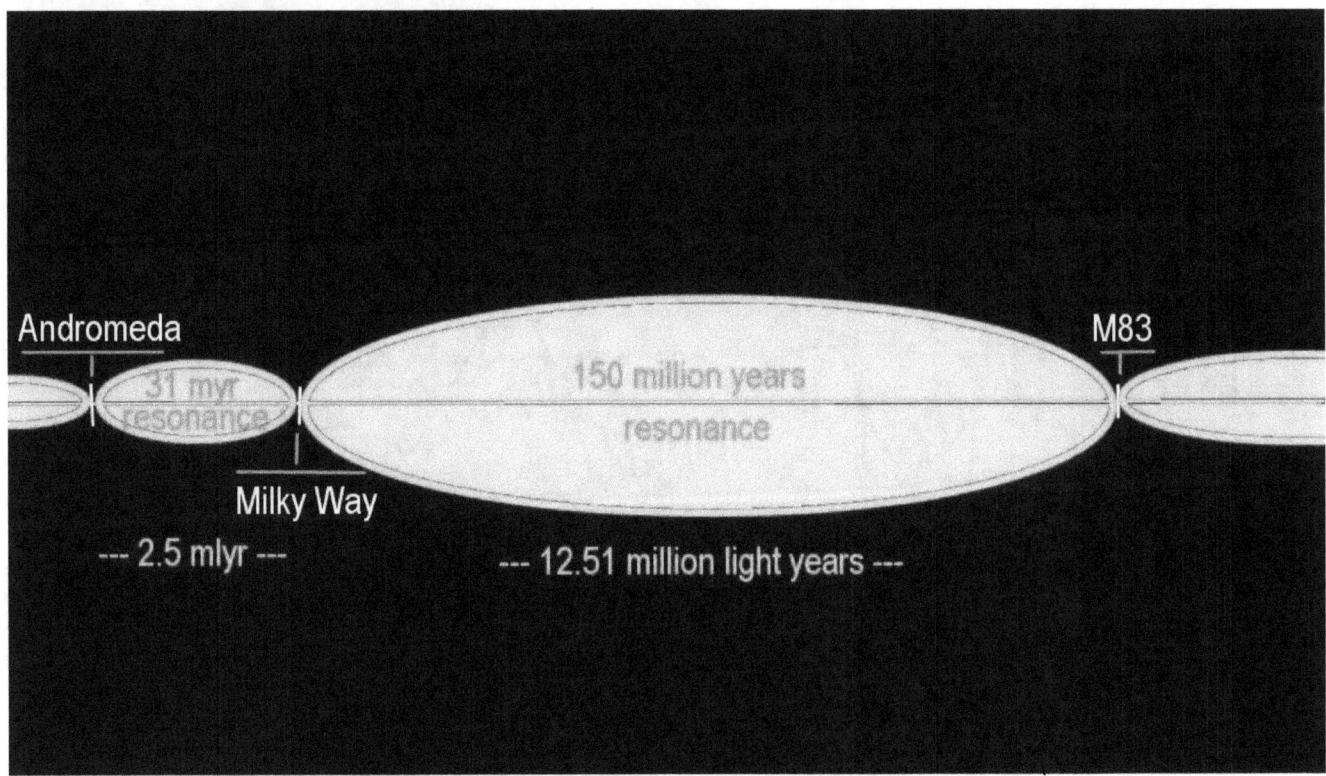

For our galaxy, the Milky Way Galaxy, two potential plasma streams to nearby galaxies come into view that form intergalactic node points. The potential resonance cycles of these very long intergalactic plasma streams, appear to be reflected in long-term climate cycles on Earth.

The two long intergalactic resonance cycles

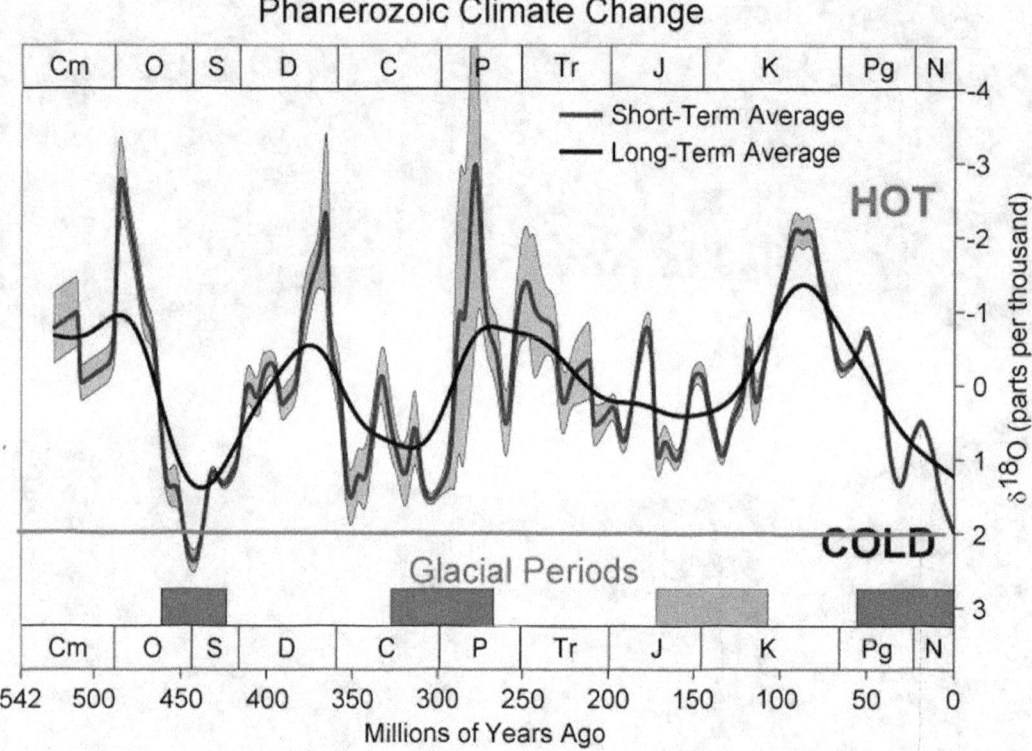

The two long intergalactic resonance cycles appear to reflect in combination the two overlapping climate cycles on Earth that we found evidence of in sediments, in the form of oxygen-18 isotope variations.

The solar system and the Earth are not fundamentally isolated entities

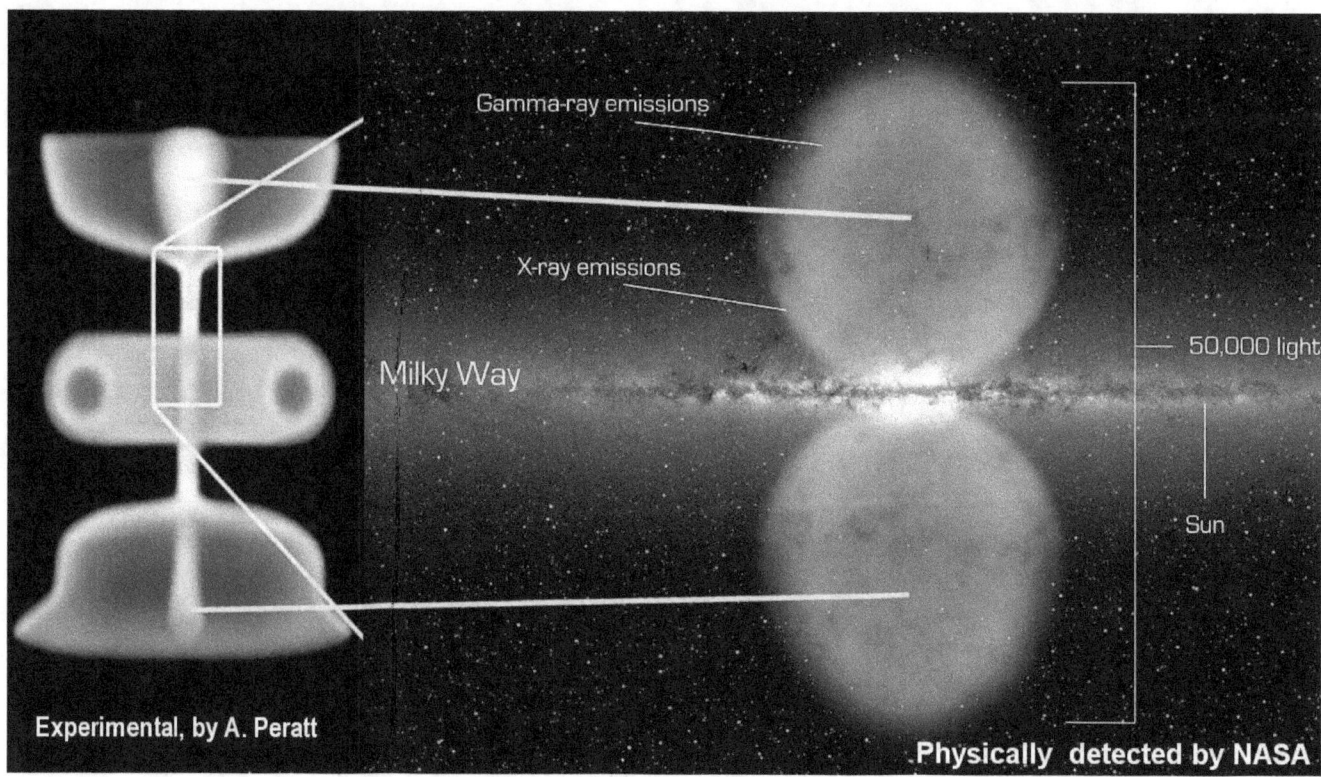

This means that the solar system and the Earth are not fundamentally isolated entities, but are intimately connected, both physically, and by the sharing of a common, universal, all-motivating principle.

This also means that the Sun is not its own master

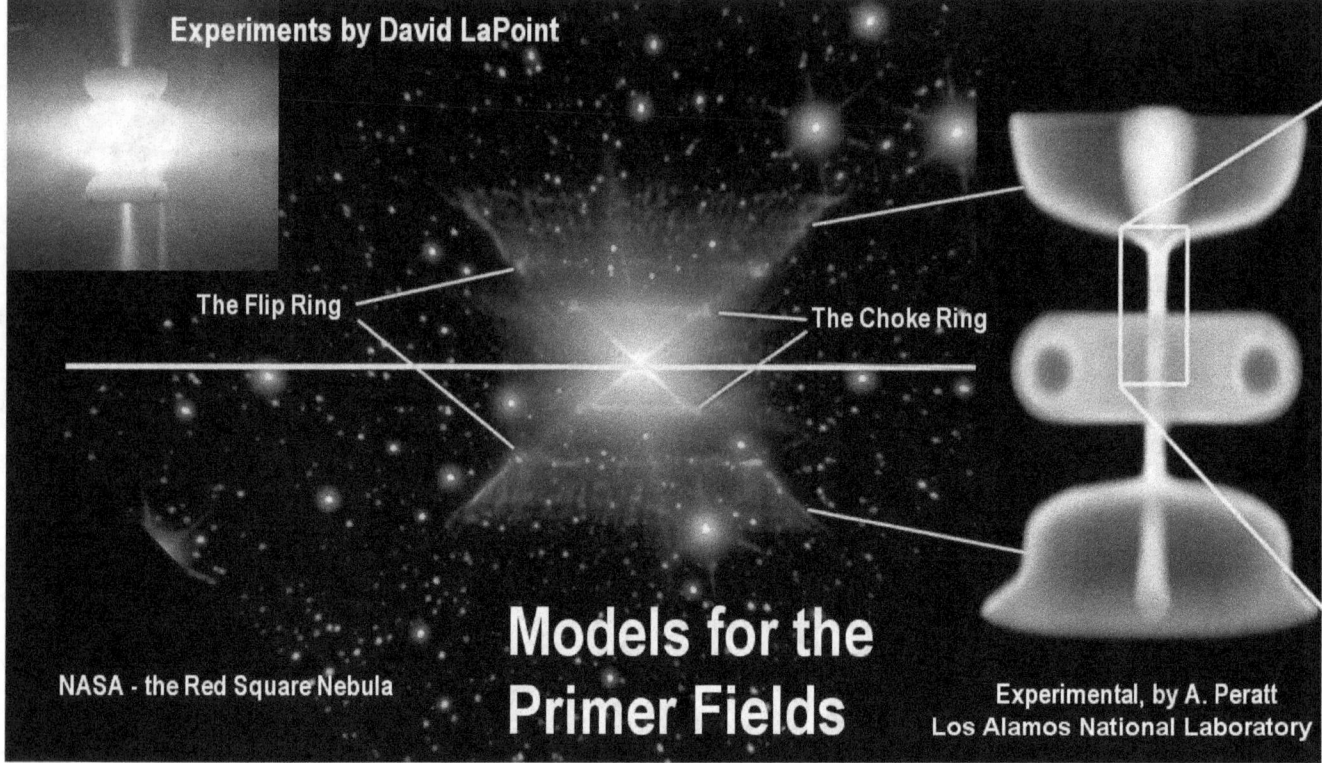

This also means that the Sun is not its own master, but exists cradled between interstellar primer fields that give it is existence as an externally powered star. This further means that the primer fields can collapse when the external conditions fail that enable the primer fields to function. When this happens, our world will change.

The phase shift in our world, onto an inactive Sun

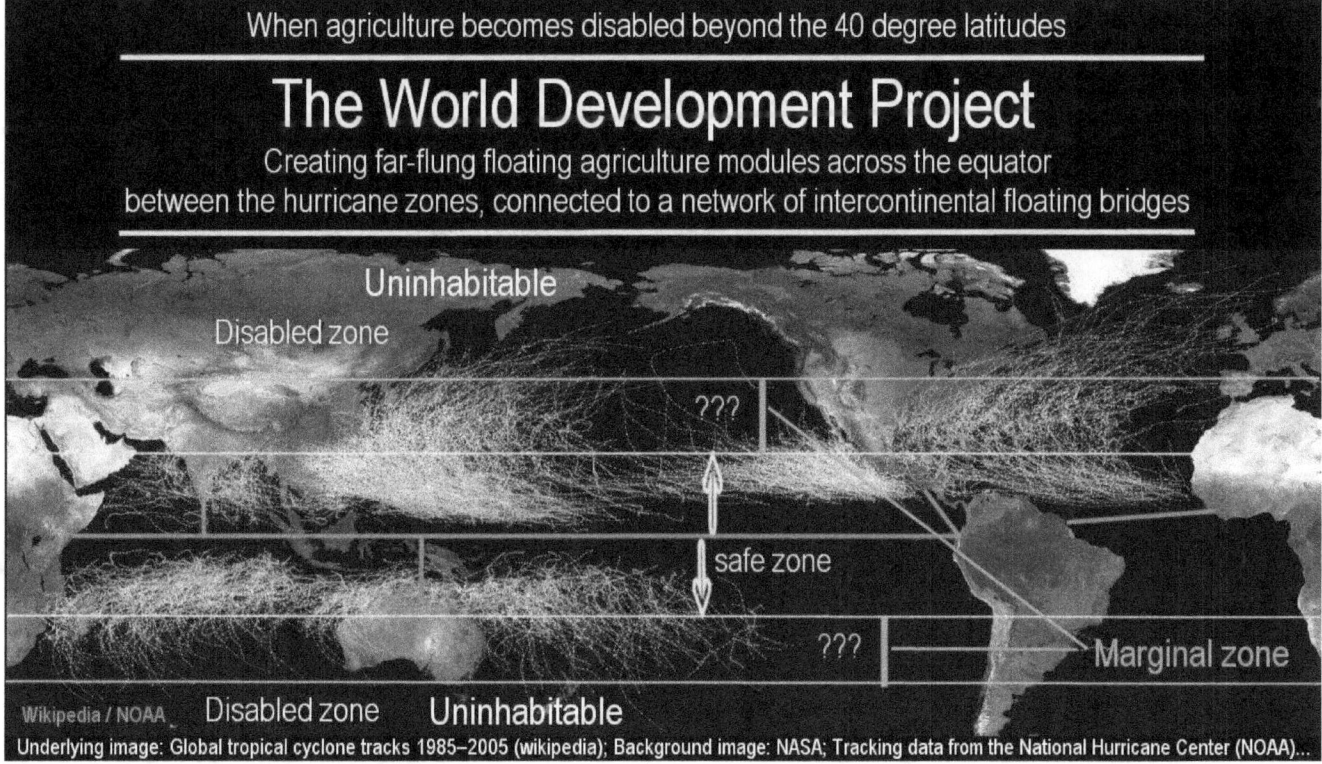

The phase shift in our world, onto an inactive Sun, which could happen in 30 years, does not imply that the end is near for mankind. It means the opposite. It means that we find ourselves impelled to utilize the scientific, technologic, economic, and cultural resources that we have developed, and meet the impending phase shift in the solar system with an equally majestic phase shift of our own, by creating the greatest renaissance of all times, in our time, against which the coming Ice Age has no sting for us.

With this we write ourselves a ticket to live and to have a future, and to live abundantly.

Part 8- The dawn of humanity, civilization, and God

The dawn of humanity, civilization, and God

Only during Ice Age environments

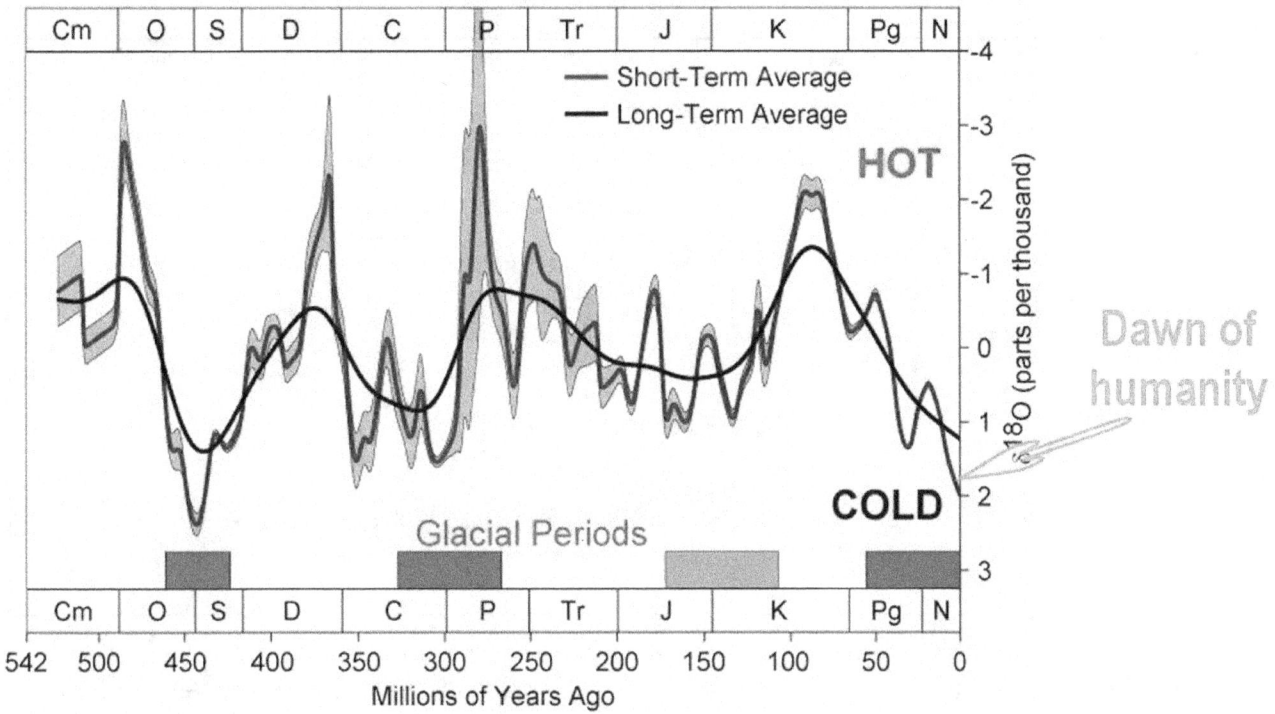

Our development into a highly advanced and capable species with creative and productive powers that no other form of life can equal, may have been made possible by one of the special conditions that exist on the Earth only during Ice Age environments.

The development of the human species didn't even begin

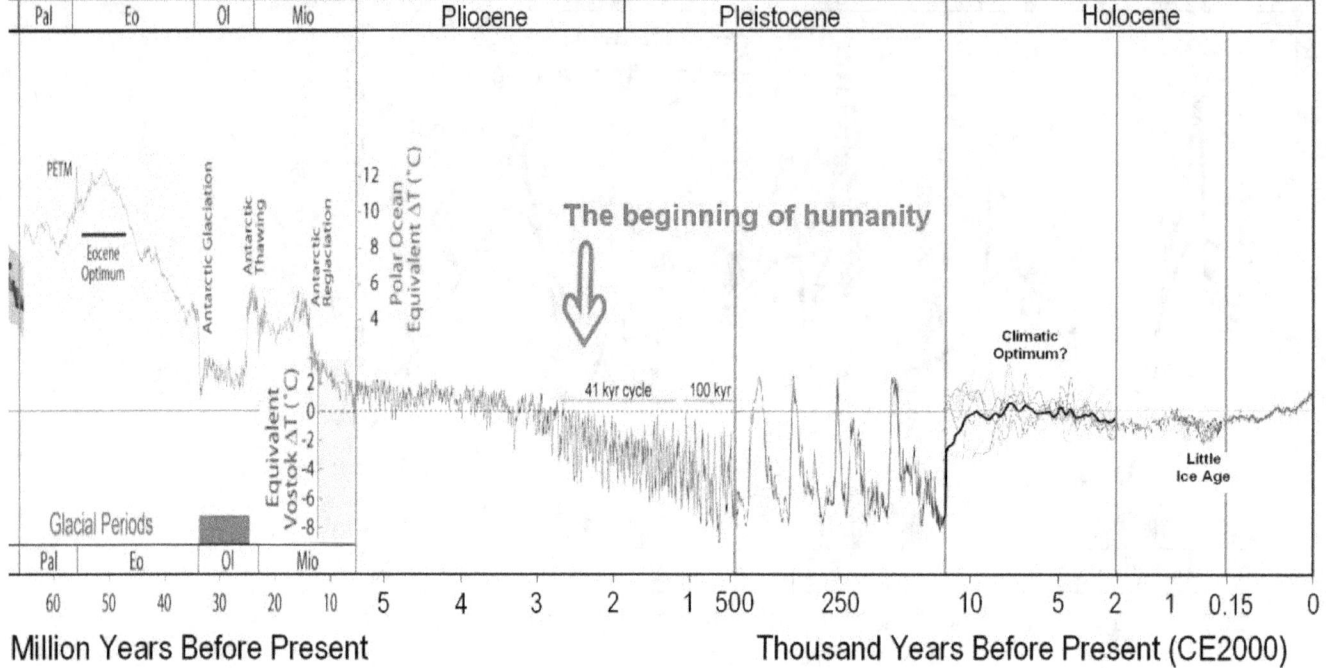

The development of the human species didn't even begin until the modern ice age epoch began roughly two million years ago.

The critical factor that dominates the timing

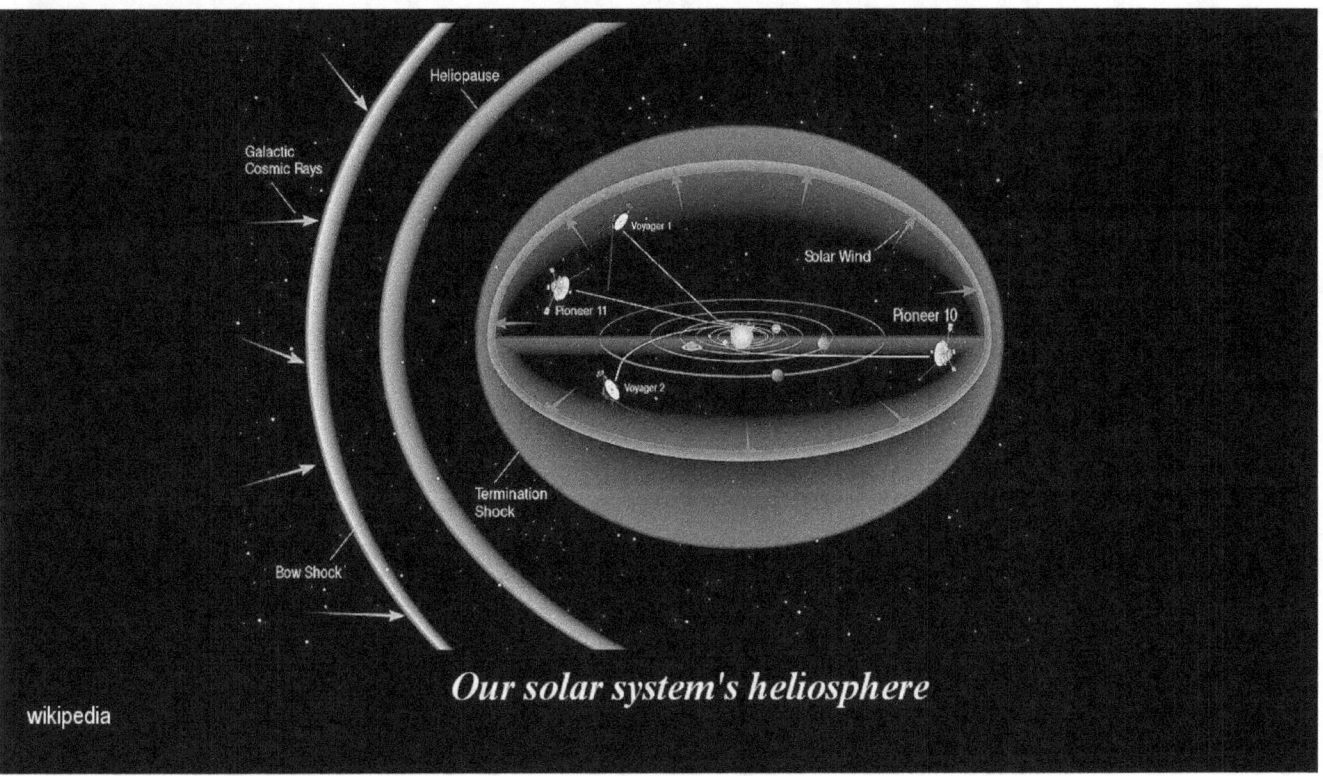

Our solar system's heliosphere

The critical factor that dominates the timing of the development of humanity, is evidently the increased galactic cosmic radiation that results during glaciation conditions when the shielding effect of the heliosphere vanishes. When the Primer Fields collapse, the spherical shell of plasma that the solar winds create around the solar system, will vanish in short order. The plasma shell is called the heliosphere. It forms when the solar winds grind to a halt, far from the Sun, whereby the plasma of the solar winds accumulates from within, into a relatively dense shell.

Without the heliospheric plasma shell surrounding the solar system, the full force and volume of the galactic cosmic-ray flux will then be able to penetrate to the Earth.

The penetration of the cosmic-ray flux, and its interaction with our biological systems. appears to have a beneficial effect.

A type of nourishment for mental development

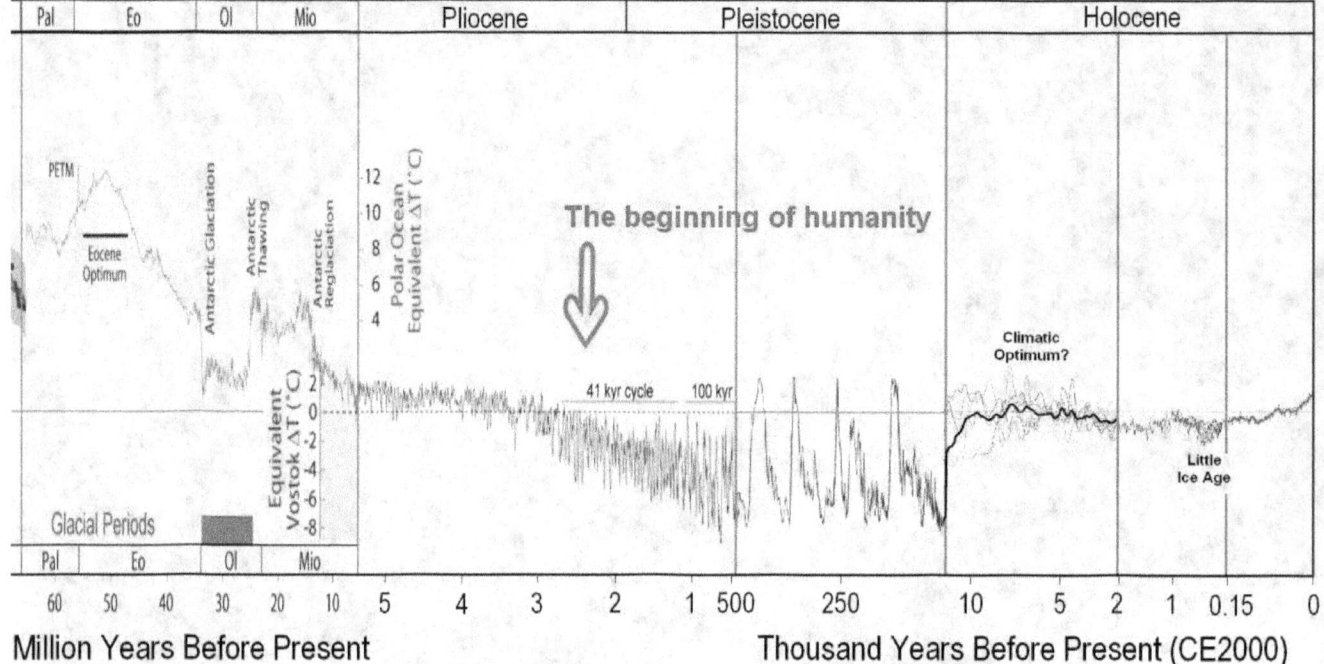

This means that the universe has given us a great gift by providing two distinct environments that in conjunction have aided the development of humanity on two critical fronts alternately. The glaciation periods furnish an environment that provides a type of nourishment for mental development, with a high rate of galactic cosmic-ray flux, while the interglacial warm periods furnish an environment for easy living with plenty of food available, under a brilliant Sun. Both aspects appear to be needed for the advance of human development. It may be that the dawn of humanity did not begin until both conditions were established.

The critical conditions simply didn't exist until the ice ages began.

Cosmic-ray particles

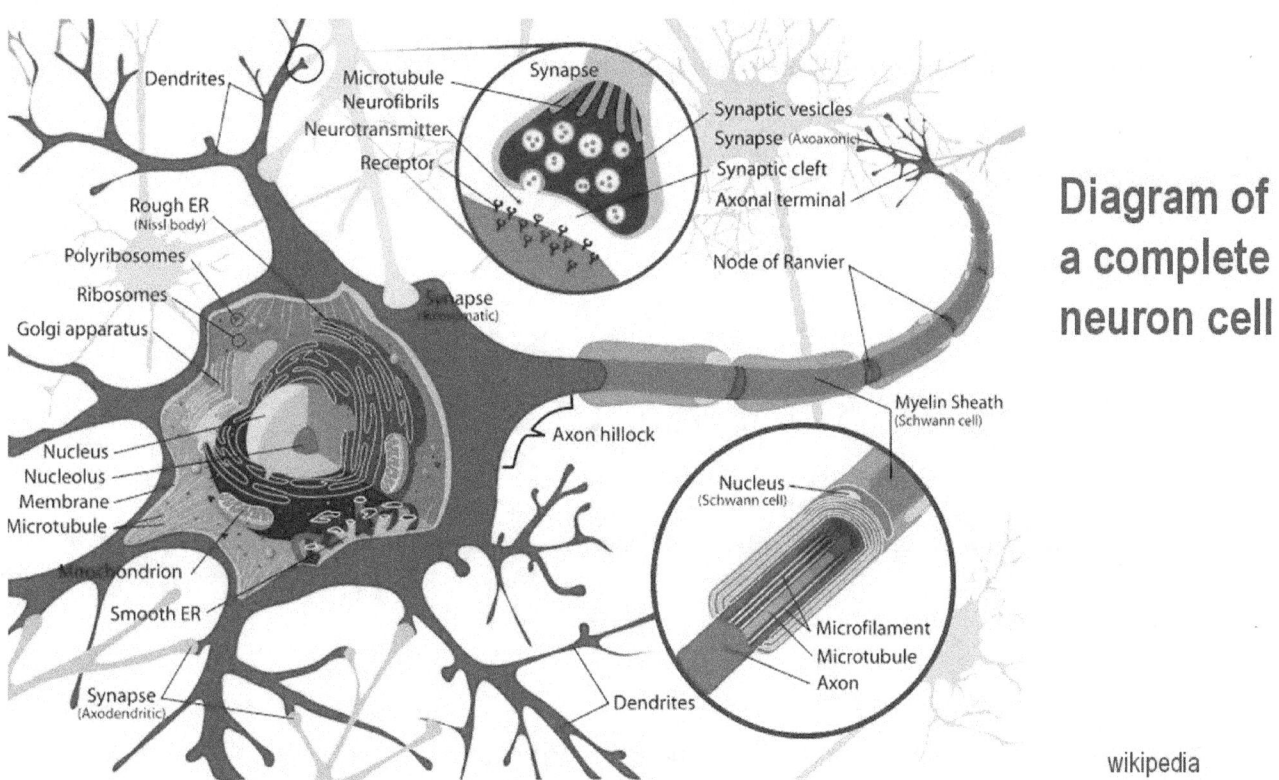

Diagram of a complete neuron cell

wikipedia

As I said before 'cosmic rays' are not 'rays' in the standard sense, like rays of light, but are high-energy electrons and protons in motion at upwards to the speed of light. With the cosmic-ray particles being electrically charged, they typically pass right through our biological systems without damaging anything. However, as they path through the biological system, they generate electric currents by induction that appear to be beneficial for the high-level neurological functioning that governs the complex human biology. They tend to facilitate critical functions that apparently might not occur otherwise.

Our biological and neurological systems, are extremely complex electric systems, which evidently benefit from increased electric activity, especially so when it is extended over long periods of time. Our cognitive, scientific, and what may be called, spiritual powers, appear to be derived from this high-level-type of electric nourishment. The cosmic-ray interaction may have enabled cognitive powers, and then aided the continuing development of them, by which we became a high-level species.

The interaction between cosmic-ray flux and human development

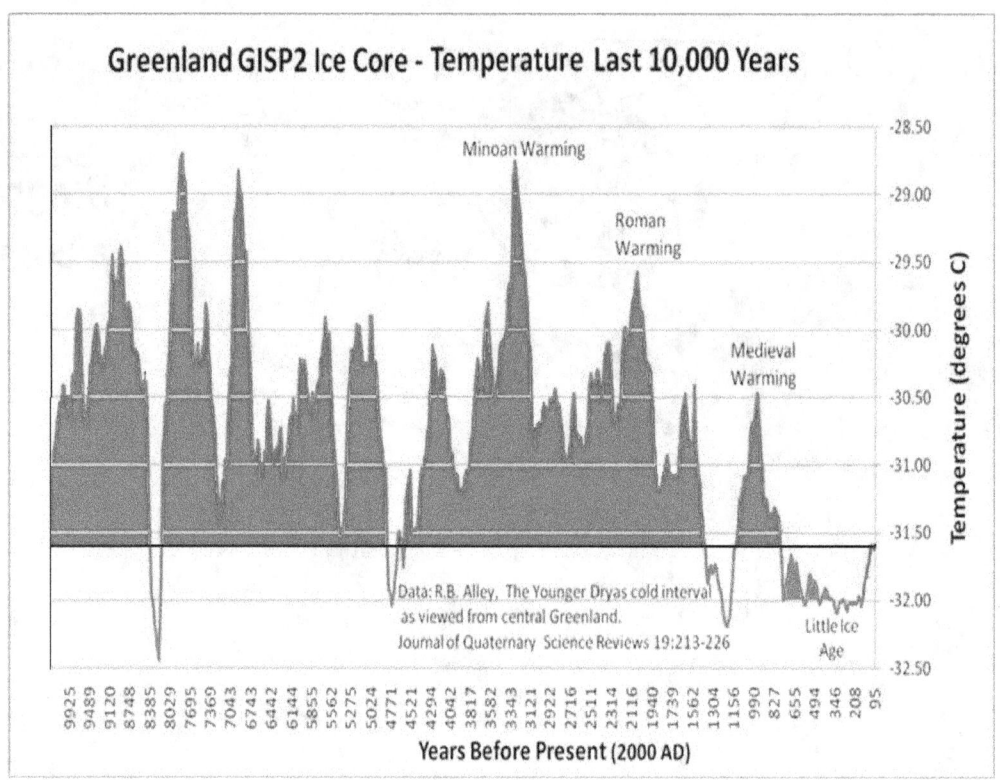

The interaction between cosmic-ray flux and human development is surprisingly evident when we look back into the history of civilization as it developed in the current interglacial period where some cosmic-ray flux is received from the Sun, primarily during the cold periods.

Since it became possible with Carbon-14 measurements

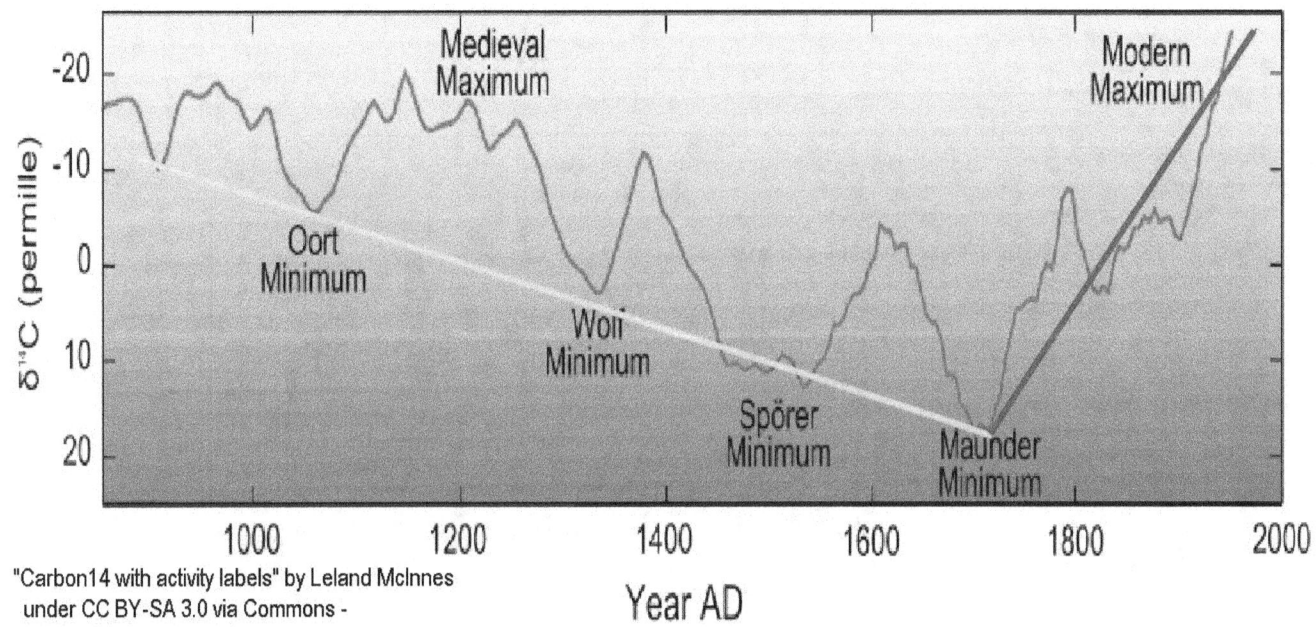

Since it became possible with Carbon-14 measurements, to measure the historic volumes of the solar cosmic-ray flux, and it became self-evident that this volume varies dramatically with changing solar activity, which is also mirrored in temperature changes, it becomes possible to correlate cultural effects with changes in solar cosmic-ray flux.

Great progressive developments in civilization

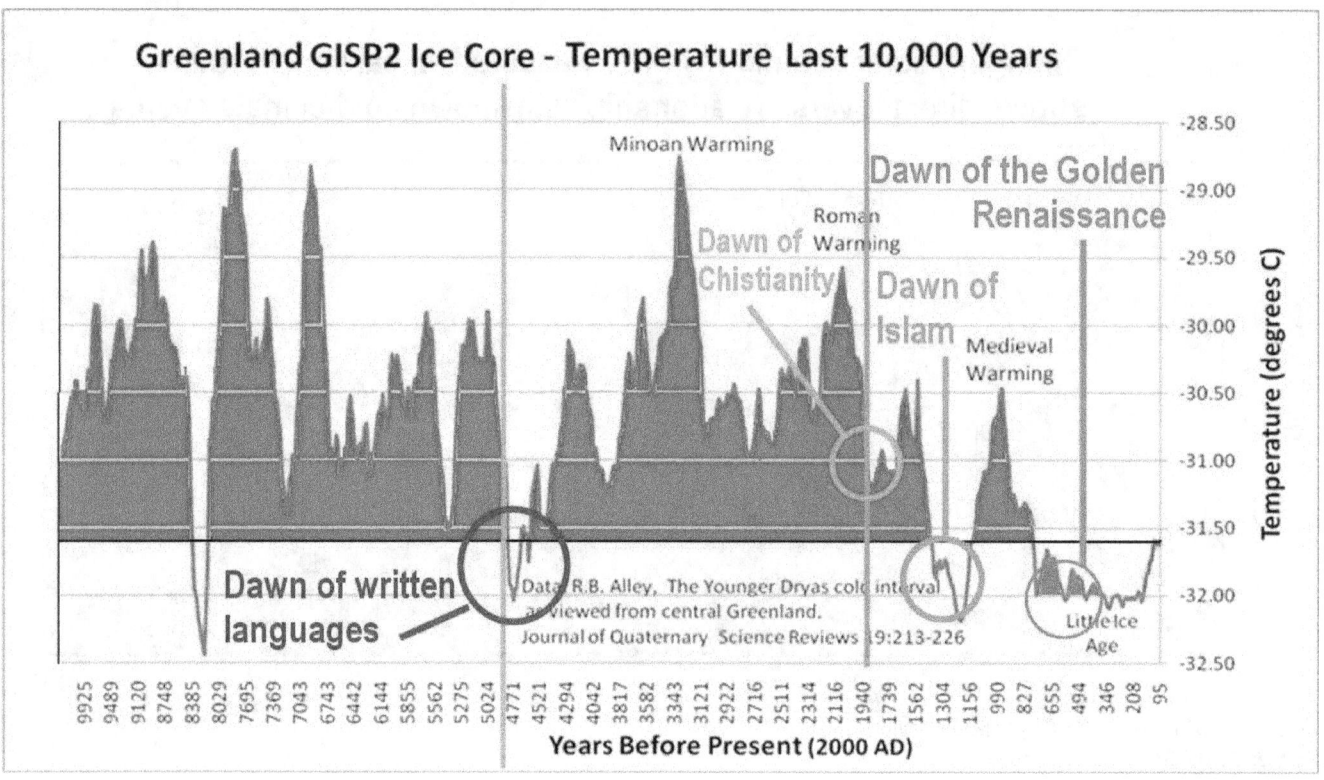

As we make the comparisons, it becomes surprisingly evident that the great progressive developments in civilization all occurred during the cold periods, which are periods of high volumes of solar cosmic-ray flux.

The Maunder Minimum

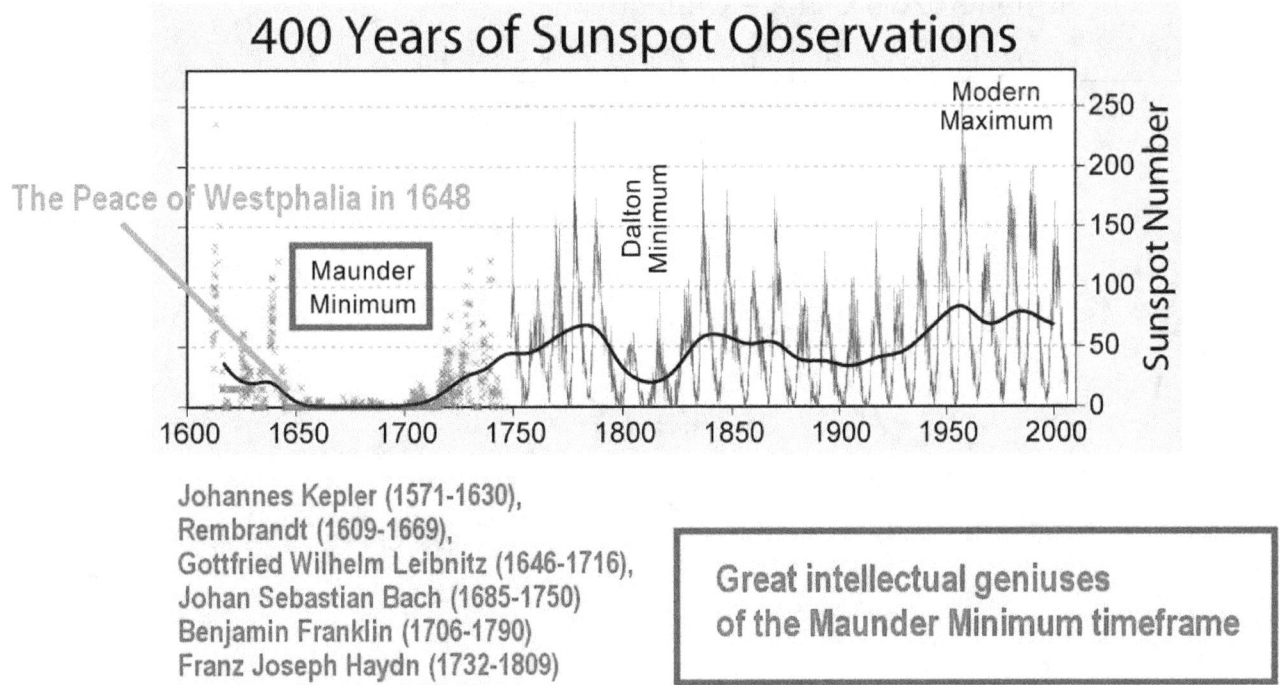

Johannes Kepler (1571-1630),
Rembrandt (1609-1669),
Gottfried Wilhelm Leibnitz (1646-1716),
Johan Sebastian Bach (1685-1750)
Benjamin Franklin (1706-1790)
Franz Joseph Haydn (1732-1809)
Amadeus Mozart (1756-1791)

Great intellectual geniuses of the Maunder Minimum timeframe

For the same reason was the period of the Maunder Minimum of low solar activity - where no sunspots have occurred - which is coincident with the Little Ice Age - a period in which enormous scientific and cultural development has occurred. During the Maunder Minimum, the extremely high solar cosmic-ray flux of that period gave us the greatest peace treaty of all times, the Treaty of Westphalia that still stands as the foundation for civilization. The period also gave us the great musical geniuses and scientific geniuses that are still admired today.

In Palaeozoic history

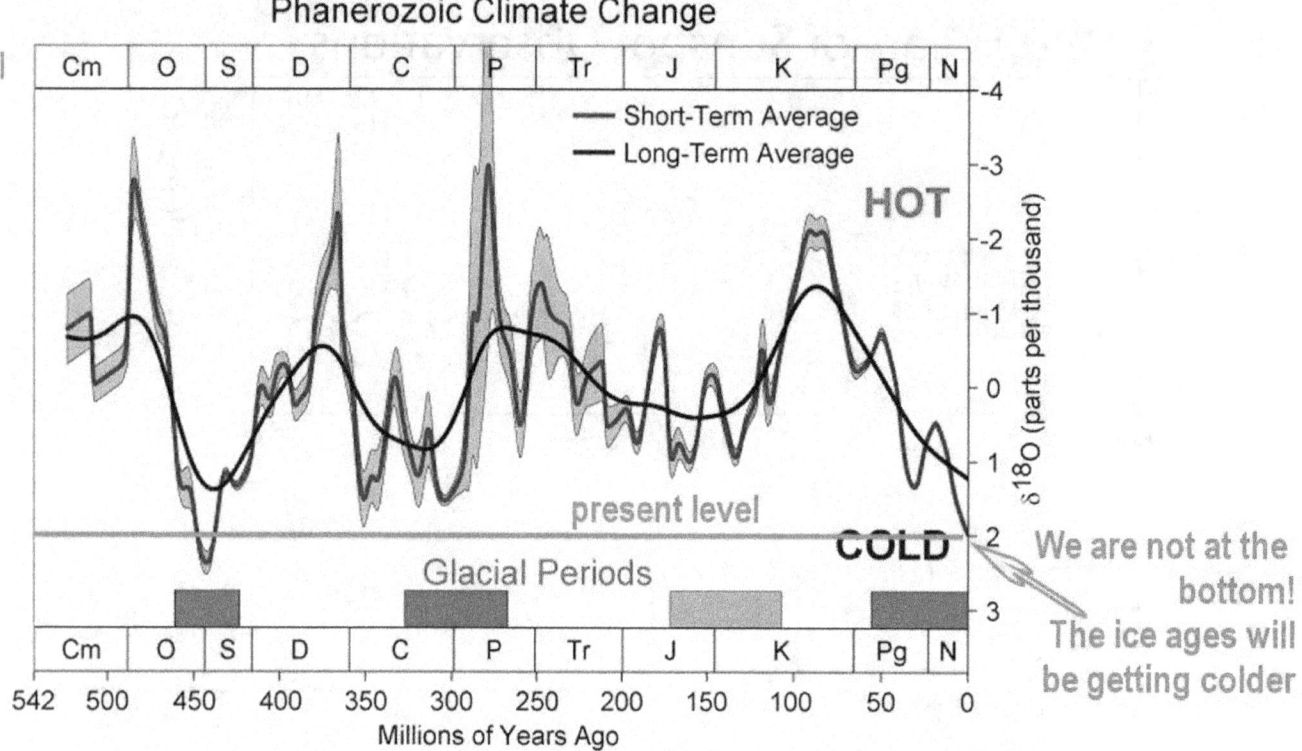

As I said before, the conditions in Palaeozoic history, when the Primer Fields for the Sun collapse and the Sun turns off, are rare. They occurred only four times, as we have records of them, with the earliest having occurred roughly 450 million years ago.

There my have been earlier, deeper glaciation periods between 650 and 750 million years ago, in which the entire Earth froze up into a giant snowball and remained frozen for tens of millions years of until the Sun became reactivated again. While the snowball-earth theory is controversial, it is well within the range of normal possibility in the context of the Primer Fields dynamics.

Rare as the great glacial periods may be, our existence is linked to them. The coincidence of the dawn of man with the modern Pleistocene ice epoch is significant, as it indicates that our very existence as a highly developed species may be the direct result of the potentially very high cosmic-ray flux density that occurs in times when the solar heliosphere does not exist.

Since the normal 'rich' conditioning for human development gets interrupted by the interglacial periods, which is all that we have known, we really don't know then what 'normal' living is like, even as we are about to become drawn into it again.

Increases in cosmic-ray flux,

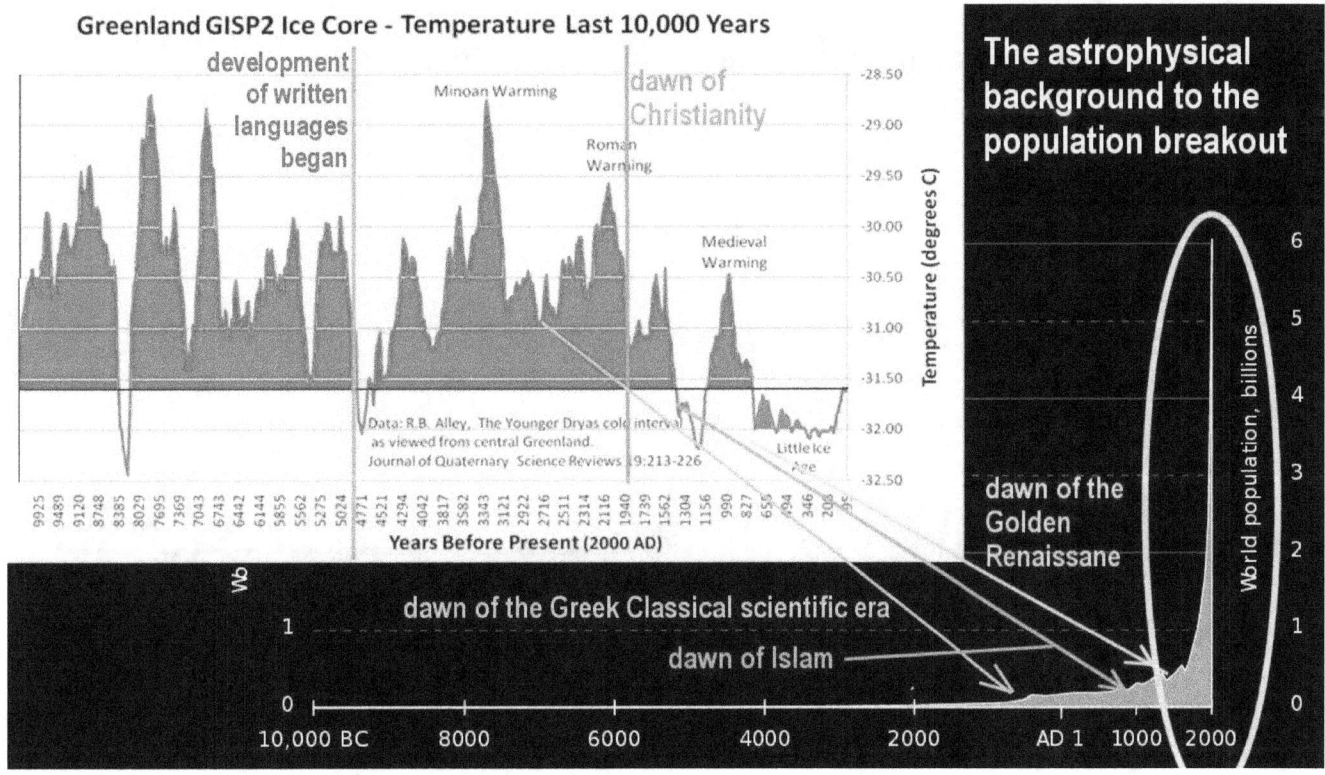

During the lean period of the interglacial, we have seen only occasional increases in cosmic-ray flux, this time coming from the Sun. History tells us that in the few occasions of high volumes of solar cosmic-ray flux, amazing cultural developments occurred. Almost all of the great cultural breakthroughs were made in these types of times. The development of written languages, for example, occurred in one of the deep cold times with high volumes of solar cosmic-ray flux.

History also tells us that the great developments that did occur in these cosmic-ray rich times, were typically cognitive and scientific in nature, which renders them to be forms of spiritual development that may be termed the pinnacle in the mental realm.

In a very real and powerful way, the rise and fall of civilization follows the ebb and flow of the recognition of spiritual values in society.

The rise and fall of spiritual recognition

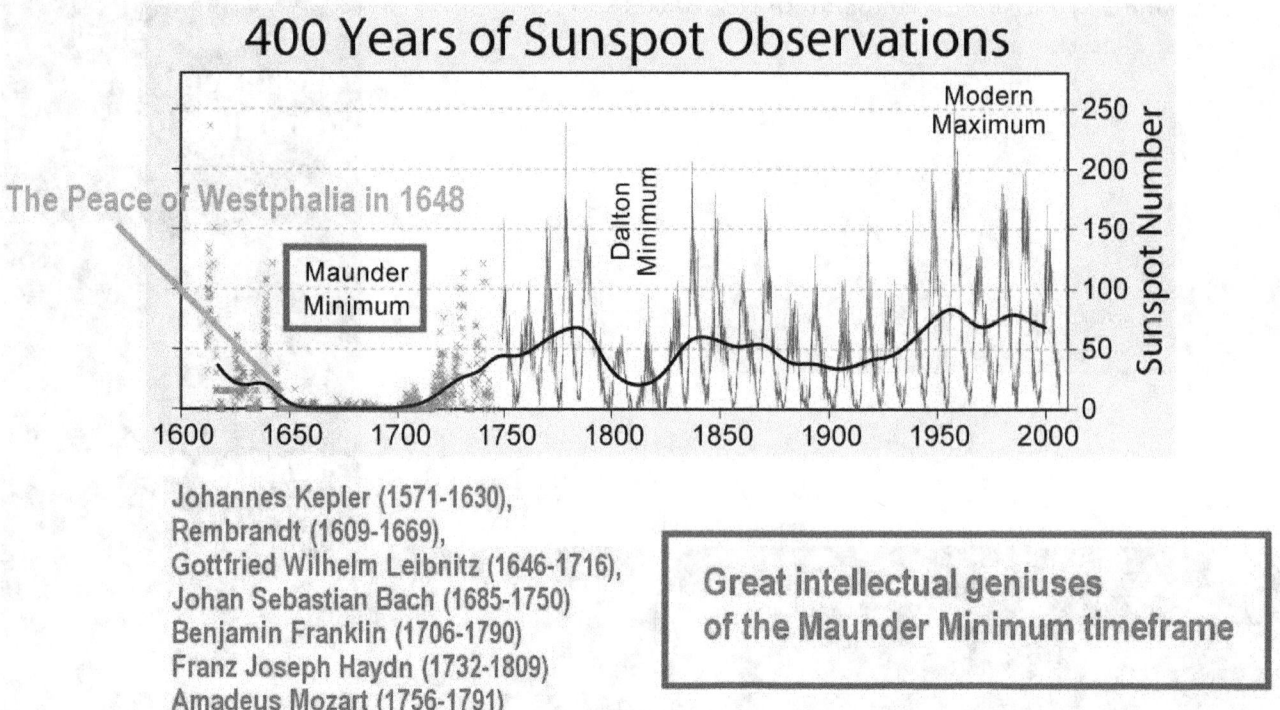

The rise and fall of spiritual recognition in turn, follows the historic rise and fall of the cosmic-ray flux reaching the Earth, which is typically the inverse trend of solar activity.

The warm periods stand out as periods of cultural destruction

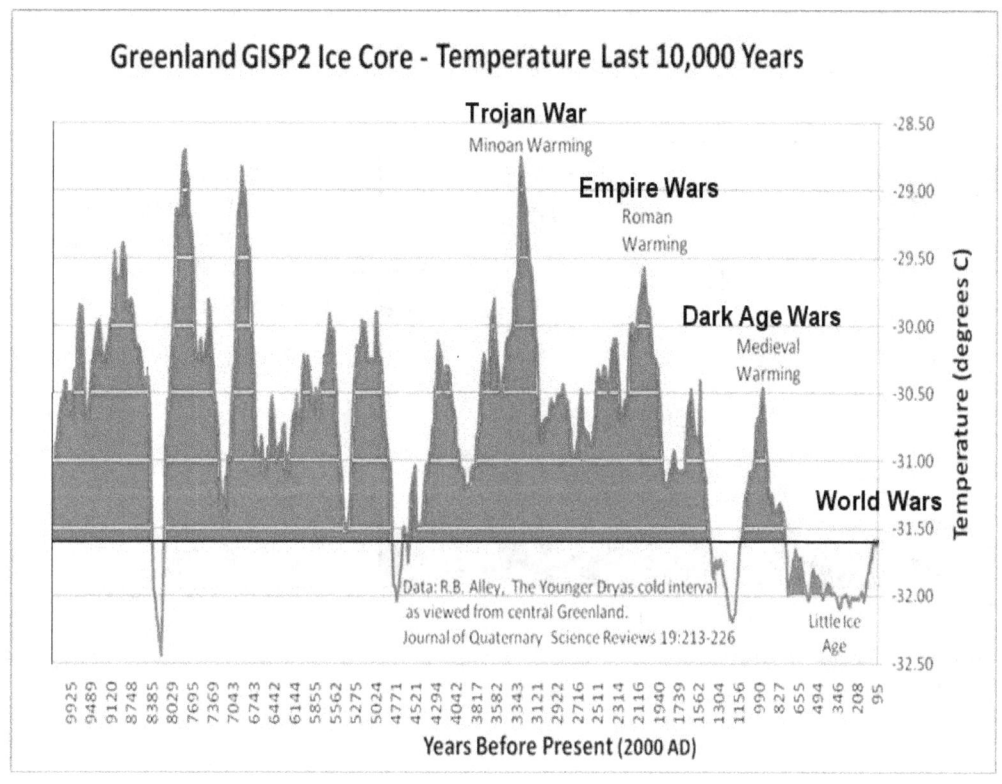

The inverse also proves the principle further. Just as the cold periods were periods of great cultural achievements, the warm periods stand out in the long sweep of history as periods of cultural destruction, periods of war, from the Trojan War, to the Roman Wars, to the Colonial Wars, to the World Wars and so on.

With the higher-resolution Carbon-14 measurements

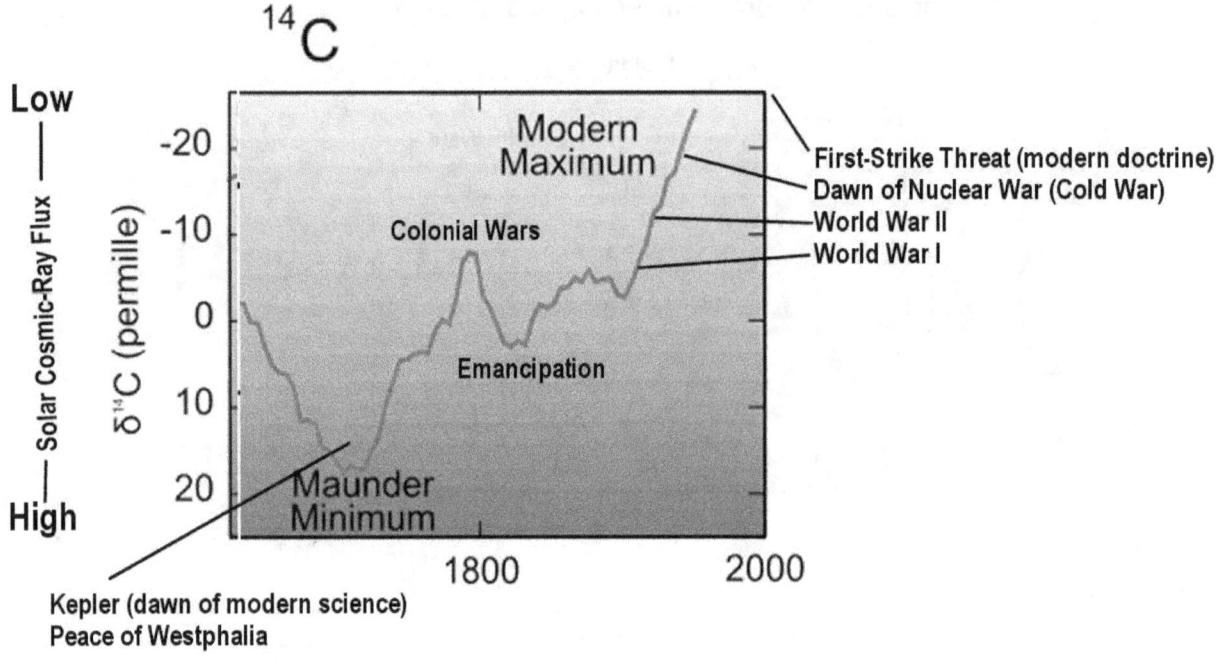

The same correlation is also evident in more recent times, in the context with the higher-resolution Carbon-14 measurements of solar cosmic-ray flux. The diminishing solar cosmic-ray flux from the time of the Maunder Minimum to the strong solar time near the year 2000, has been a period of accelerating collapse in civilization, a period of increasing insanity that became reflected in world wars, nuclear-war terror, and the modern hair-trigger stand-off towards the now fully prepared for, first-strike thermonuclear destruction of all life on Earth, for which the decision time on warning has been shrunk to roughly one minute, and the execution time to roughly one hour.

That's the present state. If this isn't utter insanity, what would qualify for the term?

Beyond the insanity of the Nazi holocaust

We have drifted far beyond the insanity of the Nazi holocaust that had murdered 6 million people in six years of fascist madness. Modern society exceeds this 100-fold. It is murdering upwards to 100 million people per year in the biofuels holocaust, quietly, unseen, with the sword of starvation forged by the mass-burning of food in a starving world. The amount of diverted agricultural resources that goes into the project to be burned, would normally nourish 400 million people. This adds up to supreme genocide in a world that has a billion people living in chronic starvation.

The Sleep of Reason Produces Monsters

The Spanish painter Francisco Goya illustrated the connection in his etching 'The Sleep of Reason Produces Monsters', of the Caprichos series of 1798.

It appears, by what we see happening all around us, that the pinnacle of our humanity, our cognitive and spiritual capability, and also the lack thereof as the case may be, is deeply affected by the every-changing cosmic-ray flux that the Primer Fields stand in the background to as an element of the larger scene.

The future of humanity is not inherently bound to the cosmic default

Francisco Goya (1746–1828) - Disasters of War series - Wikipedia

However, the future of humanity is not inherently bound to the cosmic default for the mental environment that the cosmic-ray environment dishes up. The sleep of reason is not our inevitable fate. The sleep can be ended. The age of reason can be restored.

Humanity has in its path become a highly developed species

The School of Athens - fresco by Raffaello Sanzio (1511) at the Vatican - Wikipedia

Humanity has in its path become a highly developed species with far greater capabilities than society gives itself credit for.

Of course it takes some active commitment by society to raise itself above the 'sod,' and consciously become that human factor in the world that it has the potential to be. There is no need for a society of human beings to allow events to grind it down, when it has the power to generate its own events. Nor is there a need for humanity to wait for changing conditions to raise it up, when it has the potential to raise itself up, to create its own conditions, and to make this happen decisively.

This means that no huge feat is required for humanity to free itself

Castle Bravo - the first U.S. test of a dry fuel thermonuclear hydrogen bomb - March 1, 1954 at Bikini Atoll, Marshall Islands

This means that no huge feat is required for humanity to free itself from the threat of thermonuclear annihilation, that is now fully prepared for. The feat to end this impending Armageddon can be accomplished in a week. It would take less than a day to remove the trigger-happy rulers from power who wield the club of war and terror, and own the button for Armageddon. After that is done, it wouldn't take more than a week to physically disable and dismantle the entire nuclear-war machine anywhere in the world. To accomplish this is not a huge physical task, and by taking these steps, humanity would write itself a ticket to have a future, which presently isn't even a concept anymore.

The biofuels holocaust can be stopped in a similar manner

The biofuels holocaust can be stopped in a similar manner, and just as fast. All the doctrines that stand behind the holocaust are built on lies, from the manmade global warming doctrine to the depopulation doctrine. The mass murder in the world can be completely ended in a month and sanity be restored again. No gigantic feat is required for society to rebuilt its humanity that way. On this path, society would gain more than just a little self-respect.

Only the Ice Age Challenge cannot be so easily met

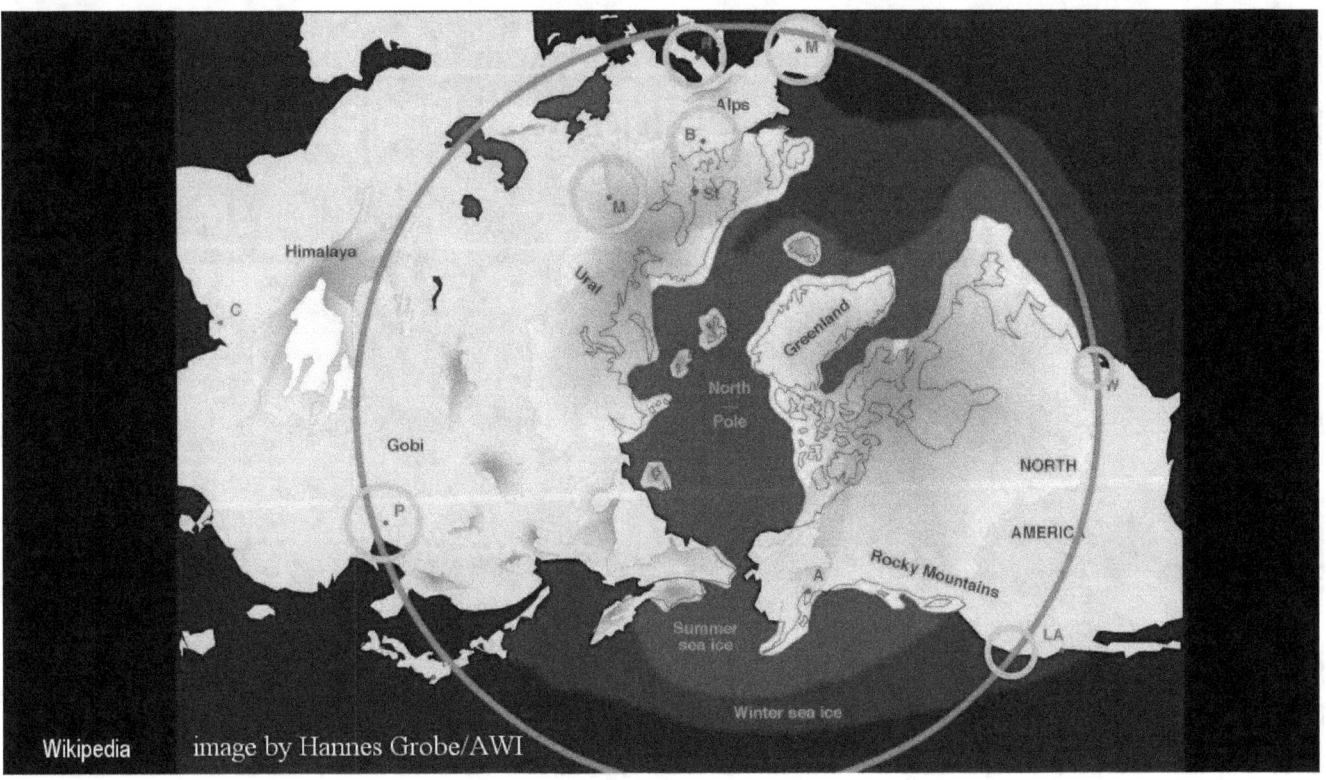

Only the Ice Age Challenge cannot be so easily met. It will take 30 years for society to build the infrastructures to relocate all countries outside the tropics, into the tropics, before their territories become uninhabitable.

The critical breakthrough can be made in a short time nevertheless

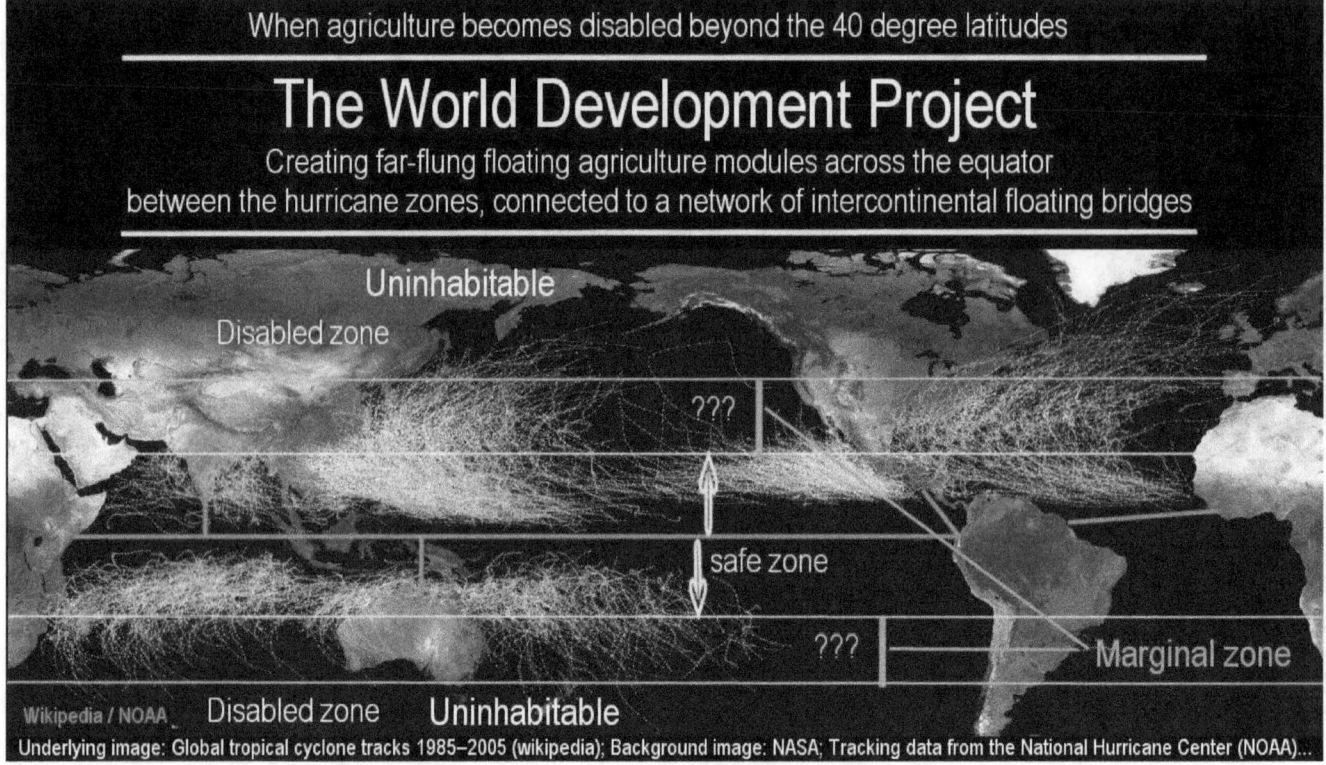

However, the critical breakthrough that is needed to get the ball rolling, can be made in a short time nevertheless. No miracles are needed. The evidence is plain. The imperative is hugely impelling. The power to meet the challenge exists in society. In taking up the challenge and moving with its imperative, humanity would give itself and its children a chance for certain survival. The concept, presently, appears to be almost non-existing.

The truth is dead

As Goya had put it in his three-part etching, "The truth is dead"

But will she rise again?

"But will she rise again?" he asks.

This is the truth

"This is the truth," he says. She is the truth that defines us all, and this truth is alive. It is us, potentially. More than this we do not need.

Part 9 - The Giza pyramids and Stone Henge

The Giza pyramids and Stone Henge

Grasping the fire

The development of humanity appears to have always been intertwined with 'grasping the fire,' both mentally and physically. On this road humanity wrought many great wonders throughout the long history of its self-development.

Some of the wonders stand as giant structures

Some of the wonders stand as giant structures that still puzzle us today. How were they built? What where they built for?

In asking these questions

In asking these questions, some of the ancient structures that we marvel at, ironically, also come to light to serve as a link to the future. They give us a hint of what the ancient builders may have seen in the sky, which can no longer be seen, because the modern age unfolds in an electrically collapsing solar environment that is trending towards its impending solar cut-off point.

The builders of the pyramids

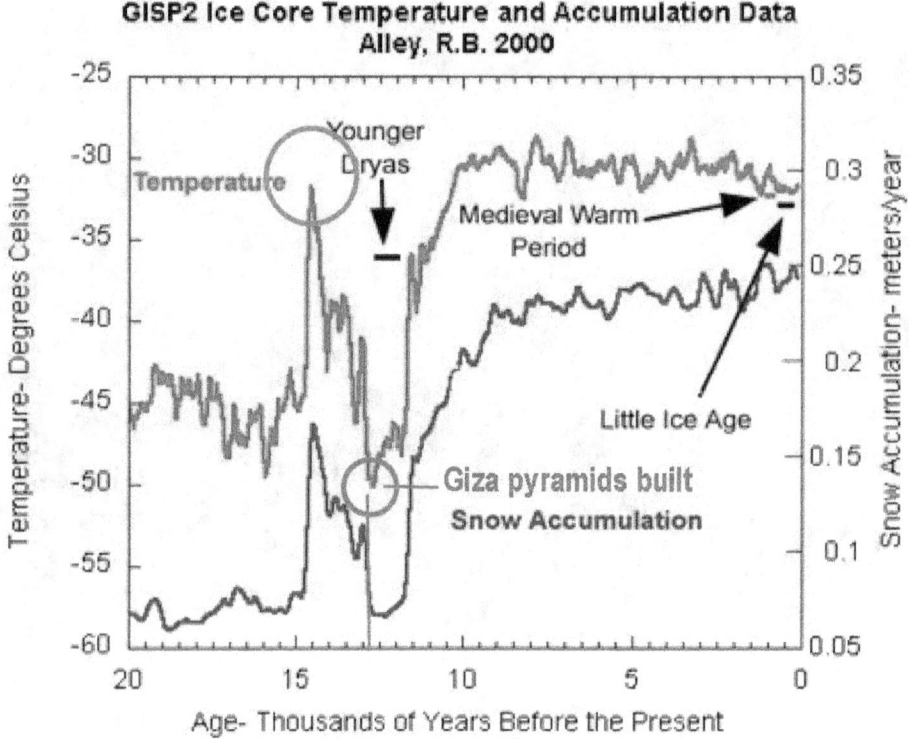

The builders of the pyramids may have experienced both of the solar extremes: A more brilliant Sun than we have today; and the inactive Sun that would have created the Younger Dryas period of renewed glaciation, in which researchers suggest the Giza pyramids have been built.

Built 12,800 years ago, and not as tombs

Contrary to assertions by Egyptology, the Giza pyramids were likely built 12,800 years ago, and not as tombs, for which they were later used. The distant date is derived by running the astrophysical clock backwards until a point is reached at which the Sphinx of the Giza complex is seeing its own image in the sky on a solstice sunrise.

The complete absence of inscriptions

Grand gallery of the Great Pyramid of Giza

The suggested date seems to be confirmed by the complete absence of inscriptions inside the pyramids. Inscriptions became customary much later, during the time of the pharaohs. Numerous other features likewise place the Giza pyramids far outside the timeframe of the pharaohs.

One of these features is the quality of their design, and the precision in construction. These place the Giza complex distinctly into a category of its own that nothing which the Egyptians had produced, and had been capable of at the time, comes even close to.

Built around astrophysical phenomena

Assuming that the pyramids were built 12,800 years before the present, which is the most reasonable estimate to date, what would have been the motive for the people at the time to build the giant structures that they built?

It appears that the motive may have been religious in a sense, while it was built around astrophysical phenomena.

Images of the Primer Fields in action

During the great warming period, 2000 years previous to the building of the pyramids, a very large Dansgaard-Oeschger event broke the deep chill of the last ice age. At this time when the Sun became extremely active, the people may have seen with their naked eye the plasma images of the Primer Fields in action, centered on the Sun.

The type of image that we see today in the Red Square nebula

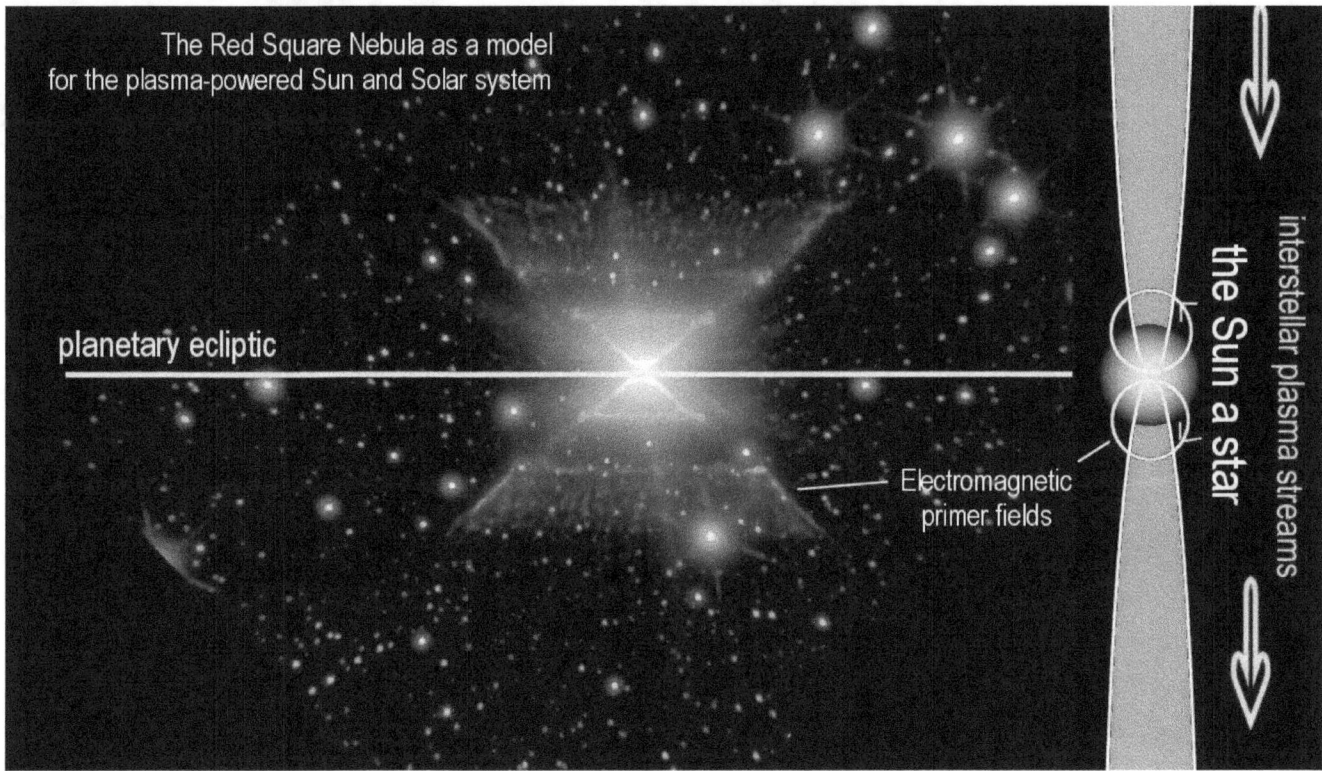

The type of image that we see today in the Red Square nebula may have been seen 'bright' and clear in the ancient skies. The full image may have been seen in the time of a solar eclipse, while a part of the image may have been 'blazing' in the night skies from dusk to dawn.

The people may have seen in the sky two pyramid shapes of light, with the Sun between then. On this basis the Sun may have became an object of worship - the giver of abundant life resting on a pyramid. Then, suddenly, the Sun becomes inactive again. The idea must have emerged at this time to build a pyramid for the Sun - a house for the gods, provided on the Earth

When a thousand years later the Sun became brilliant once more and soon thereafter went inactive again, a deep crisis may have erupted that inspired the building of the greatest pyramid of all times to entice the Sun god to restore its shiny face for his people.

The famous three stars

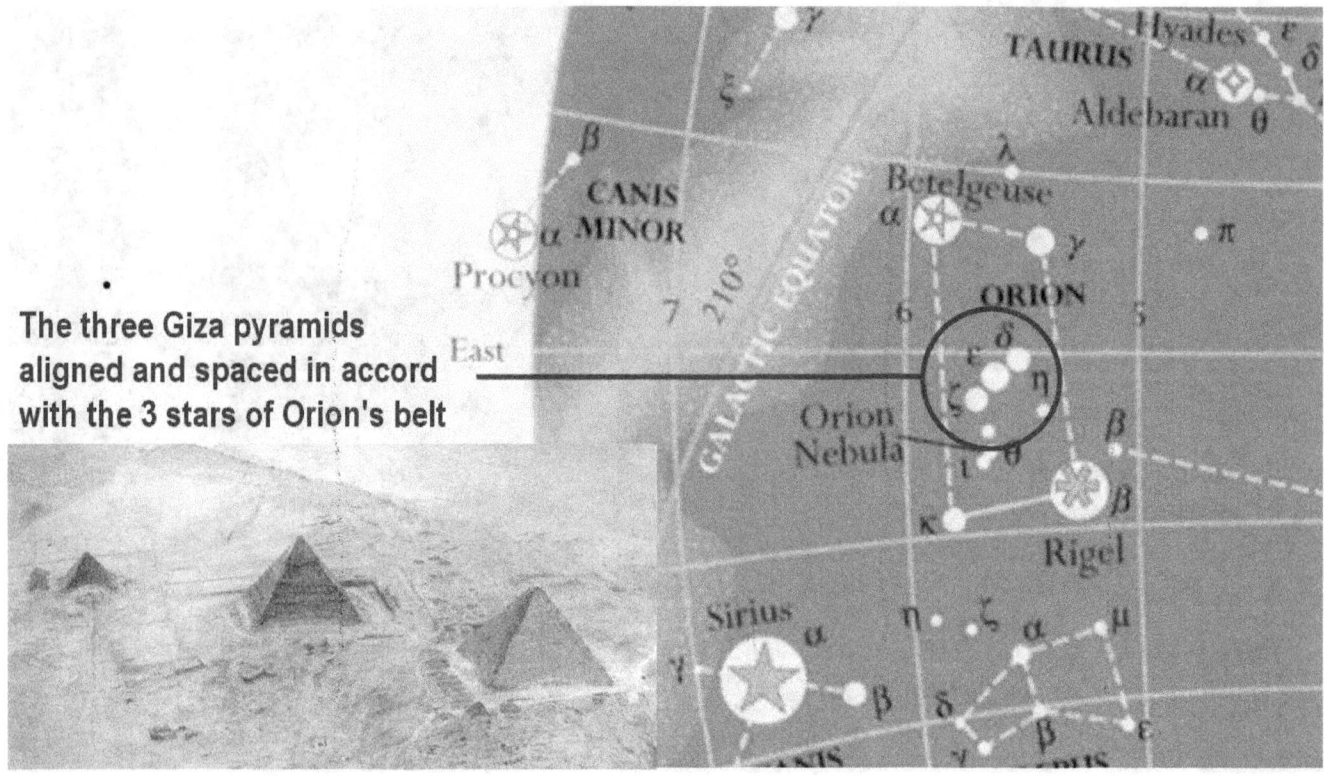

The people may also have seen among the plasma images in the sky, perhaps right at the center of them, the famous three stars of the belt of the constellation Orion.

Part 9 - Giza pyramids and Stone Henge

Three stars determined the pyramids

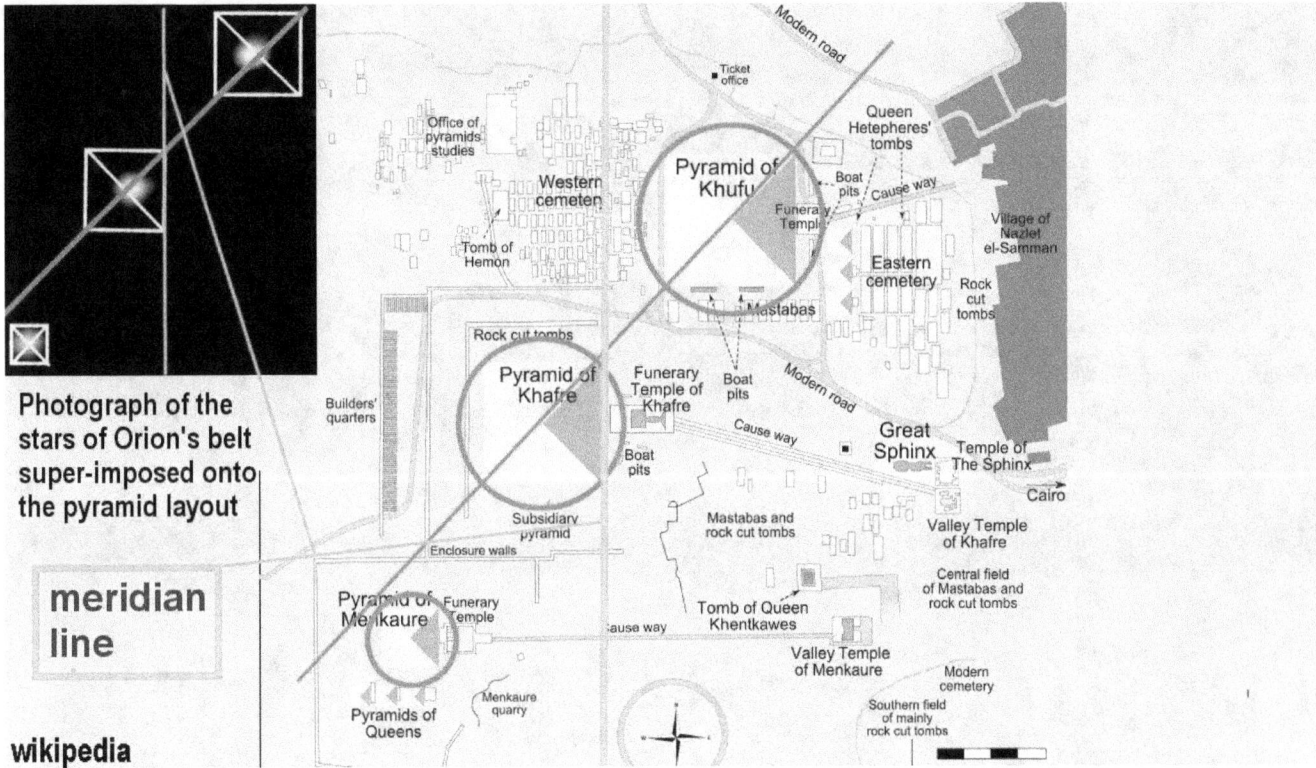

These three stars appear to have determined the positions of the pyramids, their alignment to each other, and their relative size.

Three pyramids as a single project

When the great pyramid building project was launched, it appears that the ancients did not built not just one pyramid, to inspire the gods, but have built all three pyramids together as a single project, with each pyramid being exactly aligned and proportioned in accord with the three stars of the belt of Orion.

A silent testimony

In this manner the pyramids of Giza stand for us as a silent testimony that the brilliance of the Sun is not a permanent feature, but is a fleeting phenomenon at times, which once had inspired a monumental response by a 'nation' to get the shiny face of the Sun back.

Another significant story

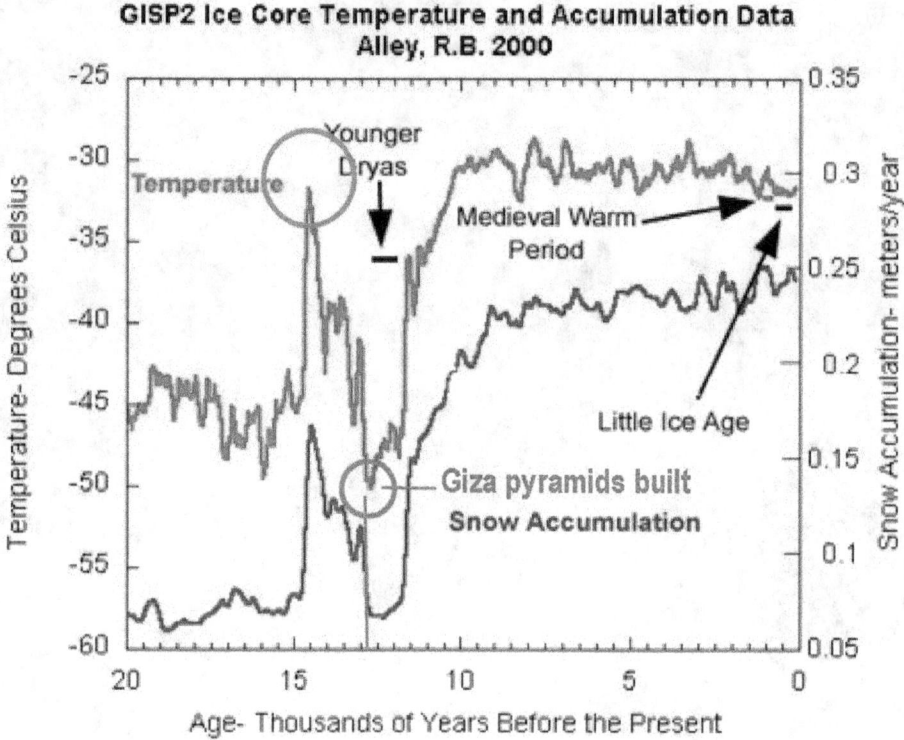

The pyramids also tell us another significant story. The evident fact that the Giza project was built 12,800 years ago is important, as it tells us that the region in which it was built was biologically sufficiently strong to support a large population.

Giza is located near Cairo

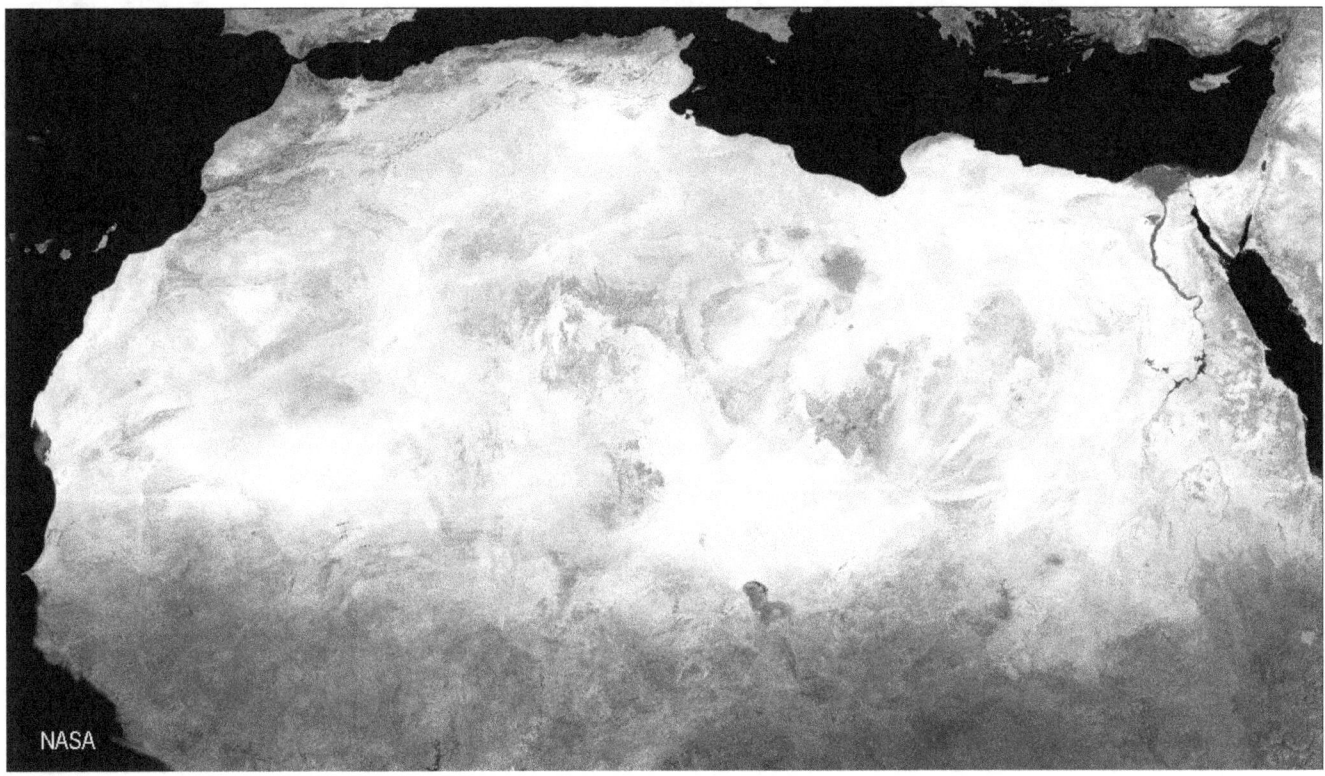

Giza is located near Cairo on the thirty degree latitude in the northern part of the Sahara.

The Sahara was a lush region

Petroglyphs indicate that the Sahara was a lush region before it became drowned in sand, perhaps by a swarm of comet fragments that became electrically fractured into sand while strafing the ionosphere, or even the atmosphere, in an event that no one lived to tell about.

Northern boundary of the ice age safe zone

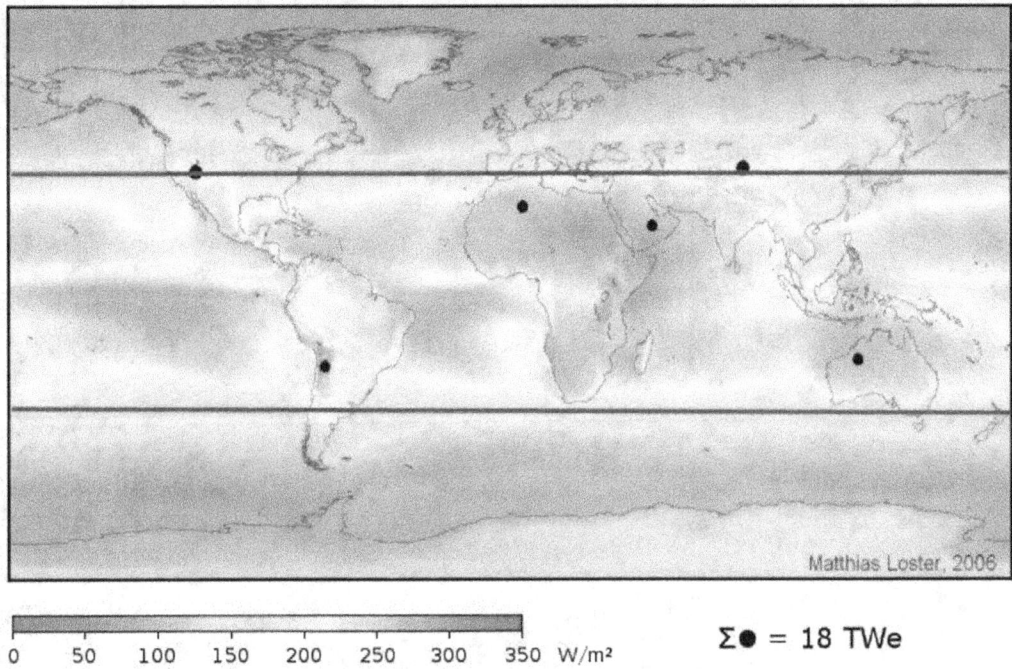

Since the Sahara was lush before this time and supported a civilization during the times of the inactive Sun, the northern boundary of the Sahara appears to be also the northern boundary of the ice age safe zone, coinciding with the thirty-five degree latitude.

The safe line

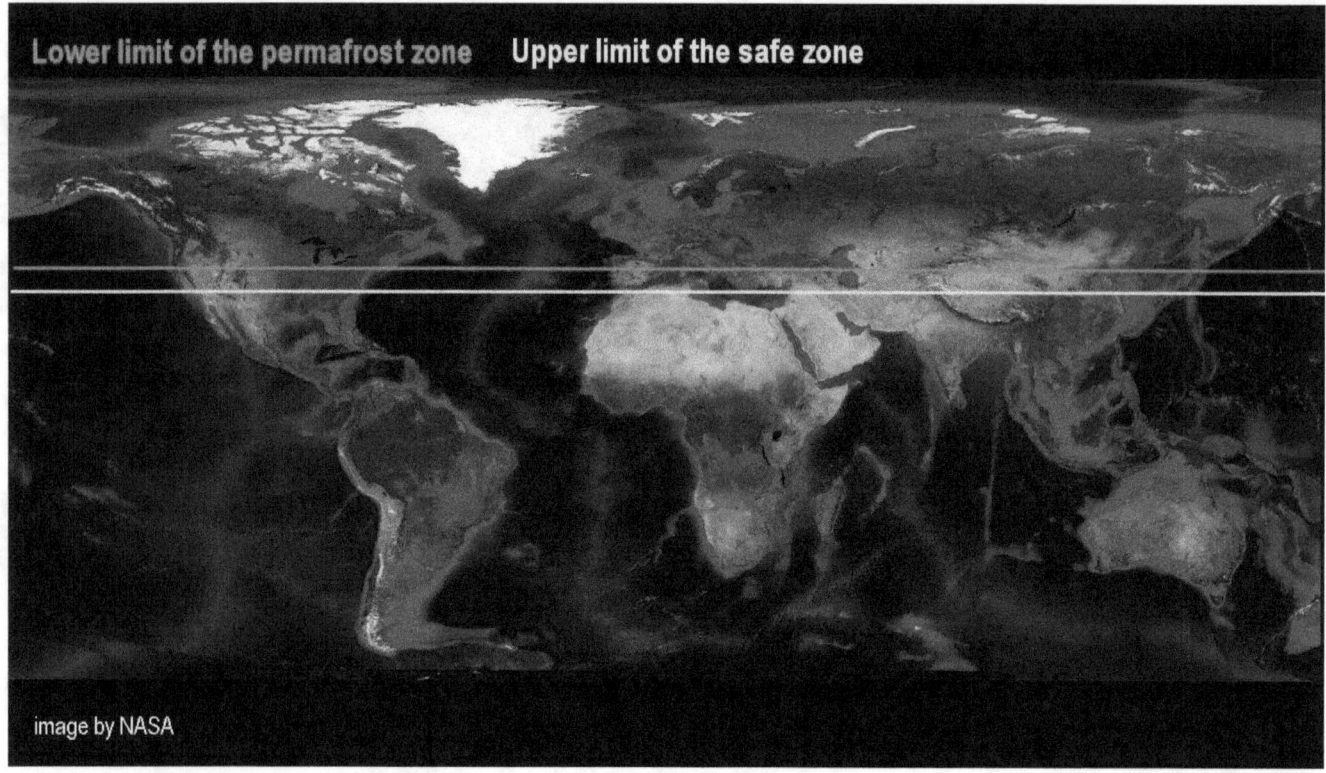

The safe line stretches across California in the USA, north of Los Angeles, and from there through the middle of the Mediterranean Sea, and in the East it cuts through the middle of China near Lianyungang, and through Japan south of Tokyo. Anything north of that, typically along the 40 degree line, from Beijing to Madrid to Philadelphia, was permafrost country during the last ice age.

This means that large portions of the Earth become 'difficult,' if not impossible, to live in during the solar inactive times. This includes primarily Russia, China, Canada, the USA, and Europe, which share a common problem, and thereby are natural partners for building the required solution.

A perfect zone free of hurricanes

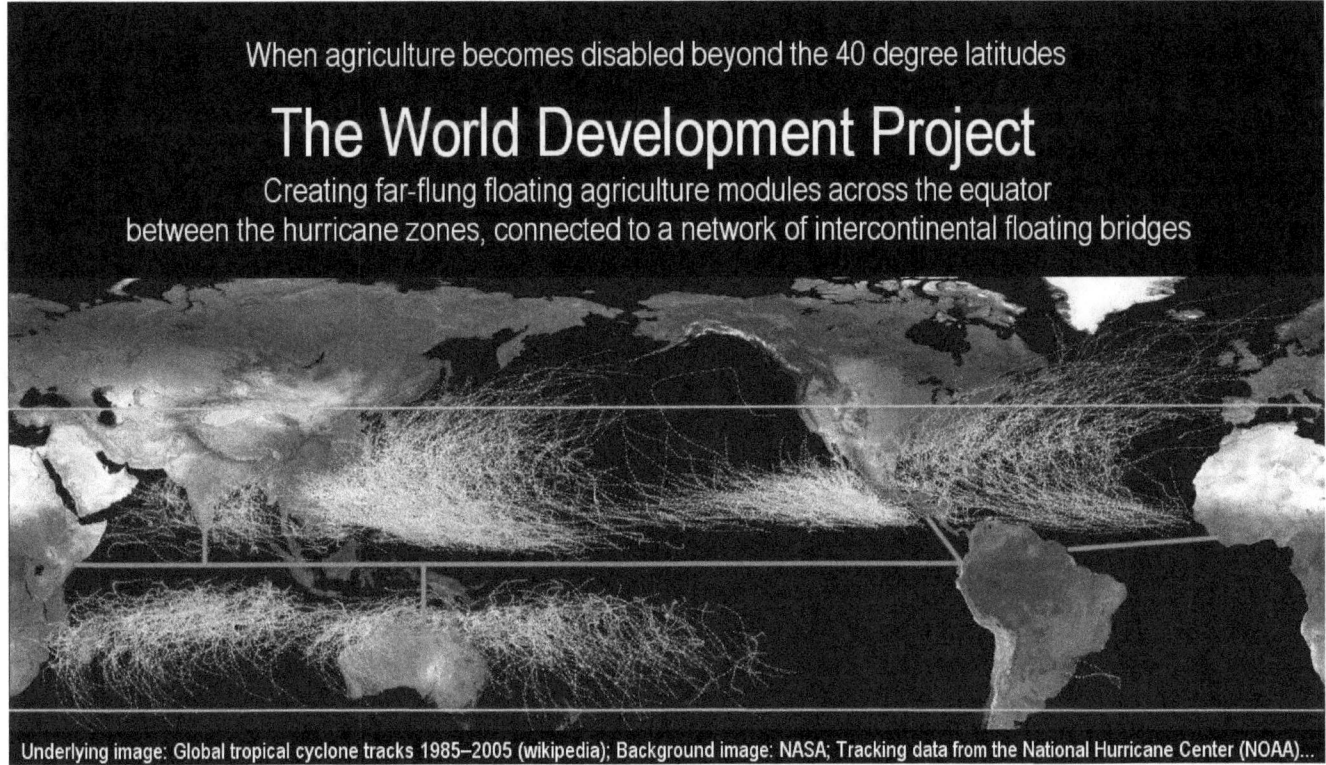

If the solution requires to place agriculture onto the sea, since land is scarce in the tropical areas, a perfect zone exists for doing this right along the equator. A narrow zone exists along the equator that is free of hurricanes. Placing agriculture onto floating modules made out of basalt and produced in nuclear powered, automated industrial processes, is probably more easily accomplished than greening the Sahara in laborious manual operations. The same is evidently also true for the automated production of complete housing-modules, as a necessary and free infrastructure to mobilize our humanity.

Some protest here, that this is too hard to do.

Floating agriculture is small stuff

If the people of a small culture, as far back as 12,800 years ago, were able to mobilize the economic resources to cut 4 million stone blocks out of the bedrock, weighing several tonnes each, transport them to the building site, and to place them with precision on a structure with a steep slope reaching 480 feet into the sky, in comparison with that, the building of bridges across the oceans with floating agriculture and floating pre-manufactured cities, is small stuff, considering the power of high-temperature automated industrial processes, motivated with nuclear power. A single 1 gigawatt plant should be able to produce 2,000 housing units an hour. We can have a whole new world coming online on this basis with comparatively little effort. No one needs to starve or perish when the Sun becomes inactive.

That's what the pyramids are telling us.

The future of Canada, Russia, and Europe

This means that the future of Canada, Russia, and Europe is logically located on the sea along the equator, and to a lesser degree the future of China and the USA. It also means that we better get busy soon, to transform our world.

The Stone Henge project

Another large construction project from ancient times tells us a similar story. The Stone Henge project, like the pyramid building project, reflects features that the ancient builders must have seen in the sky, as the features that were constructed, replicate critical aspects of high-power plasma physics that were only recently discovered in laboratory experiments.

A ring of 56 chalk pits

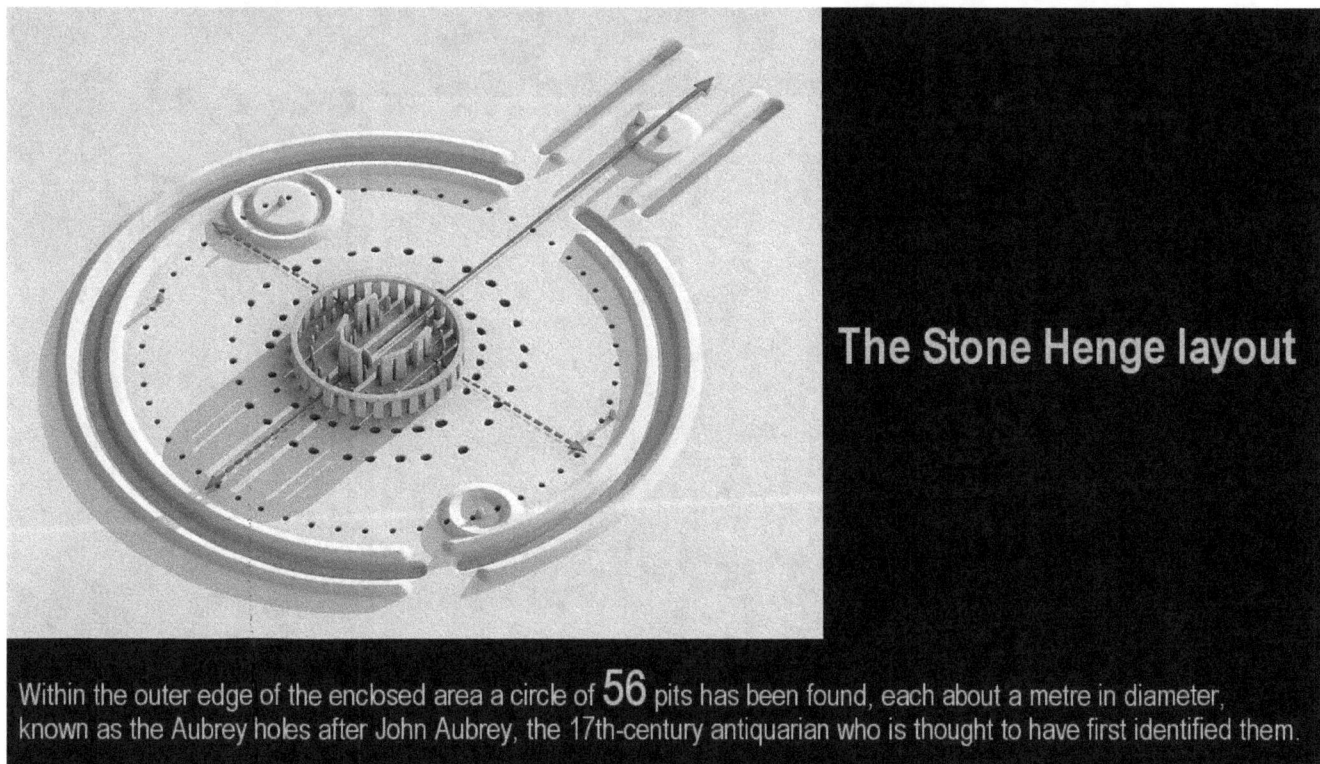

The Stone Henge layout

Within the outer edge of the enclosed area a circle of 56 pits has been found, each about a metre in diameter, known as the Aubrey holes after John Aubrey, the 17th-century antiquarian who is thought to have first identified them.

The Stone Henge monument features a ring of 56 chalk pits, a meter wide and three quarters of a meter deep, name the Aubrey holes in honour of the discoverer of them. Their purpose remains a puzzle unless one connects them with the sky.

A lab experiment, published in 2003

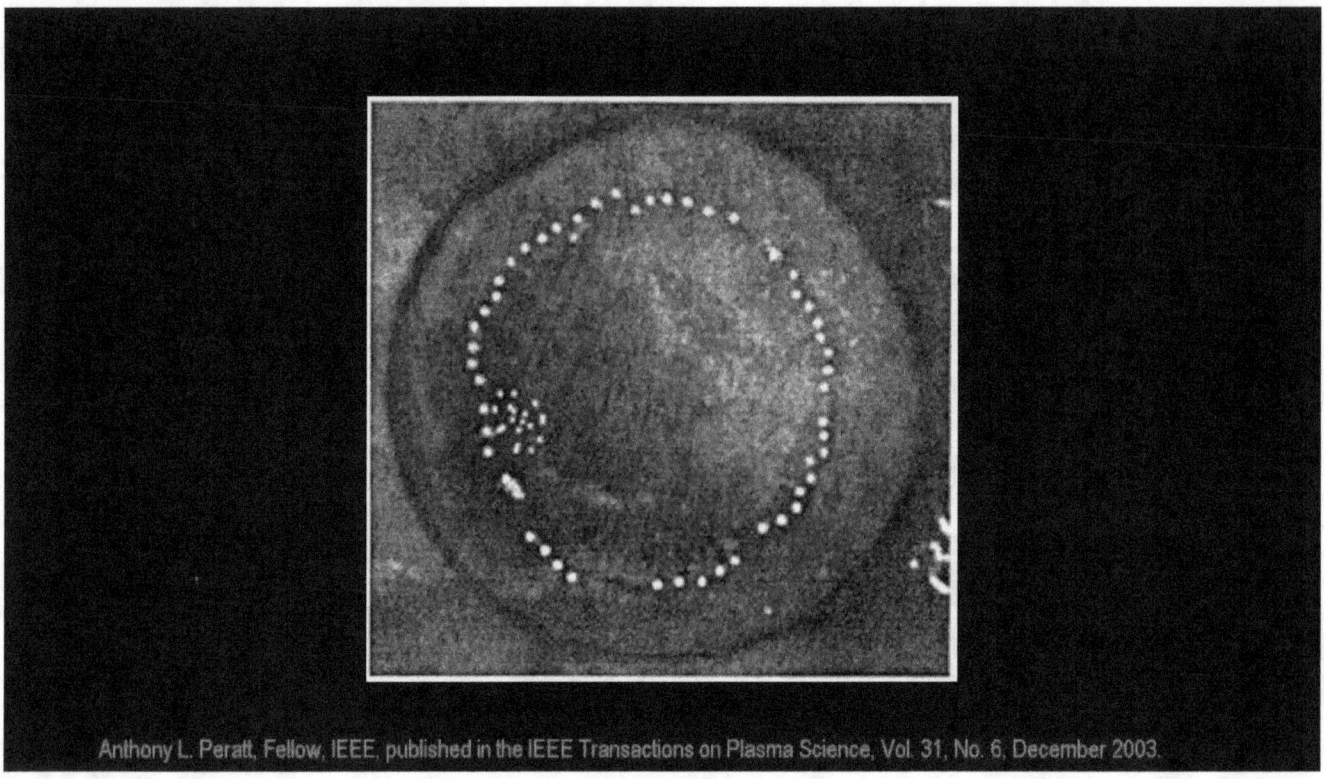

Anthony L. Peratt, Fellow, IEEE, published in the IEEE Transactions on Plasma Science, Vol. 31, No. 6, December 2003.

It was discovered in a lab experiment, published in 2003, that "A solid beam of charged particles tends to form hollow cylinders that may then filament into individual currents. When observed from below, the pattern consists of circles, circular rings of bright spots, and intense electrical discharge streamers connecting the inner structure to the outer structure." The maximum number of the filaments has been found to be 56.

A ring of 56 distinct plasma filaments

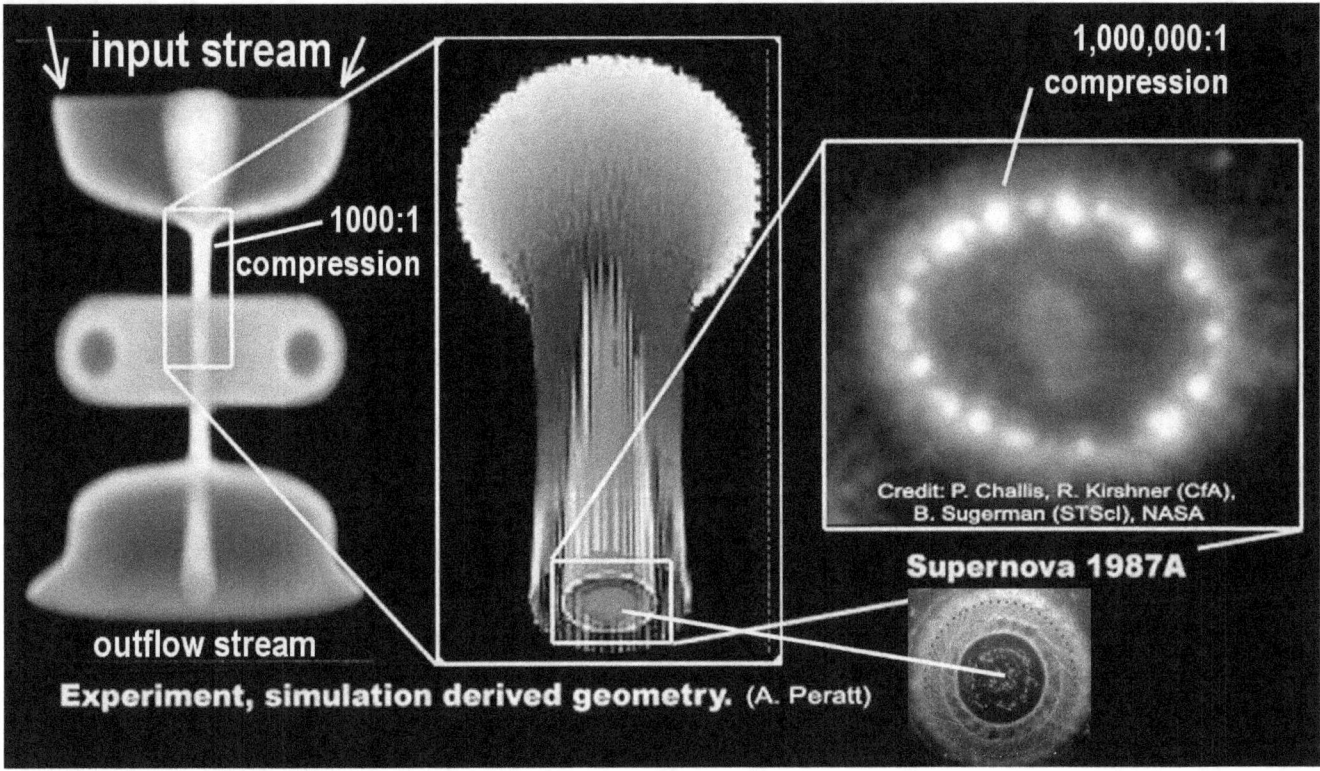

Another experiment shows that the concentrated plasma that flows between the two complementary bowl structures of the Primer Fields, forms a ring of 56 distinct plasma filaments, which in the real world, under extreme conditions, appear to have been visible in the sky.

It has been noted that over long distances the filaments merge in groups of two or three, as can be seen in this image of the Supernova 1987A.

The image shown here is giving us a cross-section view of a strong Birkeland current flowing in space.

A very-high power plasma stream

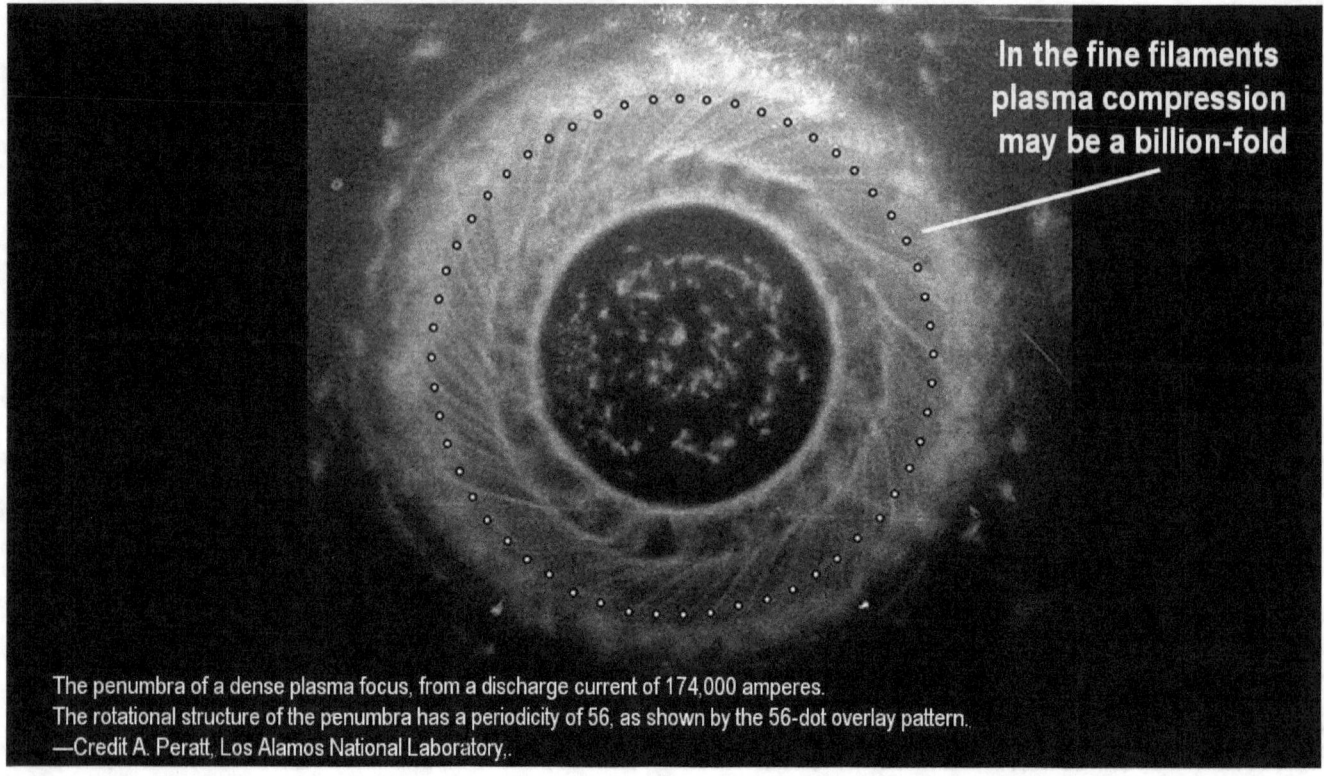

The penumbra of a dense plasma focus, from a discharge current of 174,000 amperes.
The rotational structure of the penumbra has a periodicity of 56, as shown by the 56-dot overlay pattern.
—Credit A. Peratt, Los Alamos National Laboratory,.

In the fine filaments plasma compression may be a billion-fold

In another experiment a more perfect cross section image of a very-high power plasma stream has been recorded that shows the structure of the 56 plasma filaments self-aligned into a circle, and streamers flowing into two other circles, and so on.

It is surprising to note

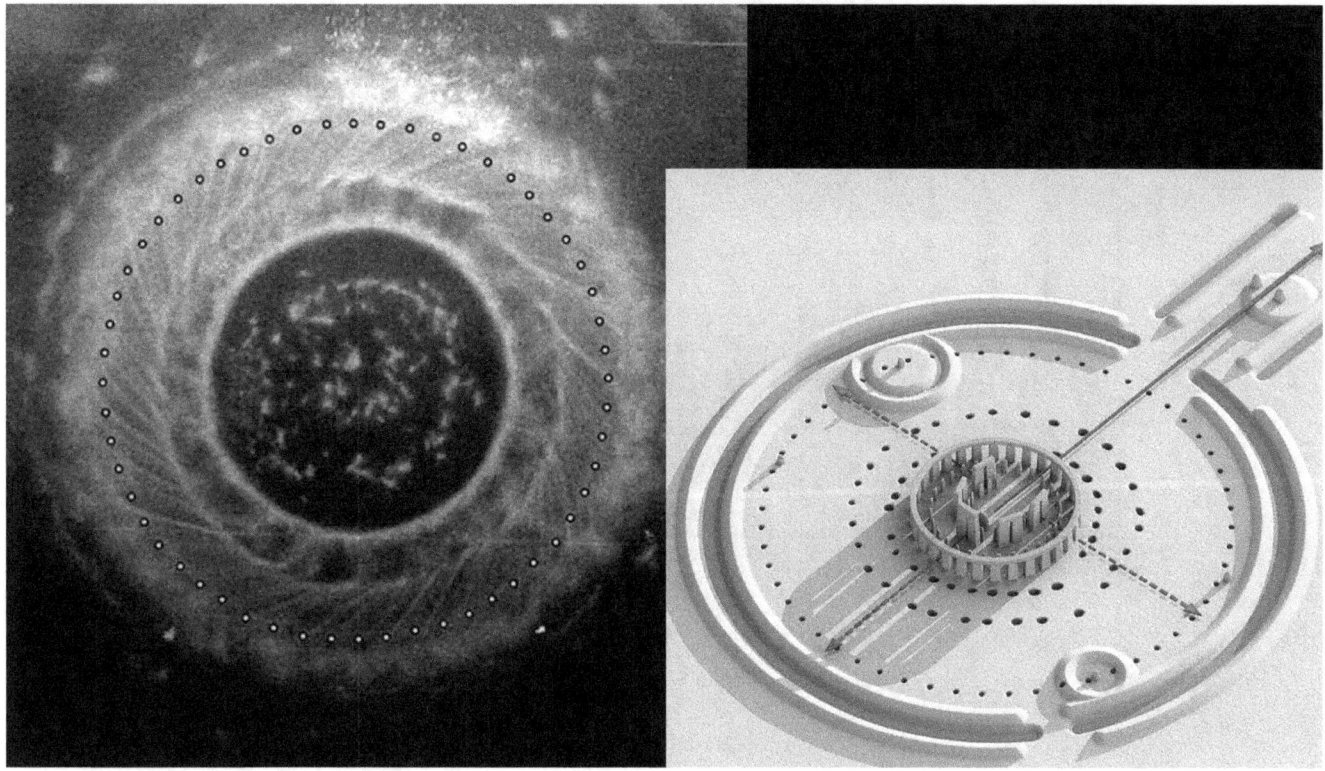

It is surprising to note how closely the layout of the Stone Henge monument, built so long ago, matches the plasma flow patterns that became visible only recently in experiments produced in the laboratory.

The surprising similarity that we find in these two cases suggests that the plasma-flow patterns had been seen in the sky in ancient times, probably seven or eight thousand years ago in times when the interglacial Sun was at it peak power level. The pattern that was seen then, which was replicated in the monument, would have been the outflow pattern of the concentrated plasma that was at the time visibly focused onto the Sun.

We no longer see these patterns

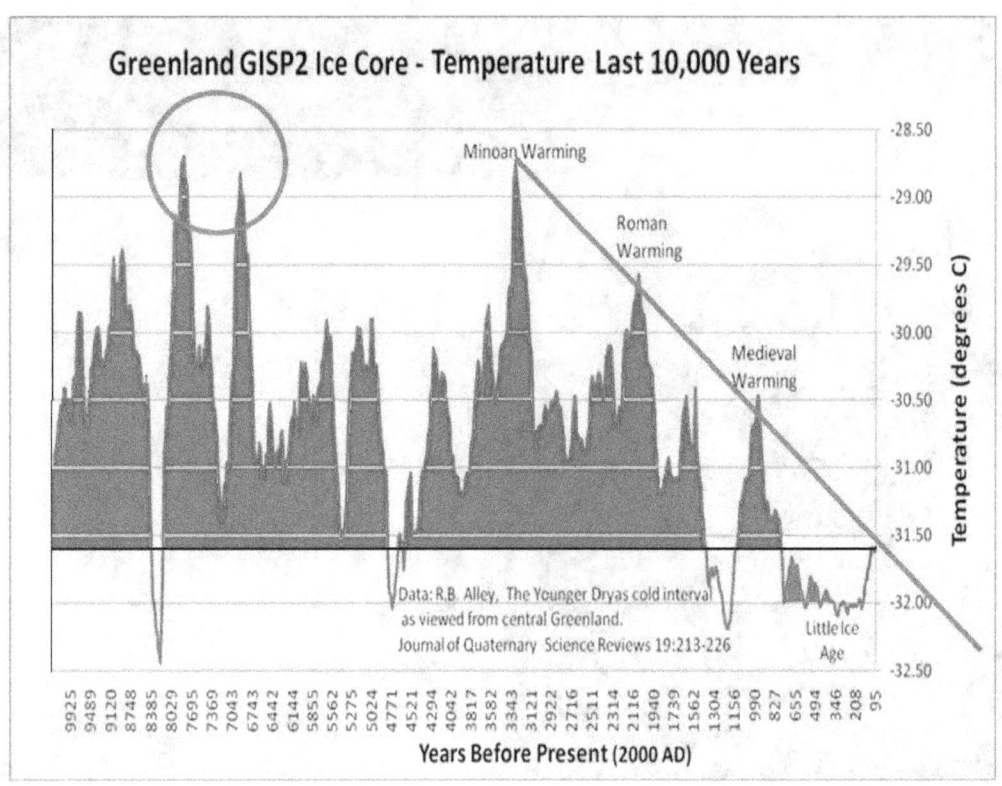

That we no longer see these types of patterns, is the natural consequence of the weak plasma environment today. There is simply not enough density left remaining for any of the plasma features forged by the Primer Fields to be visible. Soon, the Primer Fields themselves will collapse and the Sun become inactive again for long intervals, as it had been during the previous ice age period.

Part 10 - Politics versus science

Politics versus science

** The title should be: Science Uplifts Politics. But until we get there, the title remains, 'Politics versus Science.'

Humanity urgently requires the ice age challenge

Castle Bravo - the first U.S. test of a dry fuel thermonuclear hydrogen bomb - March 1, 1954 at Bikini Atoll, Marshall Islands

It may well be that humanity urgently requires the ice age challenge that now stands before us, to awake the world from its current rut before the 'sleep of reason' causes humanity to become extinct by the nuclear war that the world's leaders of empire have been building weapons for, for more than 50 years already. It would only take an hour and a half of thermonuclear war to create the conditions that causes the extinction of humanity and most forms of life with it.

A spiritual task, more than a political task

We need to raise the value of our humanity sufficiently high to take the steps required to cleanse the human landscape of the terror of nuclear war and everything that is linked to it. This is a spiritual task, more than a political task. The correlation of cosmic-ray flux with solar activity, indicates that the advance of civilization is a spiritual task. This alone should inspire us to pursue spiritual development as the highest priority objective of modern time, in order to become sufficiently spiritual minded to protect ourselves.

We wield weapons out of weakness

We wield weapons not out of strength. We wield weapons out of weakness. We wield them because of our inability to face one another as human beings. Let's hope that the ice age challenge will awake us enough to consider what human living is all about, before we throw it away without ever having really lived.

We face a choice

We face a choice therefore. The choice is between the system of empire for which all wars are fought, and the platform for freedom on which human culture and human development depends. So far humanity stands impotent and tied to the ground. It stands latched to the system of empire where it hails the weapon and throws away its humanity, as if it was already dead, unable the choose life and to choose it more abundantly.

Humanity still lives in the Roman age

The Christian Martyrs' Last Prayer by Jean-Léon Gérôme (1824–1904) - Roman Empire, Wikipedia

In many respects humanity still lives in the Roman age, reflecting the grand fascist arena of monetarism married to terror, inhumanity, and depopulation. Large segments of humanity are presently being killed under the new Roman banner, and brutally starved, torn, and despised, while much of the world looks on silently, betting on the indexes for profits.

We simply starve them to death

While we no longer burn people alive by the dozen as in the grand arenas long ago, we simply starve them to death at a rate of 300,000 a day by burning their food in automobiles in the form of biofuels.

The ice age challenge as a new paradigm

We need the ice age challenge as a new paradigm for living, in order that we may heal us of the tragedies that we have allowed to come upon us.

If we fail to become human above everything else; above politics, economics, religion, status, and sex, and fail to see ourselves as a single humanity honouring one another in love, we find ourselves unfit to meet the ice age challenge, which, technologically is not a great challenge to meet.

Only we lose by our failing

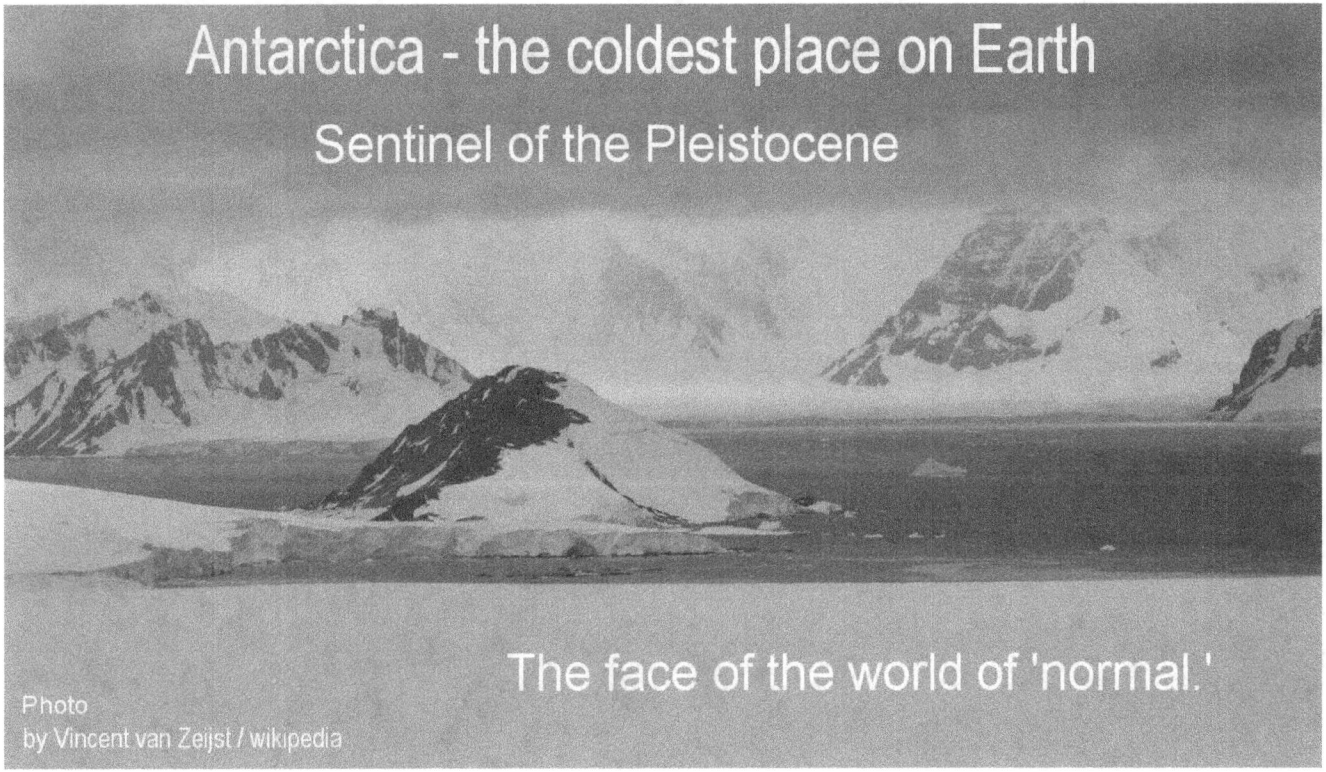

However, if we fail the ice age challenge, as we have failed in the past by not preventing the great wars, destruction, and looting carried out in the name of empires, then the coming ice age will bury us, with few exceptions. Perhaps then, in a thousand generations, or a million years, a new humanity will rise again and will not fail those challenges that we presently have no intention to even address. The universe won't be cheated by us failing ourselves. Only we lose by our failing. But why should we fail?

To rekindle that flame in the heart

We only need to rekindle that flame in the heart that rekindles our humanity and our human productive and creative power, and the beauty of our human soul and its scientific honesty. The world has become a desert landscape in all of these areas. We need to begin by turning the desert into an oasis across the world. The fading little flame of today, needs to be nourished to become a great fire in civilization. Surely, we can do this.

Not even the sky will be a limit

When we get to this point, not even the sky will be a limit for us, much less the conditions that we find on the Earth. But will we do this?

Will we dare to step up onto the wings and fly?

Will we dare to step up onto the wings and fly? Will we trust ourselves, for the great task? These remain still-open questions.

Evidently the greatest danger before us, is not the astrophysical one. This one we can deal with fairly easily and protect ourselves from the astrophysical changes.

The greatest danger that we face

The greatest danger that we face, then, is that we will do nothing, that we won't respond to the astrophysical challenge before us and die of starvation by not having created the infrastructures for our food supply that enables us to live richly in the dimmer world.

The danger is that we keep up the present course, that we continue debating global warming and manmade climate change, and that we continue to bow to the monetarism of empire till the Sun turns off and goes into its sleep mode.

Brainwashed by the choruses of the professional scoundrels

We have been brainwashed to commit ourselves to the present tragic course by the choruses of the professional scoundrels that the masters of empire have bought with their money bags to keep the dream of empire alive.

People who have sold their soul for a song

Those, who sing the empire song, are people who have sold their soul for a song and have turned civilization into a desert of starvation already, before the big challenge even begins.

That's what the global warming hoopla; the nuclear war terror; and depopulation are all about, standing against the future of humanity, aren't they?

Brainwashing to keep the internal-fusion sun theory alive

A similar type a brainwashing has also been applied to keep the internal-fusion sun theory alive at all cost, as it serves the policies of empire well. It prevents economic development, energy development, and scientific development. Humanity is being kept tied into knots by debating the internal-fusion sun theory, because the outcome of the brainwashing promises to facilitate the cherished objectives of empire, which is to eliminate 6 billion people from the face of the planet from the present 7 billion to about 1 billion. That's the officially stated policy for depopulation. Would you like to be 'depopulated?'

Children to become depopulated?

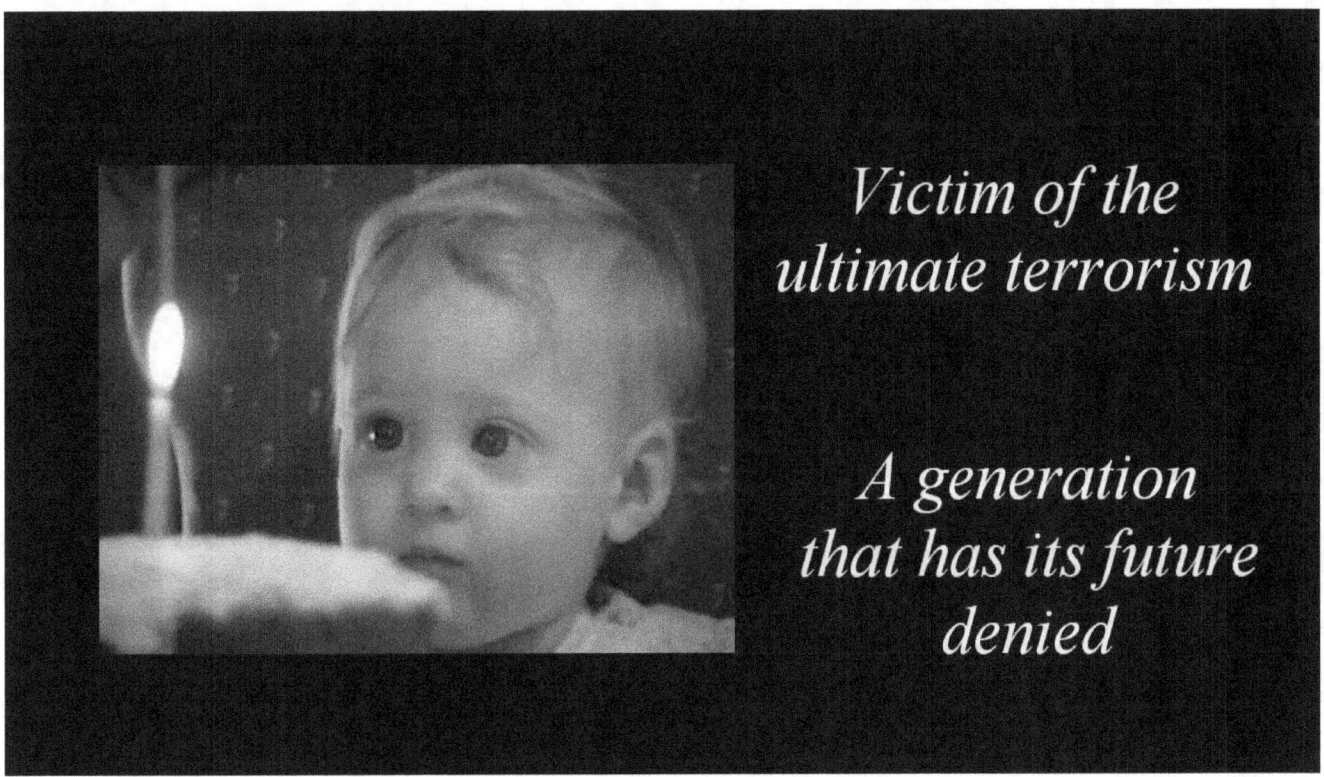

Would you like to see your children to become depopulated? This is what will happen if the ice age challenge is denied. If your answer is, yes, to the depopulation question, then sit back and do nothing, as you may do together with most of humanity, and you will have what you wish for. There is today little evidence on the horizon for a breakout from society's commitment to its small-minded thinking that has been imposed on it, such as in the context of the global warming scam and the internal-fusion sun.

The Sun is an electrically powered star

The train of the internal-fusion sun continues on track, even while the evidence is monumental that the Sun is an electrically powered star.

Behind the scene of denial

However, behind the scene of denial of the evidence that we see - a denial that is political and artificial in nature - stands the humanity that we are - a highly intelligent species that will always remain what it fundamentally is, which tends to assert itself in times of deep crisis.

The intelligence of humanity

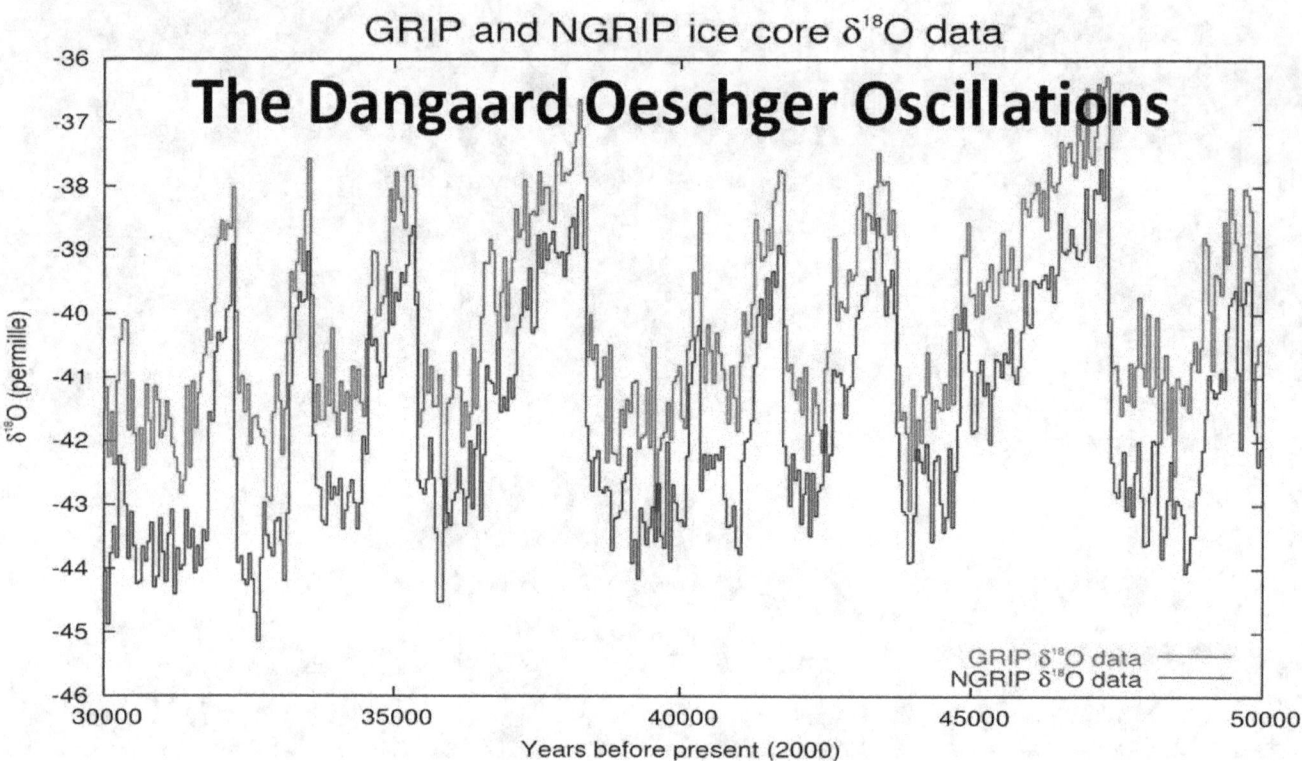

The intelligence of humanity will make its claim, slowly as this may unfold, and stand for truth, and will thereby prevent the greatest catastrophe of all times from occurring, which would result with certainty if the astrophysical processes that evidently loom before us, would remain too long ignored and not be responded to.

Break through the fog of political games

CERN - CLOUD project - Jasper Kirkby

It is inherent in the design of humanity that the leading-edge thinkers will break through the fog of political games and open their eyes to the mountains of bodies of evidence that astrophysical processes have a powerful impact on earth.

Alert scientists and truth-seekers

It is inherent in the intelligence that we embody and reflect as humanity, that the most alert scientists and truth-seekers do see the phase changes before they happen, by understanding the processes involved, and that they will inspire society to move with the requirements of the future that can be shaped with intelligent actions.

Scientific progress the hallmark of humanity

Scientific progress remains the hallmark of humanity. Even if the options are at times obscured, and the paths seem tied into knots, enormous scientific efforts have been made, and continue to be made, to push forward our understanding of the universe and its operation, and of ourselves in the unfolding process.

Unlimited energy resources

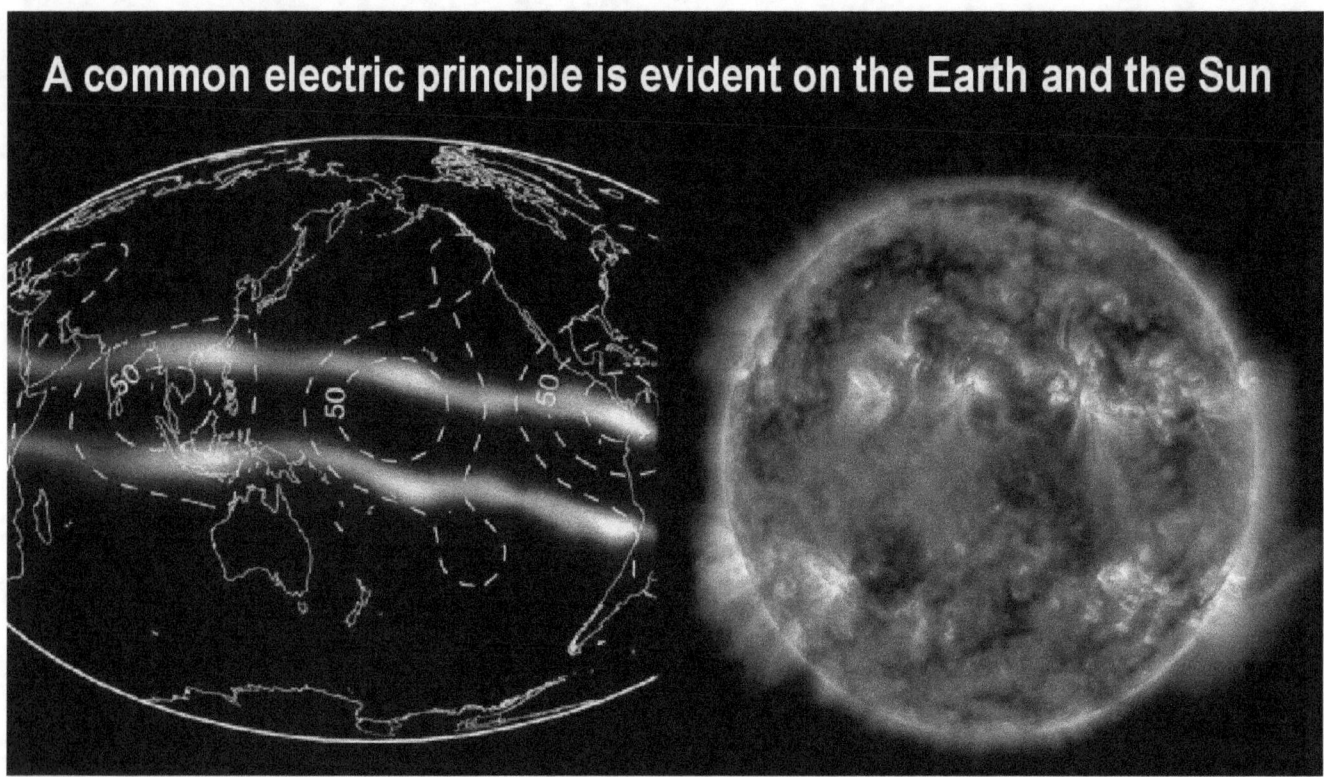

The more we move forward with our exploration on this front, the more will a great renaissance take shape before us, with unlimited energy resources that give us access to unlimited building materials and food supplies, all with the kind of density that enables us to respond to the changing solar system, which then can no longer endanger our civilization nor hinder its advances.

Potentials stand before us right now

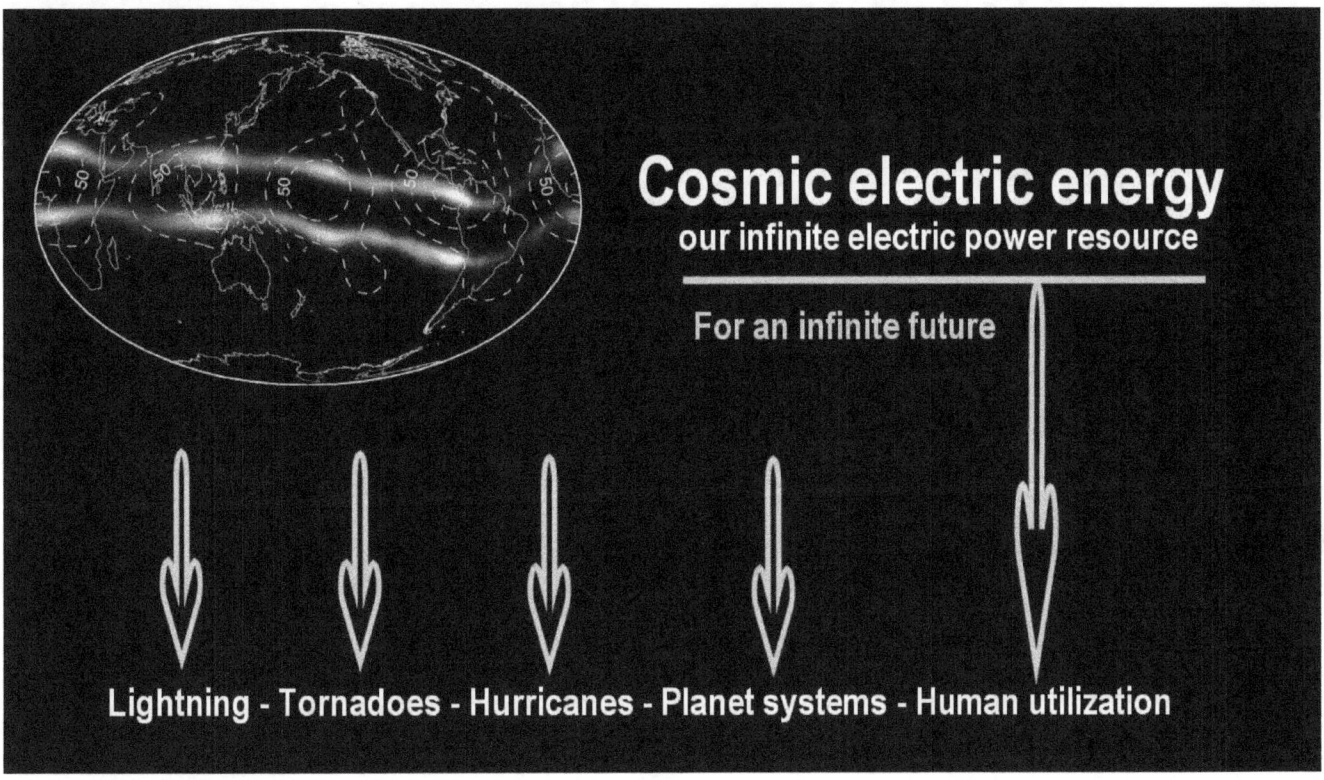

In fact these potentials stand before us right now, to be realized now, and be developed fully before the Sun reverts to its inactive state as it did during the previous ice ages.

In the previous ice ages the great potential that we have today to create a new world, did not exist, but it does exist now. As an intelligent species, the most advanced that ever existed on the Earth, we will not squander this potential.

Maybe this is what the Universe recognized and put to our credit

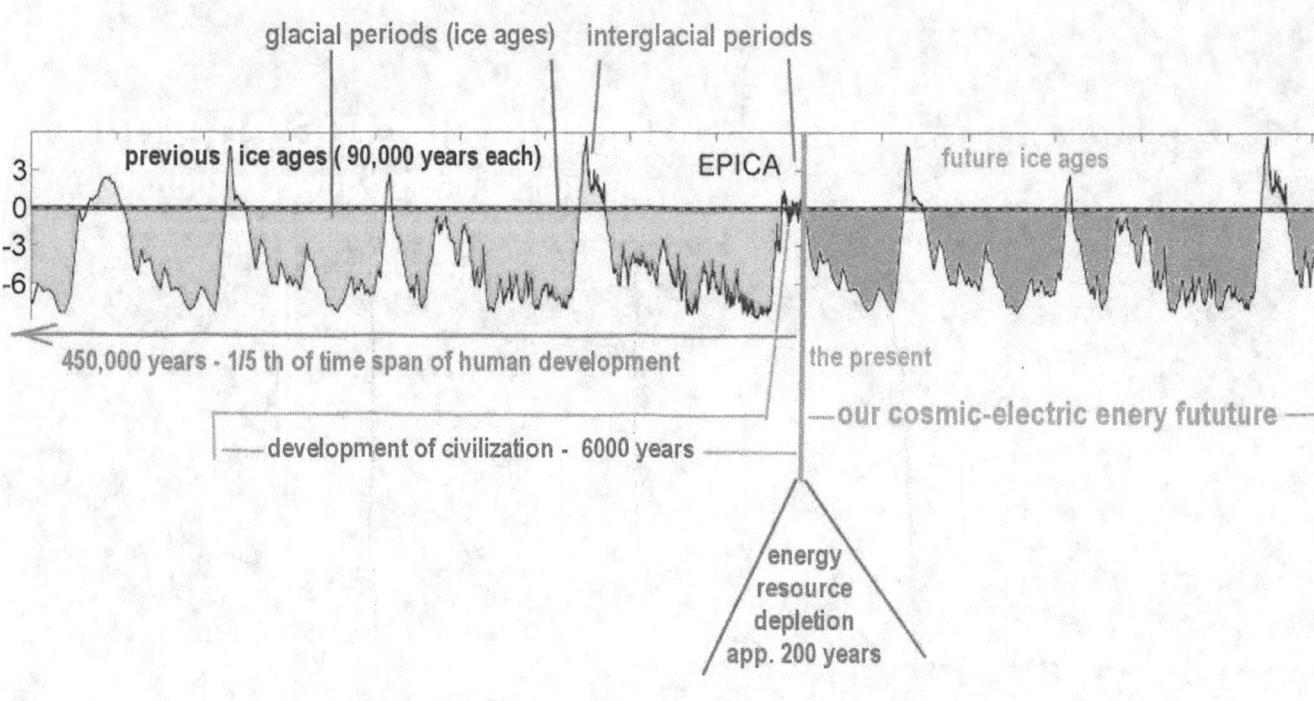

Maybe this is what the Universe recognized and put to our credit. Humanity had proven its worth before the Little Ice Age began.

To create a society without empire

During the Little Ice Age, for the first time ever, a process was set in motion to create a society without empire on the distant shores of America.

The city on a hill

Reference: "We the People" at ice-age-ahead-iaa.ca. Image from a blogger no longer online

The pilgrims' determination for the first time in history to raise the finger to empire eventually resulted in the founding of the USA as the first sovereign nation state on the planet - the city on a hill that the eyes of the world were indeed fixed upon.

The remarkable achievement that had begun, which was a radical new development in civilization - for a people to collectively take a stand against the system of empire - would likely have been lost together with much of humanity if the Little Ice Age had progressed to become the big Ice Age. But this train to near extinction was blocked with a major Dansgaard-Oeschger pulse that ended the Little Ice Age and extended the interglacial period by a few more hundred years from the 1700s on.

The greatest economic and scientific development

Ulysses was launched from the NASA Space Shuttle Discovery on Oct. 6, 1990

The extension of the interglacial was extremely critical for the development of civilization. It set the stage for the greatest economic and scientific development the world has ever seen, even though much was squandered away in recent years.

Warm climates after the Little Ice Age

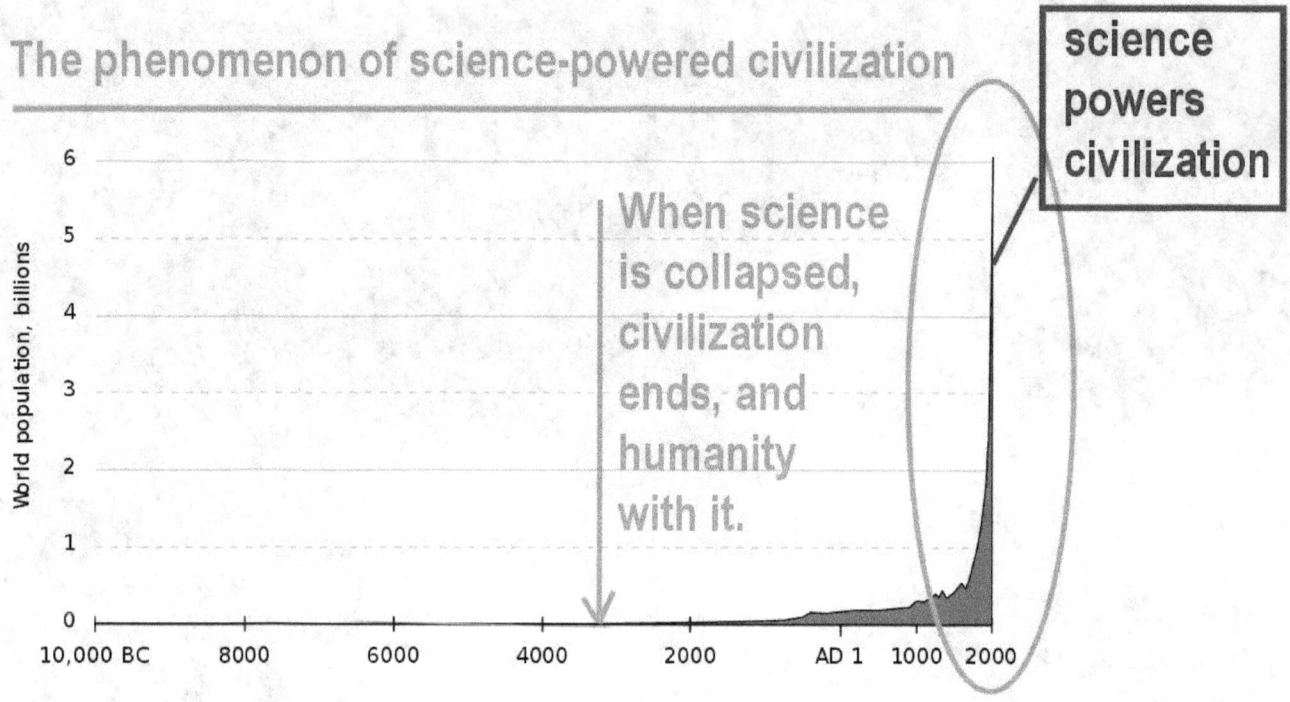

The 200-year period of warm climates after the Little Ice Age, which was by its timing evidently a major Dansgaard-Oeschger pulse, gave us the population increase that is required as an economic base for carrying out the redevelopment of the world in preparation for the coming ice age period when the Sun goes on and off for typically 90,000 years. Was this ideal coincidence, accidental? Was it a gift from heaven, to save the world? Or was it earned by humanity?

The war against empire has not yet been won

While the war against empire has not yet been won, the challenge of having to prepare the world for the environment of the dimming Sun will bring back to light the great achievements of the past and thereby swing the balance from empire and its wars and its money, to freedom and to the fulfillment of the common aims of all mankind.

This trend away from empire towards the freedom of humanity started in the 1620s with the Pilgrims that settled in Massachusetts, who came there to set up a New World on the American shores, far from empire. The Peace of Westphalia that occurred later in 1648, to end the Thirty Years War, also started a movement in the same direction in Europe, which still stands to some degree as the foundation for modern civilization.

Society's self-directed spiritual development

On this same type of path, as the power of society's self-directed spiritual development is stepped up far above the default train, the threat of nuclear war will vanish and become forgotten history as though it never existed, together with all the ugly garments that empire still parades: its monetarism, terrorism, fascism, greed, slavery, genocide, poverty, hatred, lies, and so on and on.

To meet the human need

The requirement to meet the human need in the grandest style possible and with the greatest freedoms unfolding from it, has been the basic dream and hope in every religion that ever was, which is a dream that has so far never been fully realized, but might be realized at last as we step forward in our self-directed progression from religion to the scientific realization of the grand humanity that is inherent in us all, in which we are One.

Challenge of the dimming Sun

The astrophysical challenge of the dimming Sun puts a new type of challenge before us that inspires new paradigms, since the old paradigms no longer apply.

Something to celebrate

 This is something to celebrate. It inspires us to reclaim our humanity, raise it up anew, explore it in new novels, new poetry, and new songs, on the road of our intentional development of our unlimited potentials.

This makes our future bright in spite of the dimming Sun, or because of it, for the challenge it imposes.

Harvest is Seedtime

Harvest is Seedtime

Seeds, wind-blown
Carriers of a secret still unknown
Poems in the words of nature
Sentinels of an Intelligence yet unseen
Prophets of the enduring
Apostles in an endless landscape

A poem from the novel
Endless Horizons
by Rolf A. F. Witzsche
of the series of novels
The Lodging for the Rose

Harvest is Seedtime

Seeds, wind-blown

Carriers of a secret still unknown

Poems in the words of nature

Sentinels of an Intelligence yet unseen

Prophets of the enduring

Apostles in an endless landscape

Discovering Love

Discovering Love

Seeds, wind-blown

Poems in the words of nature

Sentinels of an Intelligence yet unseen

The melody of nature - what a song!

The melody of nature - what a song!

Whispers of a splendor grander than the heavens

Like seeds are we - we whisper too

Seeds bearing gifts for the world

Gifts wrapped up in sunshine

Gems are we - unfolding a majestic song!

Lu Mountain

Lu Mountain

Whispers of a splendor grander than the heavens

Like seeds are we - we whisper too

Seeds bearing gifts for the world

Listen to the song

Listen to the song

Listen to the heart

Listen to the silence where strands of love unfold

Listen to the symphony of our humanity

In this symphony we are One

One with the Universe itself.

Flight Without Limits

Flight Without Limits

Listen to the song

Listen to the heart

Listen to the silence where strands of love unfold

Brighter than the Sun

Brighter than the Sun

Listen to the symphony of our humanity

In this symphony we are One

One with the Universe itself.

More Illustrated Science Books by Rolf A. F. Witzsche